CW01044263

Springer Texts in Statistics

Springer Texts in Statistics

For other titles published in this series, go to www.springer.com/series/417

Tze Leung Lai · Haipeng Xing

Statistical Models and Methods for Financial Markets

Tze Leung Lai
Department of Statistics
Stanford University
Stanford, CA 94305
USA
lait@stanford.edu

Haipeng Xing
Department of Statistics
Columbia University
New York, NY 10027
USA
xing@stat.columbia.edu

ISBN: 978-0-387-77826-6 e-ISBN: 978-0-387-77827-3
DOI: 10.1007/978-0-387-77827-3

Library of Congress Control Number: 2008930111

Printed on acid-free paper

9 8 7 6 5 4 3 2 1

springer.com

To Letitia and Ying

Preface

The idea of writing this book arose in 2000 when the first author was assigned to teach the required course STATS 240 (Statistical Methods in Finance) in the new M.S. program in financial mathematics at Stanford, which is an interdisciplinary program that aims to provide a master's-level education in applied mathematics, statistics, computing, finance, and economics. Students in the program had different backgrounds in statistics. Some had only taken a basic course in statistical inference, while others had taken a broad spectrum of M.S.- and Ph.D.-level statistics courses. On the other hand, all of them had already taken required core courses in investment theory and derivative pricing, and STATS 240 was supposed to link the theory and pricing formulas to real-world data and pricing or investment strategies. Besides students in the program, the course also attracted many students from other departments in the university, further increasing the heterogeneity of students, as many of them had a strong background in mathematical and statistical modeling from the mathematical, physical, and engineering sciences but no previous experience in finance. To address the diversity in background but common strong interest in the subject and in a potential career as a "quant" in the financial industry, the course material was carefully chosen not only to present basic statistical methods of importance to quantitative finance but also to summarize *domain knowledge* in finance and show how it can be combined with *statistical modeling* in financial analysis and decision making.

The course material evolved over the years, especially after the second author helped as the head TA during the years 2004 and 2005. The course also expanded to include a section offered by the Stanford Center for Professional Development, with nondegree students in the financial industry taking it on-line (`http://scpd.stanford.edu`). The steady increase in both student interest and course material led to splitting the single course into two in 2006, with STATS 240 followed by STATS 241 (Statistical Modeling in Financial Markets). Part I of this book, Basic Statistical Methods and Financial

Applications, is covered in STATS 240 and has six chapters. Chapters 1 and 2 cover linear regression, multivariate analysis, and maximum likelihood. These statistical methods are applied in Chapter 3 to a fundamental topic in quantitative finance, namely portfolio theory and investment models, for which Harry Markowitz and William Sharpe were awarded Nobel Prizes in Economics. Whereas the theory assumes the model parameters are known, in practice the parameters have to be estimated from historical data, and Chapter 3 addresses the statistical issues and describes various statistical approaches. One approach is deferred to Section 4.4 in Chapter 4, where we introduce Bayesian methods after further discussion of likelihood inference for parametric models and its applications to logistic regression and other generalized linear models that extend the linear regression models of Chapter 1 via certain "link functions." Chapter 4 also extends the least squares method in Chapter 1 to nonlinear regression models. This provides the background for the nonlinear least squares approach, which is used in various places in Part II of the book. Another important topic in quantitative finance that has attracted considerable attention in recent years, especially after the 2003 Nobel Prizes to Robert Engle and Clive Granger, is financial time series. After introducing the basic ideas and models in time series analysis in Chapter 5, Chapter 6 extends them to develop dynamic models of asset returns and their volatilities. The six chapters that constitute Part I of the book provide students in financial mathematics (or engineering) and mathematical (or computational) finance programs with basic training in statistics in a single course that also covers two fundamental topics in quantitative finance to illustrate the relevance and applications of statistical methods.

Part II of the book, Advanced Topics in Quantitative Finance, is covered in STATS 241 at Stanford. It introduces nonparametric regression in Chapter 7, which also applies the methodology to develop a substantive-empirical approach that combines domain knowledge (economic theory and market practice) with statistical modeling (via nonparametric regression). This approach provides a systematic and versatile tool to link the theory and formulas students have learned in mathematical finance courses to market data. A case in point is option pricing theory, the importance of which in financial economics led to Nobel Prizes for Robert Merton and Myron Scholes in 1997, and which is a basic topic taught in introductory mathematical finance courses. Discrepancies between the theoretical and observed option prices are revealed in certain patterns of "implied volatilities," and their statistical properties are studied in Chapter 8. Section 8.3 describes several approaches in the literature to address these discrepancies and considers in particular a substantive-empirical approach having a substantive component associated with the classical Black-Scholes formula and an empirical component that uses nonparametric regression to model market deviations from the Black-Scholes formula. Chapter 9 introduces advanced multivariate and time series methods in financial econometrics. It provides several important tools for analyzing time series data on interest rates with different maturities in Chapter 10, which also relates the statistical (empirical) analysis of real-world interest rates to stochastic process models for the valuation of interest rate derivatives in mathematical finance. The finance theories in Chapters 8 and 10 and in mathematical finance courses assume the absence of arbitrage. "Statistical arbitrage," which has become an important activity of many hedge fund managers, uses statistical learning from market prices and trading patterns to identify arbitrage opportunities in financial markets. Chapter 11 considers statistical trading strategies and related topics such as market microstructure, data-snooping checks, transaction costs, and dynamic trading. Chapter 12 applies the statistical methods in previous chapters and also describes new ones for risk management from the corporate/regulatory perspective, protecting the financial institution and its investors in case rare adverse events occur.

Since M.S. students in financial mathematics/engineering programs have strong mathematical backgrounds, the mathematical level of the book is targeted toward this audience. On the other hand, the book does not assume many prerequisites in statistics and finance. It attempts to develop the key methods and concepts, and their interrelationships, in a self-contained manner. It is also intended for analysts in the financial industry, as mentioned above in connection with the Stanford Center for Professional Development, and exposes them to modern advances in statistics by relating these advances directly to financial modeling.

Because Part I is intended for a one-semester (or one-quarter) course, it is presented in a focused and concise manner to help students study and review the material for exercises, projects, and examinations. The instructor can provide more detailed explanations of the major ideas and motivational examples in lectures, while teaching assistants can help those students who lack the background by giving tutorials and discussion sessions. Our experience has been that this system works well with the book. Another method that we have used to help students master this relatively large amount of material in a single course is to assign team projects so that they can learn together and from each other.

Besides students in the M.S. program in financial mathematics at Stanford, the course STATS 241, which uses Part II of the book, has also attracted Ph.D. students in economics, engineering, mathematics, statistics, and the Graduate School of Business. It attempts to prepare students for quantitative financial research in industry and academia. Because of the breadth and depth of the topics covered in Part II of the book, the course contents actually vary from year to year, focusing on certain topics in one year and other topics in the next. Again, team projects have proved useful for students to learn together and from each other.

After the second author joined the faculty at Columbia University in 2005, he developed a new course, W4290, in the M.S. program in statistics on the applications of statistical methods in quantitative finance. Since students in this course have already taken, or are concurrently taking, courses in statistical inference, regression, multivariate analysis, nonparametrics, and time series, the course focuses on finance applications and is based on Chapters 3, 6, and 8 and parts of Chapters 10, 11, and 12. For each finance topic covered, a review of the statistical methods to be used is first summarized, and the material from the relevant chapters of the book has been particularly useful for such a summary. Besides the core group of students in statistics, the course has also attracted many other students who are interested in finance not only to take it but also to take related statistics courses.

The Website for the book is:

`http://www.stanford.edu/~xing/statfinbook/index.html`.

The datasets for the exercises, and instructions and sample outputs of the statistical software in `R` and `MATLAB` that are mentioned in the text, can be downloaded from the Website. The Web page will be updated periodically, and corrections and supplements will be posted at the updates. We will very much appreciate feedback and suggestions from readers, who can email them to `xing@stanford.edu`.

Acknowledgments

First we want to thank all the students who took STATS 240 and STATS 241 at Stanford and W4290 at Columbia, for their interest in the subject and comments on the earlier draft of the book. We also want to express our gratitude to the teaching assistants of these courses for helping us improve the course material and develop exercises and projects. In particular, we want to thank Antje Berndt, Zehao Chen, Wei Jin, Yuxue Jin, Andreas Eckner, Noureddine El Karoui, Horel Guillaume, Jingyang Li, Wenzhi Li, Haiyan Liu, Ruixue Liu, Paul Pong, Bo Shen, Kevin Sun, Ning Sun, Zhen Wei, Chun-Yip Yau, and Qingfeng Zhang. The first author was invited to give a lecture series by the Institute of Statistical Science at Academia Sinica in December 2001. This was the first time he drafted a set of lecture notes on statistical methods in finance. He wants to thank the attendees of the lectures for their enthusiastic interest. He is particularly grateful to his coauthor on the substantive-empirical approach, Samuel Po-Shing Wong, who helped him prepare the lecture notes and delivered a lecture describing the approach. Two years later, he teamed up with Wong again to give another lecture series on statistical methods in finance at the Center of Mathematical Sciences of Zhejiang University, and he wants to thank the participants for their interest and feedback, which helped him refine the lecture notes. Comments and suggestions from friends and former students working in the financial industry have been especially helpful in shaping the contents of Part II. We are particularly grateful to Ying Chen, David Li, Viktor Spivakovsky, and Zhifeng Zhang in this connection.

Department of Statistics, Stanford University *Tze Leung Lai*
Department of Statistics, Columbia University *Haipeng Xing*
December 2007

Contents

Part I

Basic Statistical Methods and Financial Applications

The goal of Part I is to provide students in financial mathematics (or engineering) and mathematical (or computational) finance programs with a basic background in statistics. The statistical methods covered include linear regression (Chapter 1) and extensions to generalized linear models and nonlinear regression (Chapter 4), multivariate analysis (Chapter 2), likelihood inference and Bayesian methods (Chapters 2 and 4), and time series analysis (Chapter 5). Applications of these methods to quantitative finance are given in Chapter 3 (portfolio theory) and Chapter 6 (time series models of asset returns and their volatilities). The presentation attempts to strike a balance between theory and applications. Each chapter describes software implementations of the statistical methods in `R` and `MATLAB` and illustrates their applications to financial data. As mentioned in the Preface, detailed instructions and sample output of the statistical software in these illustrative examples can be downloaded from the book's Website.

As pointed out in the Preface, the treatment of the statistical methods uses mathematics that is commensurate with the background of students in mathematical finance programs. Therefore it may be at a higher mathematical level than that in master's-level courses on regression, multivariate analysis, and time series. Another difference from standard courses on these topics is that our objective is to apply the statistical methods to quantitative finance, and therefore the methodology presented should be able to handle such applications. A case in point is Chapter 1, in which we follow the standard treatment of regression analysis, assuming nonrandom regressors up to Section 1.4, and then in Section 1.5 bridging the gap between the standard theory and the applications to finance and other areas of economics, in which the regressors are typically sequentially determined input variables that depend on past outputs and inputs. The key tools to bridge this gap involve martingale theory and are summarized in Appendix A for readers interested in financial econometrics. For these readers, Appendix B summarizes a related set of tools referred to in Chapter 5.

For readers who have not taken previous courses on likelihood and Bayesian inference, regression, multivariate analysis, and time series and would like to supplement the concise treatment of these topics in Chapters 1, 2, 4, and 5 with more detailed background, the following references are recommended:

- Mongomery, D.C., Peck, E.A., and Vining, G.G. (2001). *Introduction to Linear Regression Analysis*, 3rd ed. Wiley, New York.
- Johnson, R.A. and Wichurn, D.W. (2002). *Applied Multivariate Statistical Analysis.* Prentice-Hall, Upper Saddle River, NJ.
- Rice, J.A. (2006). *Mathematical Statistics and Data Analysis*, 3rd ed. Duxbury Press, Belmont, CA.
- Tsay, R.S. (2005). *Analysis of Financial Time Series*, 2nd ed. Wiley, New York.

1

Linear Regression Models

Linear regression and the closely related linear prediction theory are widely used statistical tools in empirical finance and in many other fields. Because of the diversity of applications, introductory courses in linear regression usually focus on the mathematically simplest setting, which also occurs in many other applications. In this setting, the regressors are often assumed to be nonrandom vectors. In Sections 1.1–1.4, we follow this "standard" treatment of least squares estimates of the parameters in linear regression models with nonrandom regressors, for which the means, variances, and covariances of the least squares estimates can be easily derived by making use of matrix algebra. For nonrandom regressors, the sampling distribution of the least squares estimates can be derived by making use of linear transformations when the random errors in the regression model are independent normal, and application of the central limit theorem then yields asymptotic normality of the least squares estimates for more general independent random errors. These basic results on means, variances, and distribution theory lead to the standard procedures for constructing tests and confidence intervals described in Section 1.2 and variable selection methods and regression diagnostics in Sections 1.3 and 1.4. The assumption of nonrandom regressors, however, is violated in most applications to finance and other areas of economics, where the regressors are typically random variables some of which are sequentially determined input variables that depend on past outputs and inputs.

In Section 1.5, we bridge this gap between the standard theory and the econometric applications by extending the theory to stochastic regressors. This extension uses more advanced probability tools. Without getting into the technical details, Section 1.5 highlights two key tools from martingale theory, namely martingale strong laws and central limit theorems, given in Appendix A. (In mathematical finance, martingales are familiar objects, as they appear repeatedly in various topics.) A related point of view that Section 1.5 brings forth is the connection between least squares estimation of the regression parameters and method-of-moments estimation of the coefficients in minimum-variance linear prediction of a random variable by other random variables. This point of view and the associated linear prediction theory play a basic role in various finance applications; e.g., the minimum-variance hedge ratio for hedging futures contracts in Section 1.5.2.

An alternative to direct application of the normal distribution to evaluate confidence intervals and variances of the least squares estimates is the bootstrap methodology, which is treated in Section 1.6. An application of bootstrapping to practical implementation of Markowitz's optimal portfolio theory is given in Chapter 3. In Section 1.7, we modify the least squares criterion in ordinary least squares (OLS) estimates to account for unequal error variances at different values of the regressors, leading to generalized least squares (GLS) estimates that are often used instead of OLS in econometrics. Section 1.8 provides further implementation details and an example involving financial data in which the techniques described in Sections 1.2–1.4 are carried out to illustrate regression analysis.

1.1 Ordinary least squares (OLS)

A linear regression model relates the output (or response) y_t to q input (or predictor) variables $x_{t1}, \dots, x_{tq}$, which are also called *regressors*, via

$$y_t = a + b_1 x_{t1} + \dots + b_q x_{tq} + \epsilon_t, \tag{1.1}$$

where the ϵ_t's are unobservable random errors that are assumed to have zero means. The coefficients $a, b_1, \dots, b_q$ are unknown parameters that have to be estimated from the observed input-output vectors $(x_{t1}, \dots, x_{tq}, y_t)$, $1 \le t \le n$.

1.1.1 Residuals and their sum of squares

To fit a regression model (1.1) to the observed data, the *method of least squares* chooses $a, b_1, \dots, b_p$ to minimize the *residual sum of squares* (RSS)

$$\text{RSS} = \sum_{t=1}^{n} \{y_t - (a + b_1 x_{t1} + \dots + b_q x_{tq})\}^2 . \tag{1.2}$$

Setting to 0 the partial derivative of RSS with respect to $a, b_1, \dots, b_q$ yields $q+1$ linear equations, whose solution gives the OLS estimates. It is convenient to write a as $a x_{t0}$, where $x_{t0} = 1$. Letting $p = q + 1$ and relabeling the indices $0, 1, \dots, q$ as $1, \dots, p$, and the regression parameters $a, b_1, \dots, b_q$ as $\beta_1, \dots, \beta_p$, the regression model (1.1) can be written in matrix form as

$$\mathbf{Y} = \mathbf{X}\boldsymbol{\beta} + \boldsymbol{\epsilon}, \tag{1.3}$$

where

$$\mathbf{Y} = \begin{pmatrix} y_1 \\ \vdots \\ y_n \end{pmatrix}, \quad \boldsymbol{\beta} = \begin{pmatrix} \beta_1 \\ \vdots \\ \beta_p \end{pmatrix}, \quad \boldsymbol{\epsilon} = \begin{pmatrix} \epsilon_1 \\ \vdots \\ \epsilon_n \end{pmatrix}, \quad \mathbf{X}_j = \begin{pmatrix} x_{1j} \\ \vdots \\ x_{nj} \end{pmatrix},$$

and $\mathbf{X} = (\mathbf{X}_1, \dots, \mathbf{X}_p)$. The vector of least squares estimates of the β_i is given by

$$\widehat{\boldsymbol{\beta}} = (\mathbf{X}^T\mathbf{X})^{-1}\mathbf{X}^T\mathbf{Y}. \tag{1.4}$$

Using this matrix notation, (1.2) can be written as

$$\text{RSS} = (\mathbf{Y} - \mathbf{X}\widehat{\boldsymbol{\beta}})^T(\mathbf{Y} - \mathbf{X}\widehat{\boldsymbol{\beta}}).$$

Let $\widehat{\mathbf{Y}} = \mathbf{X}\widehat{\boldsymbol{\beta}}$. We can think of the method of least squares geometrically in terms of the projection $\widehat{\mathbf{Y}}$ of $\mathbf{Y} \in \mathbb{R}^n$ into the linear space $\mathcal{L}(\mathbf{X}_1, \dots, \mathbf{X}_p)$ spanned by $\mathbf{X}_1, \dots, \mathbf{X}_p$. Since $\widehat{\mathbf{Y}}$ is orthogonal to $\mathbf{Y} - \widehat{\mathbf{Y}}$, we have the decomposition

$$||\mathbf{Y}||^2 = ||\widehat{\mathbf{Y}}||^2 + ||\mathbf{Y} - \widehat{\mathbf{Y}}||^2, \tag{1.5}$$

where $||\mathbf{a}||$ denotes the norm $(\sum_{i=1}^n a_i^2)^{1/2}$ of a vector $\mathbf{a} = (a_1, \dots, a_n)^T$. The components $y_t - \widehat{y}_t$ of $\mathbf{Y} - \widehat{\mathbf{Y}}$ are called *residuals*, and their sum of squares is

$$||\mathbf{Y} - \widehat{\mathbf{Y}}||^2 = \sum_{t=1}^n (y_t - \widehat{y}_t)^2 = \text{RSS}.$$

Note that the projection $\widehat{\mathbf{Y}}$ of $\mathbf{Y}$ into $\mathcal{L}(\mathbf{X}_1, \dots, \mathbf{X}_p)$ is $\mathbf{X}\widehat{\boldsymbol{\beta}} = \mathbf{H}\mathbf{Y}$, where $\mathbf{H} = \mathbf{X}(\mathbf{X}^T\mathbf{X})^{-1}\mathbf{X}^T$ is the *projection matrix* (also called the *hat matrix*), which corresponds to Pythagoras' theorem for projections; see (1.5). From this geometric point of view, the symmetric matrix $\mathbf{X}^T\mathbf{X}$ does not need to be invertible. When $\mathbf{V} = \mathbf{X}^T\mathbf{X}$ is singular, we can take $(\mathbf{X}^T\mathbf{X})^{-1}$ in (1.4) as a *generalized inverse* of the matrix $\mathbf{V}$, i.e., $\mathbf{V}^{-1}$ is a $p \times p$ matrix such that given any $p \times 1$ vector $\mathbf{z}$ for which the equation $\mathbf{V}\mathbf{x} = \mathbf{z}$ is solvable, $\mathbf{x} = \mathbf{V}^{-1}\mathbf{z}$ is a solution.

1.1.2 Properties of projection matrices

A projection matrix $\mathbf{H}$ has the following properties:

(a) $\mathbf{H}$ is symmetric and idempotent; i.e., $\mathbf{H}^2 = \mathbf{H}$.

(b) If $\mathbf{H}$ is associated with projection into a linear subspace $\mathcal{L}$ (with dimension $p < n$) of $\mathbb{R}^n$, then $\mathbf{I} - \mathbf{H}$ is associated with projection into the linear space $\mathcal{L}^\perp$ consisting of vectors $\mathbf{u} \in \mathbb{R}^n$ orthogonal to $\mathcal{L}$ (i.e., $\mathbf{u}^T\mathbf{x} = 0$ for all $\mathbf{x} \in \mathcal{L}$), where $\mathbf{I}$ is the $n \times n$ identity matrix.

(c) Let $\mathbf{H} = \mathbf{X}(\mathbf{X}^T\mathbf{X})^{-1}\mathbf{X}^T$. If the column vectors $\mathbf{X}_j$ of $\mathbf{X}$ are orthogonal to each other, then $\widehat{\mathbf{Y}} = \mathbf{H}\mathbf{Y}$ is the sum of projections of $\mathbf{Y}$ into $\mathcal{L}(\mathbf{X}_j)$ over $1 \le j \le p$. Thus,

$$\widehat{\mathbf{Y}} = \sum_{j=1}^{p} (\mathbf{X}_j^T\mathbf{Y}/||\mathbf{X}_j||^2)\mathbf{X}_j = \left\{\sum_{j=1}^{p} \mathbf{X}_j\mathbf{X}_j^T \Big/ ||\mathbf{X}_j||^2\right\}\mathbf{Y} \tag{1.6}$$

since $\mathbf{X}_j^T\mathbf{Y}$ is scalar. Therefore $\mathbf{H} = \sum_{j=1}^{p} \mathbf{X}_j\mathbf{X}_j^T/||\mathbf{X}_j||^2$.

1.1.3 Properties of nonnegative definite matrices

A $p \times p$ matrix $\mathbf{V}$ is said to be *nonnegative definite* if it is symmetric and $\mathbf{a}^T\mathbf{V}\mathbf{a} \ge 0$ for all $\mathbf{a} \in \mathbb{R}^p$. If strict inequality holds for all $\mathbf{a} \ne \mathbf{0}$, then $\mathbf{V}$ is said to be *positive definite.* In particular, $\mathbf{X}^T\mathbf{X}$ is nonnegative definite since for $\mathbf{a} \in \mathbb{R}^p$

$$\mathbf{a}^T\mathbf{X}^T\mathbf{X}\mathbf{a} = (\mathbf{X}\mathbf{a})^T\mathbf{X}\mathbf{a} = \sum_{i=1}^{n} b_i^2 \ge 0, \quad \text{where } (b_1, \dots, b_n)^T = \mathbf{X}\mathbf{a}.$$

Moreover, $\mathbf{X}^T\mathbf{X}$ is positive definite if it is nonsingular.

Definition 1.1. An $n \times n$ matrix $\mathbf{Q}$ is called an *orthogonal matrix* if $\mathbf{Q}^T = \mathbf{Q}^{-1}$ or, equivalently, if the column vectors (or row vectors) are orthogonal and have unit lengths.

For a nonnegative definite matrix $\mathbf{V}$, there exists an orthogonal matrix $\mathbf{Q}$ such that $\mathbf{V} = \mathbf{Q}\mathbf{D}\mathbf{Q}^T$, where $\mathbf{D}$ is a diagonal matrix whose elements are eigenvalues of $\mathbf{V}$; see Section 2.2.2 for details. The representation $\mathbf{V} = \mathbf{Q}\mathbf{D}\mathbf{Q}^\mathbf{T}$ is called the *singular-value decomposition* of $\mathbf{V}$, which can be used to compute the inverse of $\mathbf{V}$ via $\mathbf{V}^{-1} = \mathbf{Q}\mathbf{D}^{-1}\mathbf{Q}^T$. Note that if $\mathbf{D} = \text{diag}(\lambda_1, ..., \lambda_n)$ with $\lambda_i > 0$ for all i, then $\mathbf{D}^{-1}$ is simply $\text{diag}(1/\lambda_1, ..., 1/\lambda_n)$.

Remark. To compute the least squares estimate $\widehat{\boldsymbol{\beta}} = (\mathbf{X}^T\mathbf{X})^{-1}\mathbf{X}^T\mathbf{Y}$, it is often more convenient to use instead of the singular-value decomposition of $\mathbf{X}^T\mathbf{X}$ the QR decomposition $\mathbf{X} = \mathbf{Q}\mathbf{R}$, where $\mathbf{Q}$ is an $n \times n$ orthogonal matrix and $\mathbf{R}$ is an $n \times p$ matrix with zero elements apart from the first p rows, which form a $p \times p$ upper-triangular matrix:

$$\mathbf{R} = \begin{pmatrix} \mathbf{R}_1 \\ \mathbf{0} \end{pmatrix}, \qquad \mathbf{Q} = (\mathbf{Q}_1 \ \mathbf{Q}_2), \tag{1.7}$$

in which $\mathbf{R}_1$ is $p \times p$ upper triangular and $\mathbf{Q}_1$ consists of the first p columns of $\mathbf{Q}$, so $\mathbf{X} = \mathbf{Q}\mathbf{R} = \mathbf{Q}_1\mathbf{R}_1$. Note that

$$\mathbf{X}^T\mathbf{X} = \mathbf{R}^\mathbf{T}\mathbf{Q}^\mathbf{T}\mathbf{Q}\mathbf{R} = \mathbf{R}^\mathbf{T}\mathbf{R} = \mathbf{R}_1^\mathbf{T}\mathbf{R}_1, \qquad \mathbf{X}^T\mathbf{Y} = \mathbf{R}_1^T\mathbf{Q}_1^T\mathbf{Y}. \tag{1.8}$$

Therefore the normal equations $(\mathbf{X}^T\mathbf{X})\widehat{\boldsymbol{\beta}} = \mathbf{X}^T\mathbf{Y}$ defining the OLS estimate $\widehat{\boldsymbol{\beta}}$ can be written as $\mathbf{R}_1\widehat{\boldsymbol{\beta}} = \mathbf{Q}_1^T\mathbf{Y}$, which can be solved by back-substitution (starting with the last row of $\mathbf{R}_1$) since $\mathbf{R}_1$ is upper triangular.

1.1.4 Statistical properties of OLS estimates

Consider the linear regression model (1.3) in which $n \geq p$ and

(A) x_{tj} are nonrandom constants and $\mathbf{X}$ has full rank p,
(B) ϵ_t are unobserved random disturbances with $E\epsilon_t = 0$.

Then $\widehat{\boldsymbol{\beta}}$ given by (1.4) is an unbiased estimate of $\boldsymbol{\beta}$ (i.e., $E\widehat{\beta}_j = \beta_j$). Under the additional assumption

(C) $\mathrm{Var}(\epsilon_t) = \sigma^2$ and $\mathrm{Cov}(\epsilon_i, \epsilon_j) = 0$ for $i \neq j$,

the covariance matrix of $\widehat{\boldsymbol{\beta}}$ is given by the simple formula

$$\mathrm{Cov}(\widehat{\boldsymbol{\beta}}) := \left[\mathrm{Cov}(\widehat{\beta}_i, \widehat{\beta}_j)\right]_{1\leq i,j\leq p} = \sigma^2(\mathbf{X}^T\mathbf{X})^{-1}. \tag{1.9}$$

The linear regression model with assumptions (A), (B), and (C) is often called the *Gauss-Markov model*. In this case, an unbiased estimate of σ^2 is

$$s^2 = \frac{1}{n-p}\sum_{t=1}^{n}(y_t - \widehat{y}_t)^2 = \frac{\mathrm{RSS}}{n-p}, \tag{1.10}$$

in which $n-p$ is the degrees of freedom of the model. These properties can be derived by applying the linearity of expectations of random vectors; see Section 2.1.2. In addition, we can use the multivariate normal distribution in Section 2.3 to derive distributional properties of $\widehat{\boldsymbol{\beta}}$ and s^2 when (B) and (C) are replaced by the stronger assumption that

(C*) ϵ_t are independent $N(0, \sigma^2)$,

where $N(\mu, \sigma^2)$ denotes the normal distribution with mean μ and variance σ^2.

Definition 1.2. (i) If $Z_1, \ldots, Z_n$ are independent $N(0,1)$ variables, then $U = Z_1^2 + \cdots + Z_n^2$ is said to have the *chi-square distribution* with n degrees of freedom, written $U \sim \chi_n^2$.

(ii) If $Z \sim N(0,1)$, $U \sim \chi_n^2$, and Z and U are independent, then $T = Z/\sqrt{U/n}$ is said to have the *t-distribution* with n degrees of freedom, written $T \sim t_n$.

(iii) If $U \sim \chi_m^2$, $W \sim \chi_n^2$, and U and W are independent, then $F = (U/m)/(W/n)$ is said to have the *F-distribution* with m and n degrees of freedom, written $F \sim F_{m,n}$.

Under assumptions (A) and (C*), it will be shown in Section 2.3 that

$$\widehat{\boldsymbol{\beta}} \sim N(\boldsymbol{\beta}, \sigma^2(\mathbf{X}^T\mathbf{X})^{-1}), \tag{1.11}$$

$$\sum_{t=1}^{n}(y_t - \widehat{y}_t)^2/\sigma^2 \sim \chi^2_{n-p}, \tag{1.12}$$

$$\widehat{\boldsymbol{\beta}} \text{ and } s^2 \text{ are independent.} \tag{1.13}$$

If we replace (C*) by the weaker assumption that ϵ_t is independent with mean 0 and variance σ^2 (without normality) but include additional assumptions on the x_{tj} and boundedness of higher moments of ϵ_t so that the central limit theorem is applicable, then (1.11)–(1.13) still hold asymptotically as $n \to \infty$; see Appendix A.

1.2 Statistical inference

The preceding distribution theory for OLS and its extensions given in Section 1.5 play a basic role in inference problems, described below, for the linear regression model (1.1).

1.2.1 Confidence intervals

We can use (1.11)–(1.13) to construct exact confidence intervals under assumptions (A) and (C*).

Definition 1.3. The qth *quantile* u of the probability distribution of a continuous random variable U is defined by $P(U \le u) = q$. Without assuming continuity, the qth quantile is defined as the value u satisfying $P(U \le u) \ge q$ and $P(U \ge u) \ge 1 - q$.

Confidence intervals for a regression coefficient

From (1.11), it follows that the marginal distribution of any regression coefficient $\widehat{\beta}_j$ is normal with mean β_j and variance $\sigma^2 c_{jj}$, where c_{jj} is the jth diagonal element of the matrix $\mathbf{C} = (\mathbf{X}^T\mathbf{X})^{-1}$. Combining this with (1.12) and (1.13) yields

$$\frac{\widehat{\beta}_j - \beta_j}{s\sqrt{c_{jj}}} \sim t_{n-p}, \quad j = 1, \cdots, p. \tag{1.14}$$

In view of (1.14), a $100(1-\alpha)\%$ confidence interval for β_j is

$$\widehat{\beta}_j - t_{n-p;1-\alpha/2}s\sqrt{c_{jj}} \le \beta_j \le \widehat{\beta}_j + t_{n-p;1-\alpha/2}s\sqrt{c_{jj}}, \tag{1.15}$$

in which $t_{n-p;1-\alpha/2}$ is the $(1-\alpha/2)$th quantile of the t_{n-p} distribution.

Simultaneous confidence region for all regression coefficients

From (1.11)–(1.13) and Definition 1.2(iii), it follows that

$$\frac{(\widehat{\boldsymbol{\beta}} - \boldsymbol{\beta})^T(\mathbf{X}^T\mathbf{X})(\widehat{\boldsymbol{\beta}} - \boldsymbol{\beta})/p}{s^2} \sim F_{p,n-p}, \tag{1.16}$$

which further implies that

$$P\left[\frac{(\widehat{\boldsymbol{\beta}} - \boldsymbol{\beta})^T(\mathbf{X}^T\mathbf{X})(\widehat{\boldsymbol{\beta}} - \boldsymbol{\beta})/p}{s^2} \leq F_{p,n-p;1-\alpha}\right] = 1 - \alpha, \tag{1.17}$$

where $F_{p,n-p;1-\alpha}$ is the $(1-\alpha)$th quantile of the $F_{p,n-p}$ distribution. Therefore, a $100(1-\alpha)\%$ confidence ellipsoid for $\boldsymbol{\beta}$ is

$$\left\{\boldsymbol{\beta} : \frac{(\widehat{\boldsymbol{\beta}} - \boldsymbol{\beta})^T(\mathbf{X}^T\mathbf{X})(\widehat{\boldsymbol{\beta}} - \boldsymbol{\beta})/p}{s^2} \leq F_{p,n-p;1-\alpha}\right\}. \tag{1.18}$$

Note that whereas (1.15) gives a confidence interval for the particular regression coefficient β_j, (1.18) gives a region that *simultaneously* contains all regression coefficients with probability $1 - \alpha$.

Confidence interval for the mean response

Letting $\mathbf{x}_t = (x_{t1}, \ldots, x_{tp})^T$, the rows of the matrix equation (1.3) can be written as $y_t = \boldsymbol{\beta}^T\mathbf{x}_t + \epsilon_t$, $1 \leq t \leq n$. Often one is interested in a confidence interval for the mean response $\boldsymbol{\beta}^T\mathbf{x}$ at a given regressor value $\mathbf{x}$. An unbiased estimate of $\boldsymbol{\beta}^T\mathbf{x}$ is $\widehat{\boldsymbol{\beta}}^T\mathbf{x}$, and

$$\mathrm{Var}(\widehat{\boldsymbol{\beta}}^T\mathbf{x}) = \mathrm{Var}(\mathbf{x}^T\widehat{\boldsymbol{\beta}}) = \mathbf{x}^T\mathrm{Cov}(\widehat{\boldsymbol{\beta}})\mathbf{x} = \sigma^2\mathbf{x}^T(\mathbf{X}^T\mathbf{X})^{-1}\mathbf{x}; \tag{1.19}$$

see Section 2.1. Hence a $100(1-\alpha)\%$ confidence interval for $\boldsymbol{\beta}^T\mathbf{x}$ is

$$\widehat{\boldsymbol{\beta}}^T\mathbf{x} \pm t_{n-p;1-\alpha/2}\left\{s\sqrt{\mathbf{x}^T(\mathbf{X}^T\mathbf{X})^{-1}\mathbf{x}}\right\}. \tag{1.20}$$

Prediction interval for a new observation

A similar argument can be used to construct prediction intervals for future observations. Suppose a future observation y is taken at the regressor $\mathbf{x}$. Since $y = \boldsymbol{\beta}^T\mathbf{x} + \epsilon$ with $\epsilon \sim N(0, \sigma^2)$ under assumption (C*), and since $\mathrm{Var}(y - \widehat{\boldsymbol{\beta}}^T\mathbf{x}) = \mathrm{Var}(\epsilon) + \mathrm{Var}(\widehat{\boldsymbol{\beta}}^T\mathbf{x}) = \sigma^2 + \sigma^2\mathbf{x}(\mathbf{X}^T\mathbf{X})^{-1}\mathbf{x}$, a $100(1-\alpha)\%$ prediction interval for y is

$$\widehat{\boldsymbol{\beta}}^T \mathbf{x} \pm t_{n-p;1-\alpha/2} \left\{ s\sqrt{1 + \mathbf{x}^T(\mathbf{X}^T\mathbf{X})^{-1}\mathbf{x}} \right\}. \tag{1.21}$$

1.2.2 ANOVA (analysis of variance) tests

We can apply (1.14) to test whether β_j is significantly different from 0. The null hypothesis to be tested is $H_0 : \beta_j = 0$, which means that the predictor x_j can be removed from the model in the presence of the other predictor variables. Under H_0, $\widehat{\beta}_j/(s\sqrt{c_{jj}}) \sim t_{n-p}$ by (1.14). Therefore we reject H_0 at significance level α if

$$|\widehat{\beta}_j| > t_{n-p;1-\alpha/2} s\sqrt{c_{jj}}. \tag{1.22}$$

Definition 1.4. (i) The *type I error*, or *significance level*, of a test of the null hypothesis H_0 is the probability of falsely rejecting H_0 when it is true.

(ii) The *p-value*, or *attained significance level*, of a test of the null hypothesis H_0, based on a given test statistic, is $P_{H_0}\{$test statistic exceeds the observed sample value$\}$.

In particular, the test statistic of $H_0 : \beta_j = 0$ is $|\beta_j|/(s\sqrt{c_{jj}})$, which is the absolute value of a t_{n-p} random variable under H_0. Therefore (1.22) is equivalent to $\{p\text{-value} < \alpha\}$. Most software packages report the p-value of a test instead of whether the test rejects the null hypothesis at significance level α; see Section 1.8 (Tables 1.3 and 1.5). The user can see whether, and by how much, the p-value falls below α for rejection of H_0.

Remark. If $T \sim t_{n-p}$, then $T^2 = Z^2/\{U/(n-p)\}$ by Definition 1.2(ii), where $Z \sim N(0,1)$ and $U \sim \chi^2_{n-p}$ are independent. Since $Z^2 \sim \chi^2_1$ by Definition 1.2(i), it follows from Definition 1.2(iii) that $T^2 \sim F_{1,n-p}$. Hence the t-test (1.22) can also be expressed as an F-test. Replacing $F_{1,n-p}$ with a general F-distribution leads to the following tests of more general hypotheses than the preceding null hypothesis.

F-tests of general linear hypotheses and ANOVA

Letting $E(\boldsymbol{\epsilon}) = (E\epsilon_1, \cdots, E\epsilon_n)^T$, assumption (B) implies $E(\boldsymbol{\epsilon}) = \mathbf{0}$, and therefore the regression model $\mathbf{Y} = \mathbf{X}\boldsymbol{\beta} + \boldsymbol{\epsilon}$ says that $E(\mathbf{Y})$ belongs to the linear space $\mathcal{L}$ spanned by the column vectors $\mathbf{X}_1, \ldots, \mathbf{X}_p$ of $\mathbf{X}$. Assuming $\mathbf{X}$ to be of full rank $p(\leq n)$, the dimension of this linear space is p. To test the null hypothesis $H_0 : E(\mathbf{Y}) \in \mathcal{L}_0$, where $\mathcal{L}_0$ is a linear subspace (of $\mathcal{L}$) of dimension $r < p$, we use two independent estimates of σ^2, one of which is $s^2 = ||\mathbf{Y} - \widehat{\mathbf{Y}}||^2/(n-p)$. Let $\widehat{\mathbf{Y}}_0$ be the projection of $\mathbf{Y}$ into the subspace $\mathcal{L}_0$; see Figure 1.1. Then, under H_0, $||\mathbf{Y} - \widehat{\mathbf{Y}}_0||^2/(n-r)$ is another estimate of σ^2 but is not independent of s^2. By Pythagoras' theorem for projections,

$$||\mathbf{Y} - \widehat{\mathbf{Y}}_0||^2 = ||\mathbf{Y} - \widehat{\mathbf{Y}}||^2 + ||\widehat{\mathbf{Y}} - \widehat{\mathbf{Y}}_0||^2. \tag{1.23}$$

Alongside this decomposition is the decomposition of the degrees of freedom for the residual sum of squares:

$$n - r = (n - p) + (p - r). \tag{1.24}$$

Under H_0, $||\widehat{\mathbf{Y}} - \widehat{\mathbf{Y}}_0||^2/(p-r)$ is an unbiased estimate of σ^2 and is independent of s^2; see Section 2.3 for details. Therefore

$$F := \frac{||\widehat{\mathbf{Y}} - \widehat{\mathbf{Y}}_0||^2/(p-r)}{||\mathbf{Y} - \widehat{\mathbf{Y}}||^2/(n-p)} \sim F_{p-r,n-p} \tag{1.25}$$

under H_0. The F-test rejects H_0 if F exceeds $F_{p-r,n-p;\alpha}$, so its type I error is equal to α. A particular example of H_0 is $\beta_{r+1} = \cdots = \beta_p = 0$, which means that the regression involves only r of the p input variables.

By Pythagoras' theorem, $||\widehat{\mathbf{Y}}||^2 = ||\widehat{\mathbf{Y}}_0||^2 + ||\widehat{\mathbf{Y}} - \widehat{\mathbf{Y}}_0||^2$, so the term $||\widehat{\mathbf{Y}} - \widehat{\mathbf{Y}}_0||^2$ in (1.25) can be easily computed as the difference $||\widehat{\mathbf{Y}}||^2 - ||\widehat{\mathbf{Y}}_0||^2$. Similarly, $\text{RSS} = ||\mathbf{Y} - \widehat{\mathbf{Y}}||^2$ can be computed by subtracting $||\widehat{\mathbf{Y}}||^2$ from $||\mathbf{Y}||^2$. Table 1.1 summarizes the ANOVA (analysis of variance) decomposition of the total sum of squares $||\mathbf{Y}||^2 = \sum_{i=1}^n y_i^2$.

Whereas (1.22) is used to test the significance of a particular regressor, we can make use of (1.25) to test the significance of the linear relationship (1.1). Specifically, consider the null hypothesis $H_0 : b_1 = \cdots = b_q = 0$ in (1.1), which means that the input variables $x_{t1}, \cdots, x_{tq}$ have no effect on the output y_t. Here $\mathcal{L}_0 = \mathcal{L}(\mathbf{1})$ and $\widehat{\mathbf{Y}}_0 = \bar{y}\mathbf{1}$, where $\bar{y} = n^{-1}\sum_{t=1}^n y_t$ and $\mathbf{1}$ is the vector of 1's that are associated with the intercept term a. Note that $q = p - 1, r = 1$, and $\text{RSS}(\mathcal{L}_0) = \sum_{t=1}^n (y_t - \bar{y})^2$.

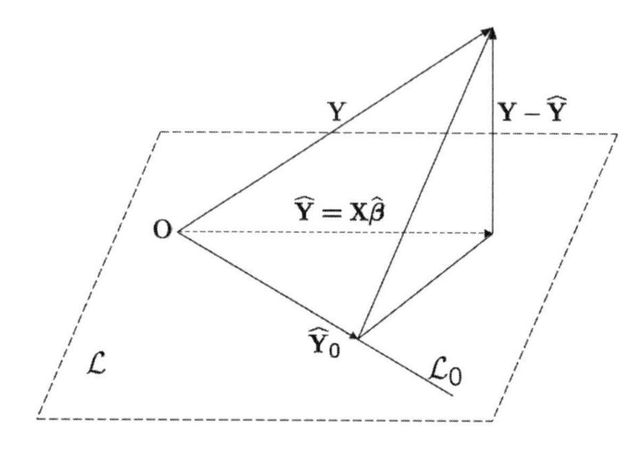

Fig. 1.1. Projections of $\mathbf{Y}$ into $\mathcal{L}$ and $\mathcal{L}_0$.

Table 1.1. ANOVA for linear hypotheses.

Source of Variation	Sum of Squares	Degrees of Freedom
Total	$\|\|\mathbf{Y}\|\|^2$	n
Regression	$\|\|\widehat{\mathbf{Y}}\|\|^2$	p
Regression ($\mathcal{L}_0$)	$\|\|\widehat{\mathbf{Y}}_0\|\|^2$	r
RSS	$\|\|\mathbf{Y} - \widehat{\mathbf{Y}}\|\|^2$	$n-p$
RSS($\mathcal{L}_0$)	$\|\|\mathbf{Y} - \widehat{\mathbf{Y}}_0\|\|^2$	$n-r$
RSS($\mathcal{L}_0$) − RSS	$\|\|\widehat{\mathbf{Y}}_0 - \widehat{\mathbf{Y}}\|\|^2$	$p-r$

1.3 Variable selection

In fitting a regression model, the more input variables (regressors) one has, the better is the fit. However, overfitting will result in deterioration of the accuracy of parameter estimates. In particular, "multicollinearity" arises when the regressors are highly correlated themselves, leading to a singular or nearly singular $\mathbf{X}^T\mathbf{X}$ and therefore a badly behaved covariance matrix $\sigma^2(\mathbf{X}^T\mathbf{X})^{-1}$ for $\widehat{\boldsymbol{\beta}}$. Letting K denote the largest possible number of regressors, we describe first some criteria for choosing regressors and then stepwise selection procedures to come up with $k(\leq K)$ regressors for inclusion in the regression model.

1.3.1 Test-based and other variable selection criteria

Partial F-statistics

A traditional test-based approach to variable selection is to test if the regression coefficient of the latest regressor entered is significantly different from 0. Specifically, letting x_j denote the latest regressor entered, perform the F-test of $H_0 : \beta_j = 0$; see Section 1.2.2 and in particular (1.25), which is called the *partial F-statistic* associated with x_j (in the presence of previously entered regressors). Stepwise inclusion of regressors terminates as soon as the latest regressor entered is not significantly different from 0. The performance of the procedure therefore depends on the choice of the significance level α, which is often chosen to be 5%. Note that since the procedure carries out a sequence of F-tests, the overall significance level can differ substantially from α. Moreover, the procedure only considers the type I error, and the type II error of accepting H_0 when $\beta_j \neq 0$ is not used by the procedure.

Multiple correlation coefficients and adjusted R^2

Besides F-tests, there are other criteria that may provide better selection procedures. Next we discuss some variable selection criteria that can be used for evaluating and comparing submodels of the full regression model. For notational convenience, let $s_p^2 = \mathrm{RSS}_p/(n-p)$, where RSS_p denotes the residual sum of squares when p input variables are included in the regression model.

The *correlation coefficient* between two random variables X and Y is defined as $\rho = \mathrm{Cov}(X,Y)/(\sigma_X \sigma_Y)$, where σ_X and σ_Y denote the standard deviations of the random variables. By the Schwarz inequality, $|\mathrm{Cov}(X,Y)| \le \sqrt{\mathrm{Var}(X)}\sqrt{\mathrm{Var}(Y)}$ and therefore $|\rho| \le 1$. The notation $\mathrm{Corr}(X,Y)$ is often used to denote the correlation coefficient between X and Y. The *method-of-moments estimate* (which replaces population moments by their sample counterparts) of ρ based on sample data $(x_1, y_1), \ldots, (x_n, y_n)$ is the sample correlation coefficient

$$r = \left\{ \sum_{i=1}^n (x_i - \bar{x})(y_i - \bar{y}) \right\} \Big/ \left\{ \sum_{i=1}^n (x_i - \bar{x})^2 \sum_{i=1}^n (y_i - \bar{y})^2 \right\}^{\frac{1}{2}}. \tag{1.26}$$

Note that

$$r^2 = \widehat{b}^2 \left\{ \sum_{i=1}^n (x_i - \bar{x})^2 \right\} \Big/ \sum_{i=1}^n (y_i - \bar{y})^2 = \sum_{i=1}^n (\widehat{y}_i - \bar{y})^2 \Big/ \sum_{i=1}^n (y_i - \bar{y})^2,$$

where $\widehat{b} = \left\{ \sum_{i=1}^n (x_i - \bar{x})(y_i - \bar{y}) \right\} / \sum_{i=1}^n (x_i - \bar{x})^2$ is the regression coefficient of the y_i on the x_i in the linear model $y_i = a + bx_i + \epsilon_i$.

In the multiple regression model $y_i = a + b_1 x_{i1} + \cdots + b_q x_{iq} + \epsilon_i$, we define similarly the *multiple correlation coefficient*

$$R^2 = \frac{\sum_{i=1}^n (\widehat{y}_i - \bar{y})^2}{\sum_{i=1}^n (y_i - \bar{y})^2} = 1 - \frac{\mathrm{RSS}}{\sum_{i=1}^n (y_i - \bar{y})^2}, \tag{1.27}$$

where $\mathrm{RSS} = \sum_{i=1}^n (y_i - \widehat{y}_i)^2$. The second equality in (1.27) is a consequence of Pythagoras' theorem; see Section 1.2.2 and particularly its last paragraph. R^2 is often used as a measure of the strength of a linear relationship, with a value of R^2 close to 1 suggesting a strong relationship. However, as will be explained in the next two paragraphs, it is not appropriate to use R^2 as a criterion for choosing the number of input variables to be included in the model, although it can be used to compare different models with the same number of input variables.

Instead of R^2, one can use for model selection the *adjusted* R^2, which is defined as

$$R^2_{adj} = 1 - \frac{s_p^2}{\sum_{i=1}^n (y_i - \bar{y})^2/(n-1)} = 1 - (1 - R^2)\frac{n-1}{n-p}. \tag{1.28}$$

One model selection procedure is to choose the model with the largest R^2_{adj}. While R^2 does not "penalize" the number p of variables used, R^2_{adj} inflates RSS_p by $(n-1)/(n-p)$ and does not necessarily increase as the number p of input variables becomes larger.

Mallows' C_p

The C_p-*statistic* introduced by Mallows is defined as

$$C_p = \frac{\mathrm{RSS}_p}{s_K^2} + 2p - n \tag{1.29}$$

when a subset of p input variables is chosen from the full set $\{x_1, \ldots, x_K\}$. Mallows (1973) proposed to choose the subset that has the smallest C_p values in view of the following considerations. First suppose that $\beta_j = 0$ whenever x_j does not belong to this set of p input variables that constitute the input vector. Let $J_p = \sum_{t=1}^n (\widehat{\boldsymbol{\beta}}^T \mathbf{x}_t - \boldsymbol{\beta}^T \mathbf{x}_t)^2/\sigma^2$, in which $\mathbf{x}_t$ is p-dimensional. Then it can be shown that

$$E(J_p) = E(\mathrm{RSS}_p)/\sigma^2 + 2p - n; \tag{1.30}$$

see Section 2.1.2. Therefore the C_p given in (1.29) is a method-of-moments estimate of (1.30). Note that the summand $2p$ in (1.29) can be interpreted as a linear penalty for adding input variables to the regression. Next consider the case of omitting important input variables x_j (with $\beta_j \neq 0$) from the regression model. Then $\widehat{\boldsymbol{\beta}}$ is biased and

$$E(\mathrm{RSS}_p) = \sum_{t=1}^n \left[E(y_t - \widehat{\boldsymbol{\beta}}^T \mathbf{x}_t)\right]^2 + (n-p)\sigma^2; \tag{1.31}$$

see Section 2.1.2. Therefore RSS_p/s_K^2 in this case tends to be substantially larger than in the regression model that includes all regressors corresponding to nonzero β_j. Therefore overfitting or underfitting will tend to increase the value of C_p, leading to Mallows' selection criterion.

Akaike's information criterion

Akaike (1973) introduced an information criterion (AIC) based on likelihood theory for parametric models. It is defined by

$$\{-2\log(\text{maximized likelihood}) + 2(\text{number of parameters})\}/n. \tag{1.32}$$

Details of maximum likelihood estimation are given in Section 2.4. Including more input variables in the regression model increases the model's "information" as measured by $2\log(\text{maximized likelihood})$, but it also increases the number of parameters to be estimated. Therefore Akaike's criterion (1.32) uses the penalty term 2(number of parameters) as in Mallows' C_p. The selection procedure is to choose the model with the smallest information criterion. Assuming (A) and (C*) in the linear regression model, the maximum likelihood estimate (MLE) of $\boldsymbol{\beta}$ is the same as the least squares estimate $\widehat{\boldsymbol{\beta}}$, while the MLE of σ^2 is $\widehat{\sigma}_p^2 = \text{RSS}_p/n$, so choosing the regression model with the smallest information criterion (1.32) is equivalent to minimizing

$$\text{AIC}(p) = \log(\widehat{\sigma}_p^2) + 2p/n, \tag{1.33}$$

in which p denotes the number of input variables; see Exercise 2.8 in Chapter 2.

Schwarz's Bayesian information criterion

Section 4.3 describes the Bayesian approach involving posterior distributions as an alternative to likelihood theory for parametric models. Using an approximation to the posterior probability of a regression with p input variables, Schwarz (1978) derived the Bayesian information criterion

$$\text{BIC}(p) = \log(\widehat{\sigma}_p^2) + (p\log n)/n. \tag{1.34}$$

The selection procedure is to choose the model with the smallest BIC. Note that whereas the AIC uses 2 as the penalty factor, the BIC penalizes the number of parameters more heavily by using $\log n$ instead of 2.

1.3.2 Stepwise variable selection

Direct application of a selection criterion to the set of all possible subsets of K variables leads to a large combinational optimization problem that can be very computationally expensive unless K is small. Stepwise methods have been developed to circumvent the computational difficulties. To fix the ideas, we use partial F-statistics for the selection criterion in describing the stepwise methods. The partial F-statistics can be replaced by Mallows' C_p-statistics or other criteria.

Forward selection procedure

To begin, we introduce *partial correlation coefficients* that are used in the forward inclusion of variables. Given response variables y_i, u_i and predictor variables $(x_{i1}, \ldots, x_{ip})$, $1 \le i \le n$, we can regress the y_i (resp. u_i) on $x_{i1}, \ldots, x_{ip}$ and denote the residuals by y_i^* (resp. u_i^*). The correlation coefficient between the y_i^* and u_i^* is called the partial correlation coefficient between y and u adjusted for $x_1, \ldots, x_p$, and is denoted by $r_{yu.1,\ldots,p}$.

To select the regressor that first enters the model, we compute the correlation coefficients r_j of $\{(y_i, x_{ij}), 1 \le i \le n\}$, $j = 1, \ldots, p$, and choose the regressor x_j with the largest r_j. After k regressors have been entered, we compute the partial correlation coefficients between y and the regressors not entered into the model and then include the regressor with the largest partial correlation coefficient. This forward stepwise procedure terminates when the partial F-statistic (see Section 1.3.1) associated with the latest regressor entered is not significantly different from 0.

Backward elimination procedure

Backward elimination begins with the full model (with all K regressors) and computes the partial F-statistic for each regressor. The smallest partial F-statistic F_L is compared with a prespecified cutoff value F^* associated with the α-quantile of the F-distribution, whose α is often chosen to be 0.05 or 0.01. If $F_L \ge F^*$, terminate the elimination procedure and choose the full model. If $F_L < F^*$, conclude that β_L is not significantly different from 0 and remove x_L from the regressor set. With the set of remaining input variables treated as the full regression model at every stage, we can carry out this backward elimination procedure inductively.

Stepwise regression

Stepwise regression procedures are hybrids of forward selection and backward elimination procedures. A computationally appealing stepwise regression procedure consists of forward selection of variables until the latest entered variable is not significantly different from 0, followed by backward elimination.

1.4 Regression diagnostics

Analysis of residuals and detection of influential observations are useful tools to check the validity of certain assumptions underlying the statistical properties of, and inference with, OLS estimates.

1.4.1 Analysis of residuals

The vector $\mathbf{e}$ of residuals $e_i = y_i - \widehat{y}_i = y_i - \widehat{\boldsymbol{\beta}}^T \mathbf{x}_i$ can be expressed as

$$\mathbf{e} = (\mathbf{I} - \mathbf{H})\mathbf{Y}.$$

Note that e_i are not independent; in fact, $\text{Cov}(\mathbf{e}) = \sigma^2(\mathbf{I} - \mathbf{H})$, i.e.,

$$\text{Var}(e_i) = \sigma^2(1 - h_{ii}), \qquad \text{Cov}(e_i, e_j) = -\sigma^2 h_{ij} \quad \text{for } j \neq i, \tag{1.35}$$

where $\mathbf{H} = (h_{ij})_{1 \leq i,j \leq n} = \mathbf{X}(\mathbf{X}^T\mathbf{X})^{-1}\mathbf{X}^T$ is the projection matrix. See Section 2.1.2 for the derivation of (1.35).

Standardized residuals

In view of (1.35), the residuals e_i can be scaled (or standardized) to yield

$$e_i' = \frac{e_i}{s\sqrt{1 - h_{ii}}}, \tag{1.36}$$

where s^2 is the unbiased estimate of σ^2 given by (1.10). The standardized residuals e_i' have zero means and approximately unit variance if the regression model indeed holds.

Jackknife (studentized) residuals

If one residual e_i is much larger than the others, it can deflate all standardized residuals by increasing the variance estimate s^2. This suggests omitting the ith observation $(\mathbf{x}_i, y_i)$ to refit the regression model, with which we can obtain the predicted response $\widehat{y}_{(-i)}$ at $\mathbf{x}_i$ and the estimate $s^2_{(-i)}$ of σ^2. Such an idea leads to the *studentized residuals*

$$e_i^* = \frac{y_i - \widehat{y}_{(-i)}}{\sqrt{\widehat{\text{Var}}(y_i - \widehat{y}_{(-i)})}}, \qquad i = 1, 2. \cdots, n, \tag{1.37}$$

where we use the subscript $(-i)$ to denote "with the ith observation deleted," and

$$\widehat{\text{Var}}(y_i - \widehat{y}_{(-i)}) = s^2_{(-i)} \left\{ 1 + \mathbf{x}_i^T \left[\mathbf{X}^T_{(-i)} \mathbf{X}_{(-i)} \right]^{-1} \mathbf{x}_i \right\}$$

is an estimate of

$$\text{Var}(y_i - \widehat{y}_{(-i)}) = \text{Var}(y_i) + \text{Var}(\widehat{y}_{(-i)}) = \sigma^2 \{ 1 + \mathbf{x}_i^T \left[\mathbf{X}^T_{(-i)} \mathbf{X}_{(-i)} \right]^{-1} \mathbf{x}_i \};$$

see (1.19). It is actually not necessary to refit the model each time that an observation is omitted since

$$e_i^* = e_i' \bigg/ \sqrt{\frac{n-p-(e_i')^2}{n-p-1}}; \tag{1.38}$$

see Exercise 1.4. As graphical analysis of residuals is an effective method of regression diagnostics, one can use the following `R` (or `S`) functions to compute and plot these residuals:

`studres` for studentized residuals,
`stdres` for standardized residuals.

(a) *Normal probability plot.* It is useful to check the assumption of normal ϵ_i that underlies the inference procedures of Section 1.2. A simple method of checking this assumption is to construct a Q-Q plot of the quantiles of studentized residuals versus standard normal quantiles.

Definition 1.5. The *quantile-quantile* (Q-Q) plot of a distribution G versus another distribution F plots the pth quantile y_p of G against the pth quantile x_p of F for $0 < p < 1$.

The `R` (or `S`) function `qqplot(x, y)` plots the quantiles of two samples `x` and `y` against each other. The function `qqnorm(x)` replaces the quantiles of one of the samples by those of a standard normal distribution. Under the normality assumption, the Q-Q plot should lie approximately on a straight line.

(b) *Plot of residuals against the fitted values.* A plot of the residuals versus the corresponding fitted values can be used to detect if the model is adequate. In general, if the plot of residuals spreads uniformly in a rectangular region around the horizontal axis, then there are no obvious model defects. However, if the plot of residuals exhibits other patterns, one can use the patterns to determine how the assumed model can be amended.

(c) *Plot of residuals against the regressors.* In multiple regression, it is also helpful to plot the residuals against each regressor. The patterns displayed in these plots can be used to assess if the assumed relationship between the response and the regressor is adequate.

1.4.2 Influence diagnostics

In fitting a regression model to data, the estimated parameters may depend much more on a few "influential" observations than on the remaining ones. It is therefore useful to identify these observations and evaluate their impact on the model. The hat matrix $\mathbf{H} = \mathbf{X}(\mathbf{X}^T\mathbf{X})^{-1}\mathbf{X}^T$ plays an important role in finding influential observations. The elements of $\mathbf{H}$ are $h_{ij} = \mathbf{x}_i^T(\mathbf{X}^T\mathbf{X})^{-1}\mathbf{x}_j$. Since $\widehat{\mathbf{Y}} = \mathbf{HY}$, $\widehat{y}_i = h_{ii}y_i + \sum_{j \neq i} h_{ij}y_j$, and h_{ii} is called the *leverage* of the ith observation. Cook (1979) proposed to use

$$D_i = \frac{(e_i')^2}{p} \frac{\mathrm{Var}(\widehat{y}_i)}{\mathrm{Var}(e_i)} = \frac{(e_i')^2}{p} \frac{h_{ii}}{1-h_{ii}}, \tag{1.39}$$

where e_i' is the standardized residual (1.36), to measure the influence of the observation y_i. This measure is called *Cook's distance*, and observations with large D_i (e.g., $D_i \geq 4/n$) are considered to be influential on the OLS estimate $\widehat{\boldsymbol{\beta}}$.

1.5 Extension to stochastic regressors

1.5.1 Minimum-variance linear predictors

Consider the problem of minimum-variance prediction of an unobserved random variable Y by a linear predictor of the form

$$\widehat{Y} = a + b_1 X_1 + \cdots + b_q X_q \tag{1.40}$$

in which $X_1, \ldots, X_q$ are observed random variables. The coefficients $b_1, \ldots, b_q$ and a in (1.40) can be determined by minimizing the mean squared error

$$S(a, b_1, \ldots, b_q) = E\big\{[Y - (a + b_1 X_1 + \cdots + b_q X_q)]^2\big\}$$

or, equivalently, by solving the "normal equations"

$$\frac{\partial S}{\partial a} = 0, \qquad \frac{\partial S}{\partial b_j} = 0 \quad (j = 1, \cdots, q), \tag{1.41}$$

which have the explicit solution

$$a = EY - (b_1 EX_1 + \cdots + b_q EX_q), \tag{1.42}$$

$$\begin{pmatrix} b_1 \\ \vdots \\ b_q \end{pmatrix} = \Big(\mathrm{Cov}(X_i, X_j)\Big)^{-1}_{1 \leq i,j \leq q} \begin{pmatrix} \mathrm{Cov}(Y, X_1) \\ \vdots \\ \mathrm{Cov}(Y, X_q) \end{pmatrix}, \tag{1.43}$$

where

$$\mathrm{Cov}(X, Y) = E\{(X - EX)(Y - EY)\} = EXY - (EX)(EY). \tag{1.44}$$

In practice, $\mathrm{Cov}(Y, X_i)$ and $\mathrm{Cov}(X_i, X_j)$ are typically unknown and have to be estimated from the data $(x_{t1}, \ldots, x_{tq}, y_t)$, $t = 1, \ldots, n$. The method of moments replaces EY, EX_j, $\mathrm{Cov}(Y, X_j)$, $\mathrm{Cov}(X_i, X_j)$ in (1.42) and (1.43) by their corresponding sample moments, yielding

$$\widehat{a} = \overline{y} - (\widehat{b}_1 \overline{x}_1 + \cdots + \widehat{b}_q \overline{x}_q), \tag{1.45}$$

$$\begin{pmatrix} \widehat{b}_1 \\ \vdots \\ \widehat{b}_q \end{pmatrix} = \left(\sum_{t=1}^{n} (x_{ti} - \overline{x}_i)(x_{tj} - \overline{x}_j) \right)^{-1}_{1 \le i,j \le q} \begin{pmatrix} \sum_{t=1}^{n} (x_{t1} - \overline{x}_1)(y_t - \overline{y}) \\ \vdots \\ \sum_{t=1}^{n} (x_{tq} - \overline{x}_q)(y_t - \overline{y}) \end{pmatrix}. \tag{1.46}$$

Note that (1.45) and (1.46) give the same estimates as OLS that minimize the residual sum of squares (1.2) in Section 1.1.1.

1.5.2 Futures markets and hedging with futures contracts

A *futures contract* is an agreement between two parties to buy (for the party taking a *long* position) or sell (for the party taking a *short* position) an asset at a future time for a certain price. Futures contracts are traded on an exchange such as the Chicago Board of Trade (CBOT) or the Chicago Mercantile Exchange (CME).

To understand the valuation of a futures contract, we begin with its *over-the-counter* (OTC) counterpart, which is a *forward contract*. The OTC market is an important alternative to exchanges, and OTC trades are carried out between two financial institutions or between a financial institution and its client, often over the phone or a computer network. A forward contract is an agreement between two parties to trade an asset at a prespecified time T in the future at a certain price K, called the *delivery price*. Let P be the price of the asset at current time t, and let r be the risk-free interest rate under continuous compounding. Then the *forward price* of the asset is $Pe^{r(T-t)}$. The *present value* of the forward contract is therefore $P - Ke^{-r(T-t)}$; see Chapter 10 for further details on interest rates and present values.

Although a futures contract is basically the same agreement between two parties as a forward contract, some modifications are needed to standardize such contracts when they are traded on an exchange. Multiple delivery prices are eliminated by revising the price of a contract before the close of each trading day. Therefore, whereas a forward contract is settled at maturity by delivery of the asset or a cash settlement, a futures contract is settled daily, with the settlement price being the price of the contract at the end of each trading day. An important device used to standardize futures contracts is *marking to market*. An investor is required to open a *margin account* with a broker, which is adjusted at the end of each trading day to reflect gains and losses and must contain at least a specified amount of cash. Delivery is made at the futures price at the delivery date, but most futures contracts are closed out prior to their delivery dates; closing out a position means entering into the opposite type of trade from the original one. When the risk-free interest rate is a known function of time, it can be shown that the futures price of a contract with a certain delivery date is the same as the forward price of a contract with the same delivery date; see Hull (2006, pp. 109–110, 127–128).

Many participants in futures markets are *hedgers* who use futures to reduce risk. We focus here on "hedge-and-forget" strategies that take a futures position at the beginning of the life of the hedge and close out the position at the end of the hedge's life. We do not consider dynamic hedging strategies that monitor the hedge closely and make frequent adjustments to eliminate as much risk as possible. The *hedge ratio* is the ratio of the size of the position taken in futures contracts to the size of the exposure. When the asset underlying the futures contract is the same as the asset being hedged, it is natural to use a hedge ratio of 1. *Cross-hedging* occurs when the asset A being hedged differs from the asset B underlying the futures contract. Let ΔP be the change in price of asset A during the life of the hedge and ΔF be the change in futures price of asset B. Let h be the optimal hedge ratio that minimizes $\mathrm{Var}(\Delta P - h\Delta F)$. Since the minimum-variance linear predictor of ΔP based on ΔF is

$$E(\Delta P) + \frac{\mathrm{Cov}(\Delta P, \Delta F)}{\mathrm{Var}(\Delta F)}\{\Delta F - E(\Delta F)\},$$

it follows that $\mathrm{Var}(\Delta P - h\Delta F)$ is minimized by

$$h = \frac{\mathrm{Cov}(\Delta P, \Delta F)}{\mathrm{Var}(\Delta F)},$$

yielding the optimal hedge ratio.

1.5.3 Inference in the case of stochastic regressors

The methods in Sections 1.1.4 and 1.2 assume that the values of the regressors x_{tj} (as in the Gauss-Markov model) are known constants. In applications to finance, however, the x_{tj} are typically random variables. As pointed out in Section 1.5.1, the OLS estimates in this case are method-of-moments estimates. They are *consistent* (i.e., converge to the parameter values in probability as the sample size approaches ∞) and *asymptotically normal* (i.e., (1.11) holds asymptotically) under certain regularity conditions. A crucial assumption is that $\mathbf{x}_t := (x_{t1}, \dots, x_{tp})^T$ is uncorrelated with ϵ_t. In fact, a condition stronger than zero correlation between $\mathbf{x}_t$ and ϵ_t has to be assumed in order to apply the limit theorems of probability theory. One simple assumption is the independence between $\{\mathbf{x}_t\}$ and $\{\epsilon_t\}$. While this assumption is satisfied in many applications, it is too strong for the regression models in financial time series. A weaker assumption is the *martingale difference* assumption

$$E(\epsilon_t | \mathbf{x}_t, \epsilon_{t-1}, \mathbf{x}_{t-1}, \dots, \epsilon_1, \mathbf{x}_1) = 0 \qquad \text{for all } t, \tag{1.47}$$

for which the assumption $E\epsilon_t^2 = \sigma^2$ is replaced by

$$\lim_{t\to 0} E(\epsilon_t^2 | \mathbf{x}_t, \epsilon_{t-1}, \mathbf{x}_{t-1}, \ldots, \epsilon_1, \mathbf{x}_1) = \sigma^2 \text{ with probability 1.} \tag{1.48}$$

If ϵ_t's are independent with $E\epsilon_t = 0$ and $E\epsilon_t^2 = \sigma^2$, and if $\{\epsilon_t\}$ and $\{\mathbf{x}_t\}$ are independent, then (1.47) and (1.48) hold. Since $\mathbf{X}^T\mathbf{X}$ is now a random matrix, statements such as $\text{Cov}(\boldsymbol{\beta}) = \sigma^2(\mathbf{X}^T\mathbf{X})^{-1}$ do not make sense. However, if we assume that

$$(\mathbf{X}^T\mathbf{X})/c_n \text{ converges in probability to a nonrandom matrix } \neq \mathbf{0} \tag{1.49}$$

for some nonrandom constants c_n such that $\lim_{n\to\infty} c_n = \infty$, then the distributional properties in Section 1.1.4 still hold asymptotically under some additional regularity conditions, details of which are given in Appendix A.

Hence we have the following *rule of thumb*: Even when the $\mathbf{x}_t$ are random, we can treat them as if they were nonrandom in constructing tests and confidence intervals for the regression parameters as in Section 1.2, by appealing to asymptotic approximations under certain regularity conditions. The assumption of nonrandom $\mathbf{x}_t$ in Section 1.1.4 has led to precise mean and variance calculations and to deriving exact distributions of the estimates when the ϵ_t are normal. Without assuming $\mathbf{x}_t$ to be nonrandom or ϵ_t to be normal, we can appeal to martingale strong laws and central limit theorems (see Appendix A) to derive similar asymptotic distributions justifying the preceding rule of thumb for approximate inference when the $\mathbf{x}_t$ are random. In Chapter 9, however, we show the importance of the condition (1.49), which basically requires that $\mathbf{X}^T\mathbf{X}$ be asymptotically nonrandom for approximate inference in the case of stochastic regressors. An example of "spurious regression" is also given when such a condition fails to hold.

1.6 Bootstrapping in regression

An alternative to inference based on approximate normal distribution theory for the least squares estimates is to make use of bootstrap resampling. We first describe the bootstrap approach in the case of i.i.d. observations $y_1, \ldots, y_n$ and then extend it to the regression setting in which the observations are $(\mathbf{x}_i, y_i), 1 \leq i \leq n$.

1.6.1 The plug-in principle and bootstrap resampling

Let $y_1, \ldots, y_n$ be independent random variables with common distribution function F. The *empirical distribution* puts probability mass $1/n$ at each y_i, or equivalently the empirical distribution function is given by $\widehat{F}(y) =$

$n^{-1}\sum_{i=1}^{n} 1_{\{y_i \le y\}}$. The "plug-in estimate" of $\theta = g(F)$ is $\widehat{\theta} = g(\widehat{F})$, where g is a functional of F. For example, the method of moments estimates the kth moment of F, for which $g(F) = \int y^k dF(y)$, by the kth sample moment $\int y^k d\widehat{F}(y) = n^{-1}\sum_{i=1}^{n} y_i^k$. For notational simplicity, let $\mathbf{y} = (y_1, \ldots, y_n)$ and also denote $g(\widehat{F})$ by $g(\mathbf{y})$.

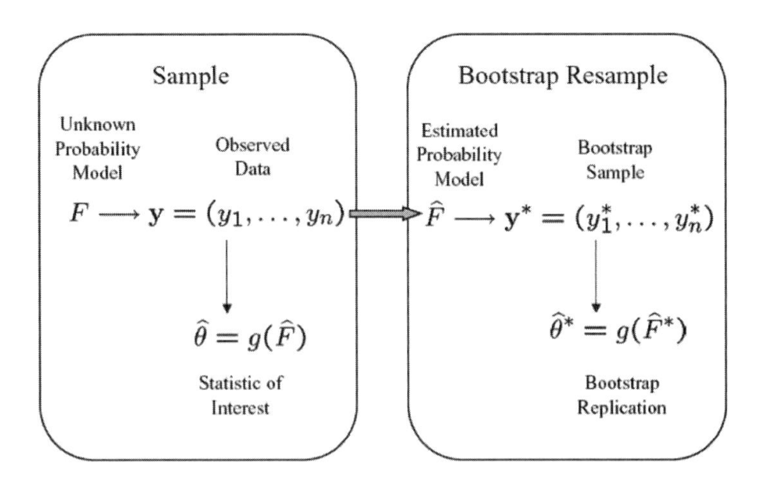

Fig. 1.2. From the observed sample to the bootstrap sample.

Given the empirical distribution function $\widehat{F}$, a *bootstrap sample* $\mathbf{y}^* = (y_1^*, \ldots, y_n^*)$ is obtained by sampling with replacement from $\widehat{F}$ so that the y_i^* are independent and have a common distribution function $\widehat{F}$. This bootstrap sample is used to form a *bootstrap replicate* of $\widehat{\theta}$ via $\widehat{\theta}^* = g(\mathbf{y}^*)$; see Figure 1.2. The sampling distribution of $\widehat{\theta}$ can be estimated by Monte Carlo simulations involving a large number of bootstrap replicates generated from $\widehat{F}$. In particular, the standard deviation of this sampling distribution, called the *standard error* of $\widehat{\theta}$ and denoted by $\text{se}(\widehat{\theta})$, can be estimated as follows:

1. Draw B independent bootstrap samples $\mathbf{y}_1^*, \ldots, \mathbf{y}_B^*$, each consisting of n independent observations from $\widehat{F}$.
2. Evaluate $\widehat{\theta}_b^* = g(\mathbf{y}_b^*)$, $b = 1, \ldots, B$.
3. Compute the average $\bar{\theta}^* = B^{-1}\sum_{b=1}^{B} \widehat{\theta}_b^*$ of the bootstrap replicates $\widehat{\theta}_1^*, \ldots, \widehat{\theta}_B^*$, and estimate the standard error $\text{se}(\widehat{\theta}) = \left(\text{Var}(\widehat{\theta})\right)^{1/2}$ by

$$\widehat{\text{se}}(\widehat{\theta}) = \left\{ \frac{1}{B-1} \sum_{b=1}^{B} (\widehat{\theta}_b^* - \bar{\theta}^*)^2 \right\}^{1/2}. \tag{1.50}$$

The procedure above, which is represented schematically in Figure 1.3, can be extended to other functionals of the sampling distribution of $\widehat{\theta}$. For example, the bias of $\widehat{\theta}$ is $b(\widehat{\theta}) = E(\widehat{\theta}) - \theta$, and the bootstrap estimate of the bias is $\widehat{b}(\widehat{\theta}) = \bar{\theta}^* - \widehat{\theta}$.

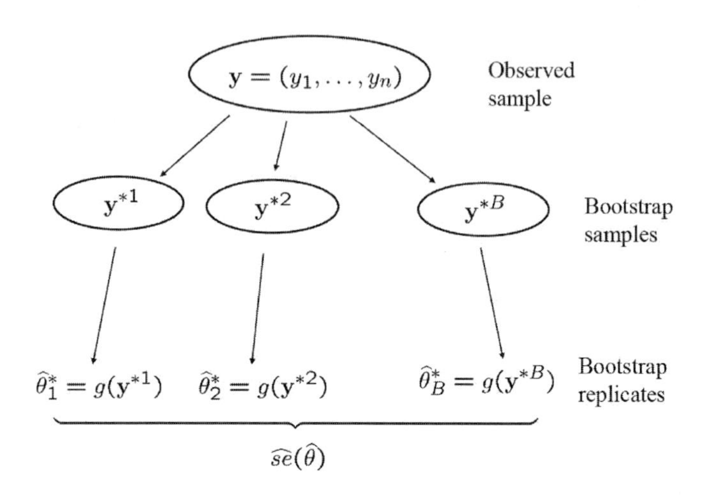

Fig. 1.3. Bootstrap estimate of standard error.

1.6.2 Bootstrapping regression models

For the linear regression model $y_i = \boldsymbol{\beta}^T \mathbf{x}_i + \epsilon_i$ $(i = 1, \ldots, n)$, there are two bootstrap approaches. In the first approach, which is called *bootstrapping residuals*, the regressors $\mathbf{x}_i = (x_{i1}, \ldots, x_{ip})^T$ are assumed to be fixed (i.e., they do not vary across samples), and the sampling variability of the least squares estimate is due to the random disturbances ϵ_i, which are assumed to be independent and have the same distribution F with mean 0. The distribution F is estimated by the empirical distribution $\widehat{F}$ of the centered residuals $\widehat{\epsilon}_i = e_i - \bar{e}$, where $e_i = y_i - \widehat{y}_i$ and $\bar{e} = n^{-1} \sum_{i=1}^n e_i$. Note that $\bar{e} = 0$ if the regression model has an intercept term. Bootstrapping residuals involves (i) defining $y_i^* = \widehat{\boldsymbol{\beta}}^T \mathbf{x}_i + \epsilon_i^*$ $(i = 1, \ldots, n)$ from a bootstrap sample $(\epsilon_1^*, \ldots, \epsilon_n^*)$ drawn with replacement from $\widehat{F}$ and (ii) computing the OLS estimate $\widehat{\boldsymbol{\beta}}^*$ based on $(\mathbf{x}_1, y_1^*), \ldots, (\mathbf{x}_n, y_n^*)$. It uses the empirical distribution of B bootstrap replicates $\widehat{\boldsymbol{\beta}}_1^*, \ldots, \widehat{\boldsymbol{\beta}}_B^*$ to estimate the sampling distribution of $\widehat{\boldsymbol{\beta}}$.

Another bootstrap approach, which is called *bootstrapping pairs*, assumes that $\mathbf{x}_i$ are independent and identically distributed (i.i.d.). Since the ϵ_i are assumed to be i.i.d., the pairs $(\mathbf{x}_i, y_i)$ are also i.i.d. Their common distribution Ψ can be estimated by the empirical distribution $\widehat{\Psi}$. Bootstrapping pairs involves drawing B bootstrap samples $\{(\mathbf{x}_{i,b}^*, y_{i,b}) : 1 \le i \le n\}$ from $\widehat{\Psi}$ and computing the OLS estimate $\widehat{\boldsymbol{\beta}}_b^*$ from the bth bootstrap sample, $1 \le b \le B$.

1.6.3 Bootstrap confidence intervals

A function of the data for which the sampling distribution does not depend on the unknown F is called a *pivot*. As shown in Appendix A, an approximate pivot is $(\widehat{\theta}-\theta)/\mathrm{se}(\widehat{\theta})$, which is asymptotically standard normal (as $n \to \infty$) but may deviate substantially from normality for the finite sample size actually used. Typically $\mathrm{se}(\widehat{\theta})$ involves unknown parameters and needs to be estimated. Let $\widehat{\mathrm{se}}$ be a consistent estimate of the standard error so that $(\widehat{\theta}-\theta)/\widehat{\mathrm{se}}$ is also an approximate pivot. From B bootstrap samples $\mathbf{y}_1^*, \ldots, \mathbf{y}_B^*$, we can compute the quantiles of

$$Z_b^* = (\widehat{\theta}_b^* - \widehat{\theta})/\widehat{\mathrm{se}}_b^*, \qquad b = 1, \ldots, B,$$

where $\widehat{\mathrm{se}}_b^*$ is the estimated standard error of $\widehat{\theta}_b^*$ based on the bootstrap sample $\mathbf{y}_b^*$. Let $\widehat{t}_\alpha$ and $\widehat{t}_{1-\alpha}$ be the αth and $(1-\alpha)$th quantiles of $\{Z_b^*, 1 \le b \le B\}$. The *bootstrap-t* interval, with confidence level $1-2\alpha$, is

$$(\widehat{\theta} - \widehat{t}_{1-\alpha}\widehat{\mathrm{se}}, \widehat{\theta} - \widehat{t}_\alpha \widehat{\mathrm{se}}). \tag{1.51}$$

1.7 Generalized least squares

The assumption that the random disturbances ϵ_i have the same variance σ^2 in Section 1.1.4 may be too restrictive in econometric studies. For example, if Y denotes a firm's profit and X is some measure of the firm's size, $\mathrm{Var}(Y)$ is likely to increase with X. Random disturbances with nonconstant variances are called *heteroskedastic* and often arise in cross-sectional studies in which one only has access to data that have been averaged within groups of different sizes n_i $(i = 1, \cdots, m)$. Besides heteroskedasticity, the assumption of uncorrelated ϵ_i may also be untenable when the Y_t are computed via moving averages, as in the case where the rate of wage change is determined by $Y_t = (w_t - w_{t-4})/w_{t-4}$ in quarter t from the wage indices w_t and w_{t-4}. These considerations have led to replacing the assumption $\mathrm{Cov}(\mathbf{Y}) = \sigma^2\mathbf{I}$ in Section 1.1.4 by

$$\mathrm{Cov}(\mathbf{Y}) = \mathbf{V}, \tag{1.52}$$

where $\mathbf{V}$ is a symmetric and positive definite matrix.

Whereas the OLS estimate $\widehat{\boldsymbol{\beta}} = (\mathbf{X}^T\mathbf{X})^{-1}\mathbf{X}^T\mathbf{Y}$ does not involve the covariance structure of $\mathbf{Y}$, we can incorporate (1.52) by modifying OLS as

$$\widehat{\boldsymbol{\beta}}_{\mathrm{GLS}} = (\mathbf{X}^T\mathbf{V}^{-1}\mathbf{X})^{-1}\mathbf{X}^T\mathbf{V}^{-1}\mathbf{Y}, \tag{1.53}$$

which is called the *generalized least squares* (GLS) estimator. Moreover, $\widehat{\boldsymbol{\beta}}_{\text{GLS}}$ is unbiased and

$$\text{Cov}(\widehat{\boldsymbol{\beta}}_{\text{GLS}}) = (\mathbf{X}^T\mathbf{V}^{-1}\mathbf{X})^{-1} \tag{1.54}$$

under assumption (1.52). Note that for the special case $\mathbf{V} = \sigma^2\mathbf{I}$, the right-hand side of (1.54) reduces to $\sigma^2(\mathbf{X}^T\mathbf{X})^{-1}$, which agrees with (1.9).

To prove (1.53) and (1.54), we use a result from Section 2.2: For a symmetric and positive definite matrix $\mathbf{V}$, there exists a symmetric and positive definite matrix $\mathbf{P}$ such that $\mathbf{PP} = \mathbf{V}$; i.e., $\mathbf{P} = \mathbf{V}^{1/2}$. Multiplying the regression model $\mathbf{Y} = \mathbf{X}\boldsymbol{\beta} + \boldsymbol{\epsilon}$ by $\mathbf{P}^{-1}$ yields

$$\mathbf{P}^{-1}\mathbf{Y} = \mathbf{P}^{-1}\mathbf{X}\boldsymbol{\beta} + \mathbf{u}, \tag{1.55}$$

where $\mathbf{u} = \mathbf{P}^{-1}\boldsymbol{\epsilon}$ has covariance matrix $\mathbf{P}^{-1}\text{Cov}(\boldsymbol{\epsilon})\mathbf{P}^{-1} = \mathbf{P}^{-1}\mathbf{PPP}^{-1} = \mathbf{I}$. Thus the model (1.55) has $\text{Cov}(\mathbf{u}) = \mathbf{I}$, for which the OLS estimate is of the form

$$[(\mathbf{P}^{-1}\mathbf{X})^T(\mathbf{P}^{-1}\mathbf{X})]^{-1}(\mathbf{P}^{-1}\mathbf{X})^T\mathbf{P}^{-1}\mathbf{Y} = (\mathbf{X}^T\mathbf{P}^{-1}\mathbf{P}^{-1}\mathbf{X})^{-1}\mathbf{X}^T\mathbf{P}^{-1}\mathbf{P}^{-1}\mathbf{Y},$$

which is the same as $\widehat{\boldsymbol{\beta}}_{\text{GLS}}$ in (1.53) since $\mathbf{P}^{-1}\mathbf{P}^{-1} = (\mathbf{PP})^{-1} = \mathbf{V}^{-1}$. Therefore, using the transformation (1.55), GLS can be transformed to OLS and thus shares the same properties of OLS after we replace $\mathbf{X}$ by $\mathbf{P}^{-1}\mathbf{X}$. In particular, (1.54) follows from (1.9) after this transformation.

The specification of $\mathbf{V}$ may involve background knowledge of the data and the empirical study, as noted in the first paragraph of this section. It may also arise from examination of the residuals in situations where there are multiple observations $(\mathbf{x}_t, y_t)$ at distinct input values. In this case, heteroskedasticity can be revealed from the residual plots and unbiased estimates of $\text{Var}(y_t)$ at each distinct input value can be obtained from the multiple observations there. The form of $\mathbf{V}$ may also arise as part of the model, as in time series modeling of asset returns and their volatilities in Chapter 6, where Section 6.4 uses maximum likelihood to estimate both the regression parameters and the parameters of $\mathbf{V}$.

1.8 Implementation and illustration

To implement the methods described in this chapter, one can use the following functions in R or Splus:

```
lm(formula, data, weights, subset, na.action)
predict.lm(object, newdata, type)
bootstrap(data, statistic, B)
step(object, scope, scale, direction)
```

For details, see Venables and Ripley (2002, pp. 139–163). Alternatively one can use the Matlab function `regress`. We illustrate the application of these methods in a case study that relates the daily log returns of the stock of Microsoft Corporation to those of several computer and software companies; see Section 3.1 for the definition and other details of asset returns.

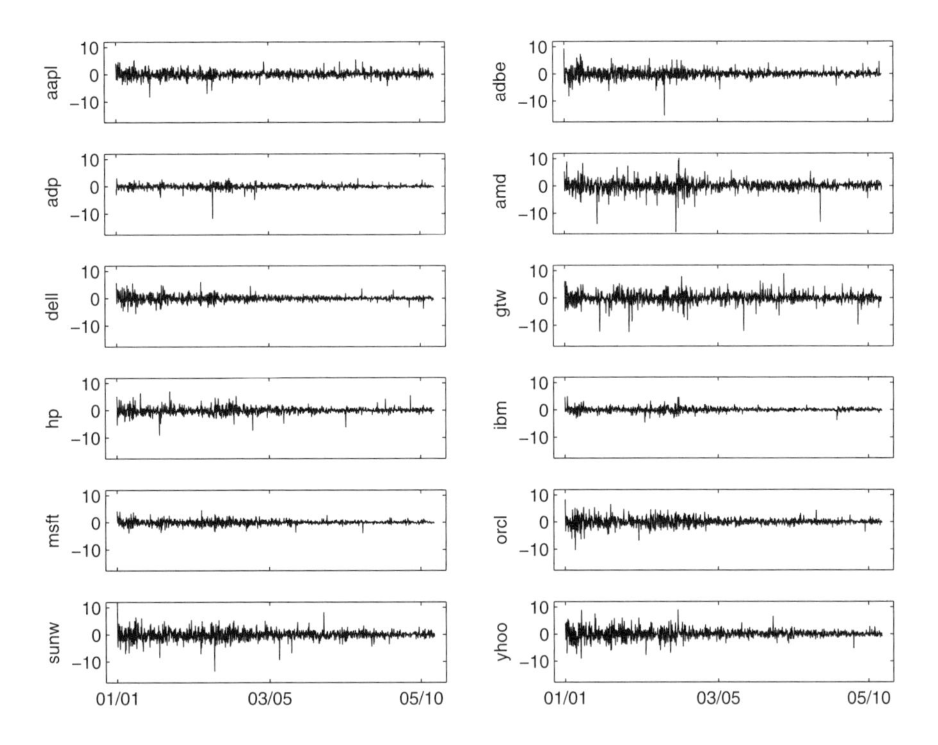

Fig. 1.4. The daily log returns of Apple Computer, Adobe Systems, Automatic Data Processing, Advanced Micro Devices, Dell, Gateway, Hewlett-Packard Company, International Business Machines Corp., Microsoft Corp., Orcale Corp., Sun Microsystems, and Yahoo! stocks from January 3, 2001 to December 30, 2005.

Microsoft Corp. (`msft`) is a leader in the application software industry. We collected a set of firms whose stock returns may be strongly correlated with `msft`. These firms include Apple Computer (`aapl`), Adobe Systems (`adbe`), Automatic Data Processing (`adp`), Dell (`dell`), Gateway (`gtw`), Hewlett-Packard Company (`hp`), International Business Machines Corp. (`ibm`), Orcale Corp. (`orcl`), Sun Microsystems (`sunw`), and Yahoo! (`yhoo`). Figure 1.4 plots the daily log returns of the stocks of these firms from January 3, 2001 to December 30, 2005. The sample size is $n = 1255$.

The correlation matrix of the daily log returns in Table 1.2 shows substantial correlations among these stocks. Figure 1.5 gives a matrix of scatterplots showing all pairs of stocks. The scatterplots also reveal a strong relationship between msft and each predictor and a strong pairwise relationship among the predictors.

Table 1.2. Pairwise correlation coefficients of daily log returns.

	aapl	adbe	adp	amd	dell	gtw	hp	ibm	msft	orcl	sunw
adbe	.387										
adp	.285	.305									
amd	.448	.382	.310								
dell	.515	.486	.316	.470							
gtw	.355	.266	.195	.347	.368						
hp	.431	.425	.342	.429	.528	.380					
ibm	.430	.451	.387	.445	.532	.282	.477				
msft	.479	.526	.366	.457	.620	.375	.493	.603			
orcl	.407	.453	.319	.385	.510	.284	.446	.539	.587		
sunw	.425	.406	.273	.440	.511	.369	.472	.472	.455	.517	
yhoo	.422	.497	.322	.422	.499	.279	.422	.441	.493	.469	.414

Regression for full model

We use OLS to fit a linear regression model to the returns data. The regression coefficient estimates and their standard errors, t-statistics, and p-values are shown in Table 1.3. The p-values in Table 1.3 measure the effect of dropping that stock from the regression model; a p-value less than 0.01 shows the regression coefficient of the corresponding stock to be significantly nonzero by the t-test in (1.22). We can use Table 1.3 together with (1.11) to construct confidence intervals of the regression coefficients. For example, a 95% confidence interval of the regression coefficient for dell is $0.1771 \pm t_{1243,0.025} 0.0213$.

Variable selection

Starting with the full model, we find that the stocks hp and sunw, with relatively small partial F-statistics, are not significant at the 5% significance

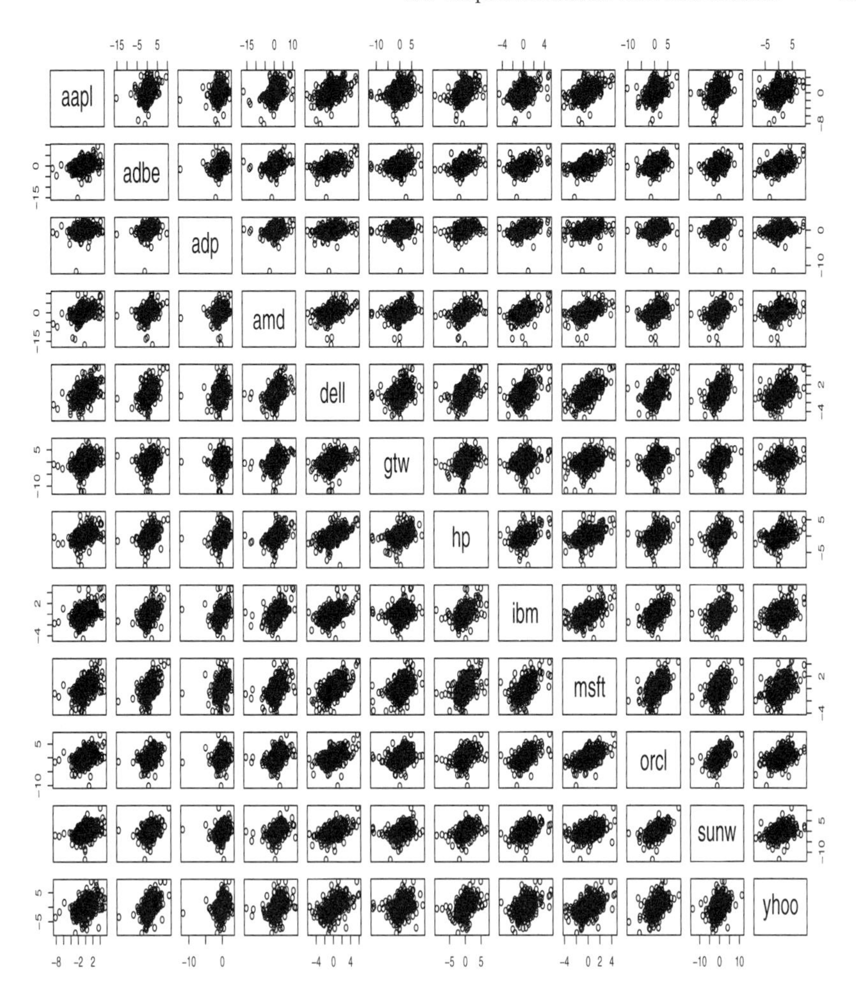

Fig. 1.5. Scatterplot matrix of daily log returns.

level. If we set the cutoff value at $F^* = 10$, which corresponds to a significance level< 0.01, then `hp` and `sunw` are removed from the set of predictors after the first step. We then refit the model with the remaining predictors and repeat the backward elimination procedure with the cutoff value $F^* = 10$ for the partial F-statistics. Proceeding stepwise in this way, the procedure terminates with six predictor variables: `aapl`, `adbe`, `dell`, `gtw`, `ibm`, `orcl`. The variables that are removed in this stepwise procedure are summarized in Table 1.4.

Table 1.5 gives the regression coefficients of the selected model

$$\texttt{msft} = \beta_0 + \beta_1\texttt{aapl} + \beta_2\texttt{adbe} + \beta_3\texttt{dell} + \beta_4\texttt{gtw} + \beta_5\texttt{ibm} + \beta_6\texttt{orcl} + \epsilon. \quad (1.56)$$

Table 1.3. Regression coefficients of the full model

Term	Estimate	Std. Error	t-statistic	p-value
Intercept	0.0028	0.0163	0.170	0.8649
aapl	0.0418	0.0162	2.573	0.0102
adbe	0.0782	0.0139	5.639	0.0000
adp	0.0489	0.0227	2.158	0.0311
amd	0.0192	0.0107	1.796	0.0727
dell	0.1771	0.0213	8.325	0.0000
gtw	0.0396	0.0099	3.991	0.0000
hp	0.0270	0.0177	1.526	0.1273
ibm	0.2302	0.0275	8.360	0.0000
orcl	0.1264	0.0154	8.204	0.0000
sunw	−0.0206	0.0121	−1.696	0.0902
yhoo	0.0258	0.0124	2.085	0.0372

RSS = 411.45, d.f.= 1243, $s = 0.575$, adjusted $R^2 = 56.9\%$.

Table 1.4. Stepwise variable selection.

Step	1	2	3
Variables removed	sunw, hp	amd	adp, yhoo

The selected model shows that, in the collection of stocks we studied, the `msft` daily log return is strongly influenced by those of its competitors. Instead of employing backward elimination using partial F-statistics, we can proceed with stepwise forward selection using the AIC. Applying the `R` or `Splus` function `stepAIC` with 6 as the maximum number of variables to be included, `stepAIC` also chooses the regression model (1.56).

Regression diagnostics

For the selected model (1.56), Figure 1.6 shows diagnostic plots from `R`. The top panels give the plots of residuals versus fitted values and the Q-Q plot. The

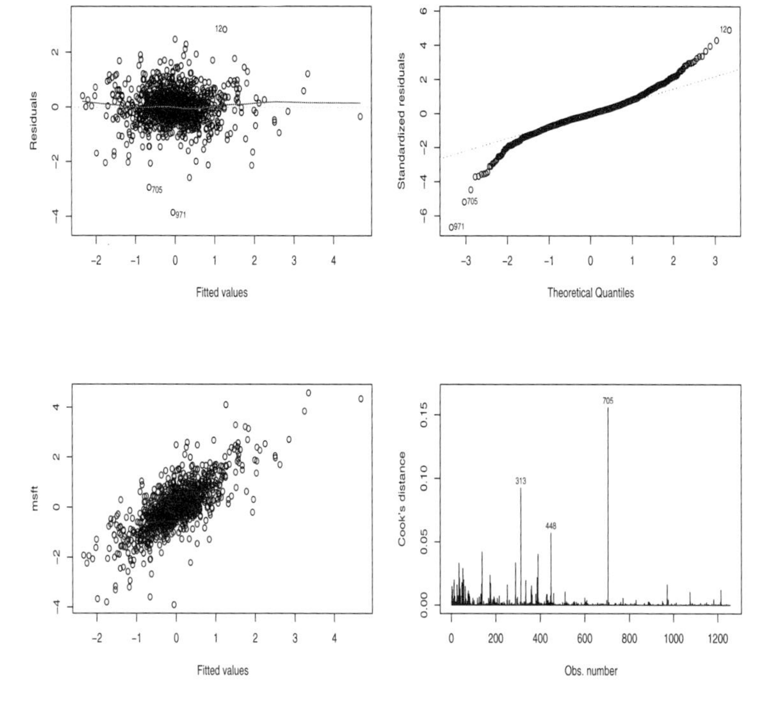

Fig. 1.6. Diagnostic plots of the fitted regression model (1.56).

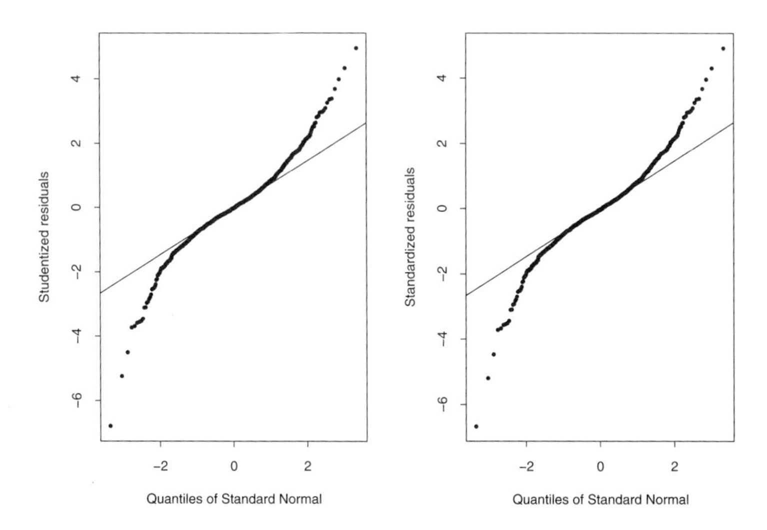

Fig. 1.7. Q-Q plots of studentized (left panel) and standardized (right panel) residuals.

Table 1.5. Regression coefficients of the selected regression model.

Term	Estimate	Std. Error	t-statistics	p-value
Intercept	0.0040	0.0164	0.2406	0.8099
aapl	0.0536	0.0159	3.3713	0.0008
adbe	0.0911	0.0134	6.7822	0.0000
dell	0.1920	0.0205	9.3789	0.0000
gtw	0.0437	0.0097	4.5218	0.0000
ibm	0.2546	0.0266	9.5698	0.0000
orcl	0.1315	0.0149	8.8461	0.0000

RSS $= 418.20$, d.f. $= 1248$, $s = 0.579$, adjusted $R^2 = 56.6\%$.

bottom panels plot (i) the `msft` log returns versus fitted values and (ii) Cook's distance for each observation, which identifies the 313th (April 8, 2002) and the 705th (October 23, 2003) observations as outliers that have substantial influence on the estimated regression coefficients.

Figure 1.7 replaces the residuals in the Q-Q plot (the left panel in the second row) of Figure 1.6 by studentized and standardized residuals, respectively. The Q-Q plots of the studentized and standardized residuals are quite similar. The tails in these Q-Q plots show substantial departures from the assumption of normal errors. The nonnormality of stock returns will be discussed further in Section 6.1.

Exercises

1.1. Given observations (x_i, y_i), $i = 1, \cdots n$, fit a quadratic polynomial regression model

$$y_i = f(x_i) + \epsilon_i, \quad \text{where } f(x) = \alpha + \beta_1 x + \beta_2 x^2.$$

Construct a 95% confidence interval for $f(x)$ at a given x.

1.2. Consider the linear regression model

$$y_i = \beta_0 + \beta_1 x_{i1} + \beta_2 x_{i2} + \beta_3 x_{i3} + \epsilon_i, \quad 1 \le i \le n.$$

Using the F-test of a general linear hypothesis, show how to test the following hypotheses on the regression coefficients:

(a) $H_0 : \beta_0 = \beta_2$.
(b) $H_0 : \beta_1 + \beta_2 = 3\beta_3$.

1.3. In the Gauss-Markov model $y_i = \alpha + \beta x_i + \epsilon_i$ $(i = 1, \cdots, n)$, let $\widehat{\alpha}$ and $\widehat{\beta}$ be the OLS estimates of α and β, and let $\bar{y} = n^{-1} \sum_{i=1}^{n} y_i$.
(a) Show that $\bar{y}$ and $\widehat{\beta}$ are uncorrelated.
(b) If $\widehat{y}_0 = \widehat{\alpha} + \widehat{\beta} x_0$, show that

$$\mathrm{Var}(\widehat{y}_0) = \left(\frac{1}{n} + \frac{(x_0 - \bar{x})^2}{\sum_{i=1}^{n} (x_i - \bar{x})^2} \right) \sigma^2.$$

1.4. Using the same notation as in Section 1.4.1 on studentized residuals, prove (1.38) by proceeding in the following steps.
(a) Let $\mathbf{A}$ be a $p \times p$ nonsingular matrix and $\mathbf{U}$ and $\mathbf{V}$ be $p \times 1$ vectors. The matrix inversion lemma represents $(\mathbf{A} + \mathbf{U}\mathbf{V}^T)^{-1}$ in terms of $\mathbf{A}^{-1}$ as follows:

$$(\mathbf{A} + \mathbf{U}\mathbf{V}^T)^{-1} = \mathbf{A}^{-1} - \mathbf{A}^{-1}\mathbf{U}(\mathbf{I} + \mathbf{V}^T\mathbf{A}^{-1}\mathbf{U})^{-1}\mathbf{V}^T\mathbf{A}^{-1}.$$

Use the matrix inversion lemma to prove

$$(\mathbf{X}_{(-i)}^T \mathbf{X}_{(-i)})^{-1} = (\mathbf{X}^T\mathbf{X})^{-1} + \frac{(\mathbf{X}^T\mathbf{X})^{-1}\mathbf{x}_i\mathbf{x}_i^T(\mathbf{X}^T\mathbf{X})^{-1}}{1 - h_{ii}},$$

$$\mathbf{x}_i^T (\mathbf{X}_{(-i)}^T \mathbf{X}_{(-i)})^{-1} \mathbf{x}_i = \frac{h_{ii}}{1 - h_{ii}}.$$

(b) Show that, after the deletion of $(\mathbf{x}_i, y_i)$, we have

$$\widehat{\boldsymbol{\beta}}_{(-i)} = \widehat{\boldsymbol{\beta}} - \frac{(\mathbf{X}^T\mathbf{X})^{-1}\mathbf{x}_i}{1 - h_{ii}} e_i,$$

where $e_i = y_i - \mathbf{x}_i^T \widehat{\boldsymbol{\beta}}$. Letting $\widehat{e}_{(-i)} = y_i - \mathbf{x}_i^T \widehat{\boldsymbol{\beta}}_{(-i)}$ and $e_i' = e_i / (s\sqrt{1 - h_{ii}})$, derive the representations

$$\widehat{e}_{(-i)} = \frac{e_i}{1 - h_{ii}} = \frac{e_i'}{\sqrt{1 - h_{ii}}} s, \qquad s_{(-i)}^2 = \frac{n - p - (e_i')^2}{n - p - 1} s^2.$$

(c) Show that $\widehat{\mathrm{Var}}(\widehat{e}_{(-i)}) = s_{(-i)}^2 / (1 - h_{ii})$, and hence prove (1.38).

1.5. The log return of a stock at week t is defined as $r_t = \log(P_t / P_{t-1})$, where P_t is the stock price at week t. Consider the weekly log returns of Citigroup Inc., Pfizer Inc., and General Motors from the week of January 4, 1982 to the week of May 21, 2007 in the file `w_logret_3stocks.txt`.
(a) Plot the weekly log returns of each stock over time. A widely held assumption in mathematical models of stock prices is that the

corresponding log returns are independent and identically distributed (i.i.d.); see Section 3.1. Do your plots show departures from this assumption?

(b) For each stock, show the Q-Q plot of the weekly log returns and thereby check the assumption that they are normally distributed.

(c) For every pair of stocks, estimate the correlation coefficient of the weekly log returns from these data, and give a bootstrap estimate of its standard error.

1.6. The file `w_logret_3stocks.txt` contains the weekly log returns of Pfizer stock from the week of January 4, 1982 to the week of May 21, 2007.

(a) Plot these data over time. Does the plot show deviations from the i.i.d. assumption? Consider in particular the 897th week (which is the week of March 8, 1999). Are the returns before and after that week markedly different?

(b) Fit the following regression model to these data:

$$r_t = \mu_1 \mathbf{1}_{\{t<t_0\}} + \mu_2 \mathbf{1}_{\{t \geq t_0\}} + \epsilon_t \tag{1.57}$$

with $t_0 = 897$. Provide 95% confidence intervals for μ_1 and μ_2.

(c) Test the null hypothesis $\mu_1 = \mu_2$.

1.7. The file `w_logret_3automanu.txt` contains the weekly log returns of three auto manufacturers, General Motors Corp., Ford Motor Corp., and Toyota Motor Corp., from the week of January 3, 1994 to the week of June 25, 2007. Consider the regression model

$$\texttt{gm} = \beta_0 + \beta_1 \texttt{toyota} + \beta_2 \texttt{ford} + \epsilon.$$

(a) Estimate β_0, β_1, and β_2 by least squares, and give 95% confidence intervals for these parameters.

(b) Construct a 95% confidence region for $\boldsymbol{\beta} = (\beta_0, \beta_1, \beta_2)^T$.

(c) Test the null hypothesis $\beta_1 = 0$.

(d) Check the adequacy of the regression model with plots of the standardized and studentized residuals.

(e) Calculate the leverage and Cook's distance for each observation. What do these numbers show?

(f) Use bootstrapping to estimate the standard errors of the OLS estimates of the regression coefficients. Compare your results with those in (a).

1.8. The file `d_nasdaq_82stocks.txt` contains the daily log returns of the NASDAQ Composite Index and 82 stocks from January 3, 1990 to December 29, 2006. We want to track the returns of NASDAQ by using a small number of stocks from the given 82 stocks.

(a) Construct a full regression model.

(b) Use partial F-statistics and backward elimination to select variables from the full regression model in (a). Write down the selected model.
(c) Compare the full and selected models. Summarize your comparison in an ANOVA table.
(d) For the selected regression model in (b), perform residual diagnostics.
(e) If you can only use at most five stocks to track the daily NASDAQ log returns, describe your model selection procedure and your constructed model.

2

Multivariate Analysis and Likelihood Inference

Statistical analysis of multivariate data arises in many empirical studies in finance and econometrics. A classical example, considered in Chapter 3, is the implementation and statistical analysis of Markowitz's optimal portfolio theory based on historical data, which are multivariate, on the mean levels and the covariance matrix of different assets that are used to form the portfolios. Another important example, studied in Chapter 10, involves multivariate data of bond yields over different maturities.

A generic form of multivariate data is $\mathbf{X}_1, \ldots, \mathbf{X}_n$, in which n is the number of observations and the ith observation $\mathbf{X}_i$ is a vector $(X_{i1}, \ldots, X_{ik})^T$. For example, for bond yields over different maturities, X_{ij} is the yield of a zero-coupon bond with maturity m_j on the ith trading day in the past. Stochastic models for the data therefore involve the distribution of $\mathbf{X}_i$. Section 2.1 reviews the concept of joint and marginal distributions of components of $\mathbf{X}_i$. The special case of jointly normal random variables is considered in Section 2.3, in which the multivariate normal distribution is generalized to the k-dimensional normal distribution by replacing $(x-\mu)^2/\sigma^2$ by $(\mathbf{x}-\boldsymbol{\mu})^T\mathbf{V}^{-1}(\mathbf{x}-\boldsymbol{\mu})$ and $\sqrt{2\pi\sigma^2}$ by $(2\pi)^{-k/2}\{\det(\mathbf{V})\}^{-1/2}$, where $\mathbf{V}$ denotes the covariance matrix of the normal random vector $\mathbf{X}$. Important properties of the multivariate normal distribution are given in Section 2.3. In particular, it is shown that the marginal and conditional distributions of subvectors of a normal vector $\mathbf{X}$ are normal and that uncorrelated components of $\mathbf{X}$ are independent. Making use of these properties, we derive in Section 2.3 the distribution theory of the OLS estimates listed in Section 1.1 under conditions (A) and (C*). Whereas the chi-square distribution (1.12) is related to the unbiased estimate s^2 of σ^2, Section 2.3 also considers the multivariate analog in which σ^2 is generalized to the covariance matrix $\mathbf{V}$ and the chi-square distribution is generalized to the Wishart distribution.

Although it may appear from the preceding paragraph that multivariate statistical analysis simply extends univariate analysis by using analogous vectors and matrices, an important practical issue in multivariate analysis is dimension reduction. If the dimension k of $\mathbf{X}_i$ is not small compared with n, then even though n may be large, the data $\mathbf{X}_1, \ldots, \mathbf{X}_n$ are still relatively sparse in $\mathbb{R}^k$ unless they tend to cluster around lower-dimensional subspaces. Section 2.2 provides an important technique, called *principal component analysis*, to uncover these low-dimensional subspaces. Other dimension reduction techniques and more advanced topics in multivariate analysis are given in Chapter 9.

Section 2.4 describes a powerful and widely applicable approach to parametric estimation in multivariate statistical models. It is based on the *likelihood function*, which is the joint density function of the data regarded as a function only of the unknown parameters. Maximum likelihood estimation, likelihood ratio tests, and their asymptotic (large-sample) theory are introduced not only for the classical setting of independent observations sampled from a population distribution but also for more general stochastic models of financial data. The parametric bootstrap is also introduced as an alternative or supplement to the asymptotic methods in likelihood inference.

2.1 Joint distribution of random variables

The distribution of a random vector $\mathbf{X} = (X_1, \ldots, X_k)^T$ is characterized by its *joint density function* f such that

$$P\{X_1 \le a_1, \ldots, X_k \le a_k\} = \int_{-\infty}^{a_k} \cdots \int_{-\infty}^{a_1} f(x_1, \ldots, x_k) dx_1 \cdots dx_k \tag{2.1}$$

when $\mathbf{X}$ has a differentiable distribution function (which is the left-hand side of (2.1)), and for the case of discrete X_i,

$$f(a_1, \ldots, a_k) = P\{X_1 = a_1, \ldots, X_k = a_k\}.$$

Given the joint density function f of $\mathbf{X}$, the density function f_i of X_i, called the *marginal density function*, can be obtained from f by

$$f_i(x_i) = \int \cdots \int f(x_1, \ldots, x_k) dx_1 \cdots dx_{i-1} dx_{i+1} \cdots dx_k$$

in the continuous case and by summing over the other components of $\mathbf{X}$ in the discrete case. More generally, for any $1 \le i_1 < \cdots < i_j \le k$, we can obtain from f the joint density function $f_{i_1,\ldots,i_j}$ of the subvector $(X_{i_1}, \ldots, X_{i_j})^T$ of $\mathbf{X}$ by summing or integrating over the other components of $\mathbf{X}$. The random variables $X_1, \ldots, X_k$ are *independent* if

$$f(x_1, \ldots, x_k) = \prod_{i=1}^{k} f_i(x_i)$$

for all $x_1, \ldots, x_k$. In general,

$$f(x_1, \ldots, x_k) = f_1(x_1) \prod_{i=2}^{k} f_i(x_i | x_1, \ldots, x_{i-1}),$$

where $f_i(x_i | x_1, \ldots, x_{i-1})$ is the *conditional density function* of X_i given $(X_1, \ldots, X_{i-1}) = (x_1, \ldots, x_{i-1})$.

2.1.1 Change of variables

Suppose $\mathbf{X}$ has a differentiable distribution function with density function $f_{\mathbf{X}}$. Let $\mathbf{g} : \mathbb{R}^k \to \mathbb{R}^k$ be continuously differentiable and one-to-one. Let $\mathbf{Y} = \mathbf{g}(\mathbf{X})$. Since g is a one-to-one transformation, we can solve $\mathbf{Y} = \mathbf{g}(\mathbf{X})$ to obtain $\mathbf{X} = \mathbf{h}(\mathbf{Y})$; $\mathbf{h} = (h_1, \ldots, h_k)^T$ is the inverse of the transformation $\mathbf{g}$. The density function $f_{\mathbf{Y}}$ of $\mathbf{Y}$ is given by

$$f_{\mathbf{Y}}(\mathbf{y}) = f_{\mathbf{X}}(\mathbf{h}(\mathbf{y})) |\det(\partial \mathbf{h} / \partial \mathbf{y})|, \tag{2.2}$$

where $\partial \mathbf{h} / \partial \mathbf{y}$ is the Jacobian matrix $(\partial h_i / \partial y_j)_{1 \le i,j \le k}$. The inverse function theorem of multivariate calculus gives $\partial \mathbf{h} / \partial \mathbf{y} = (\partial \mathbf{g} / \partial \mathbf{x})^{-1}$, or equivalently $\partial \mathbf{x} / \partial \mathbf{y} = (\partial \mathbf{y} / \partial \mathbf{x})^{-1}$.

2.1.2 Mean and covariance matrix

The mean vector of $\mathbf{Y} = (Y_1, \ldots, Y_k)^T$ is $E(\mathbf{Y}) = (EY_1, \ldots, EY_k)^T$. The covariance matrix of $\mathbf{X}$ is

$$\operatorname{Cov}(\mathbf{Y}) = (\operatorname{Cov}(Y_i, Y_j))_{1 \le i,j \le k} = E\{(\mathbf{Y} - E(\mathbf{Y}))(\mathbf{Y} - E(\mathbf{Y}))^T\}.$$

The following linearity (and bilinearity) properties hold for $E\mathbf{Y}$ (and $\operatorname{Cov}(\mathbf{Y})$). For nonrandom $p \times k$ matrix $\mathbf{A}$ and $p \times 1$ vector $\mathbf{B}$,

$$E(\mathbf{A}\mathbf{Y} + \mathbf{B}) = \mathbf{A}E\mathbf{Y} + \mathbf{B}, \tag{2.3}$$

$$\operatorname{Cov}(\mathbf{A}\mathbf{Y} + \mathbf{B}) = \mathbf{A}\operatorname{Cov}(\mathbf{Y})\mathbf{A}^T. \tag{2.4}$$

We next apply (2.3) and (2.4) to derive some of the statistical properties of the OLS estimate $\widehat{\boldsymbol{\beta}} = (\mathbf{X}^T\mathbf{X})^{-1}\mathbf{X}^T\mathbf{Y}$ in the regression model $\mathbf{Y} = \mathbf{X}\boldsymbol{\beta} + \boldsymbol{\epsilon}$ of Section 1.1. Under (A) and (B), since $E\boldsymbol{\epsilon} = \mathbf{0}$ and $\widehat{\boldsymbol{\beta}} = (\mathbf{X}^T\mathbf{X})^{-1}\mathbf{X}^T\boldsymbol{\epsilon} + \boldsymbol{\beta}$, it follows from (2.3), with nonrandom $\mathbf{A} = (\mathbf{X}^T\mathbf{X})^{-1}\mathbf{X}^T$, that $E(\widehat{\boldsymbol{\beta}}) = \boldsymbol{\beta}$, showing that $\widehat{\boldsymbol{\beta}}$ is unbiased. Moreover, since $\text{Cov}(\boldsymbol{\epsilon}) = \sigma^2\mathbf{I}$, it follows from (2.4) that

$$\text{Cov}(\widehat{\boldsymbol{\beta}}) = (\mathbf{X}^T\mathbf{X})^{-1}\mathbf{X}^T(\sigma^2\mathbf{I})\mathbf{X}(\mathbf{X}^T\mathbf{X})^{-1} = \sigma^2(\mathbf{X}^T\mathbf{X})^{-1}.$$

The vector of residuals is $\mathbf{e} = \mathbf{Y} - \mathbf{X}\widehat{\boldsymbol{\beta}} = \mathbf{X}(\boldsymbol{\beta} - \widehat{\boldsymbol{\beta}}) + \mathbf{X}\boldsymbol{\epsilon}$, so $E(\mathbf{e}) = \mathbf{0}$. Moreover, as pointed out in Section 1.4.1, we can write $\mathbf{e} = (\mathbf{I} - \mathbf{H})\mathbf{Y}$, where $\mathbf{H} = \mathbf{X}(\mathbf{X}^T\mathbf{X})^{-1}\mathbf{X}^T$ is the projection matrix. Hence, by (2.4),

$$\text{Cov}(\mathbf{e}) = (\mathbf{I} - \mathbf{H})\text{Cov}(\mathbf{Y})(\mathbf{I} - \mathbf{H}) = (\mathbf{I} - \mathbf{H})\text{Cov}(\boldsymbol{\epsilon})(\mathbf{I} - \mathbf{H}) = \sigma^2(\mathbf{I} - \mathbf{H}),$$

noting that $(\mathbf{I} - \mathbf{H})^2 = \mathbf{I} - \mathbf{H}$. This proves (1.35).

The *trace* of a $k \times k$ matrix $\mathbf{A} = (a_{ij})_{1 \le i,j \le k}$ is the sum of its diagonal elements; i.e., $\text{tr}(\mathbf{A}) = \sum_{i=1}^k a_{ii}$. An important property of a trace is

$$\text{tr}(\mathbf{AB}) = \text{tr}(\mathbf{BA}). \tag{2.5}$$

Also $\text{tr}(\mathbf{I}_k) = k$, where to highlight the dimensionality of an identity matrix, we write $\mathbf{I}_k$ instead of $\mathbf{I}$. Note that the residual sum of squares can be expressed as $\text{RSS} = \sum_{i=1}^n \mathbf{e}_i^2 = \mathbf{e}^T\mathbf{e} = \text{tr}(\mathbf{e}\mathbf{e}^T)$ by (2.5). Since $E\text{tr}(\mathbf{M}) = \text{tr}(E\mathbf{M})$ for a random $k \times k$ matrix $\mathbf{M}$, it follows that

$$E(\text{RSS}) = E\{\text{tr}(\mathbf{e}\mathbf{e}^T)\} = \text{tr}(E(\mathbf{e}\mathbf{e}^T)) = \text{tr}(\text{Cov}(\mathbf{e})) = \sigma^2\text{tr}(\mathbf{I} - \mathbf{H}) = (n-p)\sigma^2, \tag{2.6}$$

noting that $\text{tr}(\mathbf{I}_n) = n$ and $\text{tr}(\mathbf{H}) = \text{tr}(\mathbf{X}^T(\mathbf{X}^T\mathbf{X})^{-1}\mathbf{X}) = p$ by (2.5).

We next make use of (2.3) and (2.5) to prove (1.30), which is related to Mallows' C_p-statistic in Section 1.3. Under the assumption $Ey_t = \boldsymbol{\beta}^T\mathbf{x}_t$,

$$\begin{aligned}
\sigma^2 E(J_p) &= \sum_{t=1}^n E\big[\mathbf{x}_t^T(\widehat{\boldsymbol{\beta}} - \boldsymbol{\beta})\big]^2 = \sum_{t=1}^n \mathbf{x}_t^T E\big[(\widehat{\boldsymbol{\beta}} - \boldsymbol{\beta})(\widehat{\boldsymbol{\beta}} - \boldsymbol{\beta})^T\big]\mathbf{x}_t \quad (\text{by } (2.4)) \\
&= \sum_{t=1}^n \mathbf{x}_t^T \text{Cov}(\widehat{\boldsymbol{\beta}})\mathbf{x}_t = \sigma^2 \sum_{t=1}^n \mathbf{x}_t^T(\mathbf{X}^T\mathbf{X})^{-1}\mathbf{x}_t \\
&= \sigma^2 \sum_{t=1}^n \text{tr}\big[\mathbf{x}_t^T(\mathbf{X}^T\mathbf{X})^{-1}\mathbf{x}_t\big] = \sigma^2\text{tr}\Big[(\mathbf{X}^T\mathbf{X})^{-1}\sum_{t=1}^n \mathbf{x}_t\mathbf{x}_t^T\Big] \quad (\text{by } (2.5)) \\
&= \sigma^2\text{tr}\big[(\mathbf{X}^T\mathbf{X})^{-1}(\mathbf{X}^T\mathbf{X})\big] = \sigma^2\text{tr}(\mathbf{I}_p) = \sigma^2 p.
\end{aligned}$$

Since $E(\text{RSS}_p) = (n-p)\sigma^2$, as shown in (2.6), it then follows that $E(\text{RSS}_p)/\sigma^2 + 2p - n = p = E(J_p)$, proving (1.30).

In the case where $\mathbf{x}_t$ omits input variables associated with nonzero β_j's, $Ey_t \ne E(\widehat{\boldsymbol{\beta}}^T\mathbf{x}_t)$ and therefore $Ee_t = E(y_t - \boldsymbol{\beta}^T\mathbf{x}_t) \ne 0$. Since $Ee_t^2 = (Ee_t)^2 + \text{Var}(e_t)$, it then follows that

$$E(\text{RSS}) = \sum_{t=1}^{n} Ee_t^2 = \sum_{t=1}^{n} \left[E(y_t - \widehat{\boldsymbol{\beta}}^T \mathbf{x}_t)\right]^2 + \text{tr}\big(\text{Cov}(\mathbf{e})\big).$$

Combining this with (2.6) yields (1.31).

2.2 Principal component analysis (PCA)

2.2.1 Basic definitions

Definition 2.1. Let $\mathbf{V}$ be a $p \times p$ matrix. A complex number λ is called an *eigenvalue* of $\mathbf{V}$ if there exists a $p \times 1$ vector $\mathbf{a} \neq \mathbf{0}$ such that $\mathbf{Va} = \lambda \mathbf{a}$. Such a vector $\mathbf{a}$ is called an *eigenvector* of $\mathbf{V}$ corresponding to the eigenvalue λ.

We can rewrite $\mathbf{Va} = \lambda \mathbf{a}$ as $(\mathbf{V} - \lambda \mathbf{I})\mathbf{a} = 0$. Since $\mathbf{a} \neq \mathbf{0}$, this implies that λ is a solution of the equation $\det(\mathbf{V} - \lambda \mathbf{I}) = 0$. Since $\det(\mathbf{V} - \lambda \mathbf{I})$ is a polynomial of degree p in λ, there are p eigenvalues, not necessarily distinct. If $\mathbf{V}$ is symmetric, then all its eigenvalues are real and can be ordered as $\lambda_1 \geq \cdots \geq \lambda_p$. Moreover,

$$\text{tr}(\mathbf{V}) = \lambda_1 + \cdots + \lambda_p, \qquad \det(\mathbf{V}) = \lambda_1 \ldots \lambda_p. \tag{2.7}$$

If $\mathbf{a}$ is an eigenvector of $\mathbf{V}$ corresponding to the eigenvalue λ, then so is $c\mathbf{a}$ for $c \neq 0$. Moreover, premultiplying $\lambda \mathbf{a} = \mathbf{Va}$ by $\mathbf{a}^T$ yields

$$\lambda = \mathbf{a}^T \mathbf{Va} / ||\mathbf{a}||^2. \tag{2.8}$$

In the rest of this section, we shall focus on the case where $\mathbf{V}$ is the covariance matrix of a random vector $\mathbf{x} = (X_1, \ldots, X_p)^T$. Not only are its eigenvalues real because $\mathbf{V}$ is symmetric, but they are also nonnegative since $\mathbf{V}$ is nonnegative definite; see Section 1.1.3. Consider the linear combination $\mathbf{a}^T\mathbf{x}$ with $||\mathbf{a}|| = 1$ that has the largest variance over all such linear combinations. To maximize $\mathbf{a}^T\mathbf{Va}(= \text{Var}(\mathbf{a}^T\mathbf{x}))$ over $\mathbf{a}$ with $||\mathbf{a}|| = 1$, introduce the Lagrange multiplier λ to obtain

$$\frac{\partial}{\partial a_i} \left\{\mathbf{a}^T\mathbf{Va} + \lambda(1 - \mathbf{a}^T\mathbf{a})\right\} = 0 \quad \text{for } i = 1, \ldots, p. \tag{2.9}$$

The p equations in (2.9) can be written as the linear system $\mathbf{Va} = \lambda \mathbf{a}$. Since $\mathbf{a} \neq \mathbf{0}$, this implies that λ is an eigenvalue of $\mathbf{V}$ and $\mathbf{a}$ is the corresponding eigenvector, and that $\lambda = \mathbf{a}^T\mathbf{Va}$ by (2.8).

Let $\lambda_1 = \max_{\mathbf{a}:||\mathbf{a}||=1} \mathbf{a}^T\mathbf{Va}$ and $\mathbf{a}_1$ be the corresponding eigenvector with $||\mathbf{a}_1|| = 1$. Next consider the linear combination $\mathbf{a}^T\mathbf{x}$ that maximizes $\text{Var}(\mathbf{a}^T\mathbf{x}) = \mathbf{a}^T\mathbf{Va}$ subject to $\mathbf{a}_1^T\mathbf{a} = 0$ and $||\mathbf{a}|| = 1$. Introducing the Lagrange multipliers λ and η, we obtain

$$\frac{\partial}{\partial a_i}\{\mathbf{a}^T\mathbf{V}\mathbf{a} + \lambda(1 - \mathbf{a}^T\mathbf{a}) + \eta\mathbf{a}_1^T\mathbf{a}\} = 0 \text{ for } i = 1, ..., p.$$

As in (2.9), this implies that the Lagrange multiplier λ is an eigenvalue of $\mathbf{V}$ with corresponding unit eigenvector $\mathbf{a}_2$ that is orthogonal to $\mathbf{a}_1$. Proceeding inductively in this way, we obtain the eigenvalue $\lambda_1 \geq \lambda_2 \geq \cdots \geq \lambda_p$ of $\mathbf{V}$ with the optimization characterization

$$\lambda_{k+1} = \max_{\mathbf{a}:\|\mathbf{a}\|=1, \mathbf{a}^T\mathbf{a}_j=0 \text{ for } 1\leq j\leq k} \mathbf{a}^T\mathbf{V}\mathbf{a}. \tag{2.10}$$

The maximizer $\mathbf{a}_{k+1}$ of the right-hand side of (2.10) is an eigenvector corresponding to the eigenvalue λ_{k+1}.

Definition 2.2. $\mathbf{a}_i^T\mathbf{x}$ is called the ith *principal component* of $\mathbf{x}$.

2.2.2 Properties of principal components

(a) From the preceding derivation, $\lambda_i = \text{Var}(\mathbf{a}_i^T\mathbf{x})$.

(b) The elements of the eigenvectors $\mathbf{a}_i$ are called *factor loadings* in PCA (principal component analysis), which can be carried out by the software package `princomp` in `R` or `Splus`. Since $\mathbf{a}_i^T\mathbf{a}_j = 0$ for $i \neq j$ and $||\mathbf{a}_i|| = 1$, $(\mathbf{a}_1, \ldots, \mathbf{a}_p)$ is an orthogonal matrix and therefore we can decompose the identity matrix $\mathbf{I}$ as

$$\mathbf{I} = (\mathbf{a}_1, \ldots, \mathbf{a}_p)(\mathbf{a}_1, \ldots, \mathbf{a}_p)^T = \mathbf{a}_1\mathbf{a}_1^T + \cdots + \mathbf{a}_p\mathbf{a}_p^T. \tag{2.11}$$

More importantly, summing $\lambda_i\mathbf{a}_i\mathbf{a}_i^T = \mathbf{V}\mathbf{a}_i\mathbf{a}_i^T$ over i and applying (2.11), we obtain the following decomposition of $\mathbf{V}$ into p rank-one matrices:

$$\mathbf{V} = \lambda_1\mathbf{a}_1\mathbf{a}_1^T + \cdots + \lambda_p\mathbf{a}_p\mathbf{a}_p^T. \tag{2.12}$$

(c) Since $\mathbf{V} = \text{Cov}(\mathbf{x})$ and $\mathbf{x} = (X_1, \ldots, X_p)^T$, $\text{tr}(\mathbf{V}) = \sum_{i=1}^p \text{Var}(X_i)$. Hence it follows from (2.7) that

$$\lambda_1 + \cdots + \lambda_p = \sum_{i=1}^p \text{Var}(X_i). \tag{2.13}$$

An important goal of PCA is to determine if the first few principal components can account for most of the overall variance $\sum_{i=1}^p \text{Var}(X_i)$. In view of (2.13), this amounts to determining whether

$$\Big(\sum_{i=1}^k \lambda_i\Big)\Big/\text{tr}(\mathbf{V}) \text{ is near 1 for some small } k. \tag{2.14}$$

The proportion in (2.14) can be evaluated by using `screeplot`, which plots $\lambda_i/\mathrm{tr}(\mathbf{V})$ for each i, in `R` or `Splus`.

Singular-value decomposition

We can write the representation of $\mathbf{V}$ in (2.12) as

$$\mathbf{V} = \mathbf{Q}\mathrm{diag}(\lambda_1, \ldots, \lambda_p)\mathbf{Q}^T, \tag{2.15}$$

where $\mathbf{Q} = (\mathbf{a}_1, \ldots, \mathbf{a}_p)$. The matrix $\mathbf{Q}$ is orthogonal, and (2.15) is called the *singular-value decomposition* of $\mathbf{V}$. We can use (2.15) to define the square root of $\mathbf{V}$ by $\mathbf{V}^{1/2} = \mathbf{Q}\mathrm{diag}(\sqrt{\lambda_1}, \ldots, \sqrt{\lambda_p})\mathbf{Q}^T$, noting that $\mathbf{V}^{1/2}\mathbf{V}^{1/2} = \mathbf{Q}\mathrm{diag}(\lambda_1, \ldots, \lambda_p)\mathbf{Q}^T = \mathbf{V}$.

PCA of sample covariance matrix and sampling theory of $\widehat{\lambda}_j$, $\widehat{\mathbf{a}}_j$

Suppose $\mathbf{x}_1, \ldots, \mathbf{x}_n$ are n independent observations from a multivariate population with mean $\boldsymbol{\mu}$ and covariance matrix $\mathbf{V}$. The mean vector $\boldsymbol{\mu}$ can be estimated by $\overline{\mathbf{x}} = \sum_{i=1}^n \mathbf{x}_i/n$, and the covariance matrix can be estimated by

$$\widehat{\mathbf{V}} = \sum_{i=1}^n (\mathbf{x}_i - \bar{\mathbf{x}})(\mathbf{x}_i - \bar{\mathbf{x}})^T/(n-1),$$

which is the sample covariance matrix. Let $\widehat{\mathbf{a}}_j = (\widehat{a}_{1j}, \ldots, \widehat{a}_{pj})^T$ be the eigenvector corresponding to the jth largest eigenvalue $\widehat{\lambda}_j$ of the sample covariance matrix $\widehat{\mathbf{V}}$. For fixed p, the following asymptotic results have been established under the assumption $\lambda_1 > \cdots > \lambda_p > 0$ as $n \to \infty$:

(i) $\sqrt{n}(\widehat{\lambda}_i - \lambda_i)$ has a limiting $N(0, 2\lambda_i^2)$ distribution. Moreover, for $i \neq j$, the limiting distribution of $\sqrt{n}(\widehat{\lambda}_i - \lambda_i, \widehat{\lambda}_j - \lambda_j)$ is that of two independent normals.

(ii) $\sqrt{n}(\widehat{\mathbf{a}}_i - \mathbf{a}_i)$ has a limiting normal distribution with mean 0 and covariance matrix

$$\sum_{h \neq i} \frac{\lambda_i \lambda_h}{(\lambda_i - \lambda_h)^2} \mathbf{a}_h \mathbf{a}_h^T.$$

Since $\mathbf{a}_i$ is not uniquely defined because multiplication by -1 still preserves the eigenvector property $\mathbf{Va} = \lambda\mathbf{a}$ with $\|\mathbf{a}\| = 1$, the preceding result means that $\widehat{\mathbf{a}}_i$ and $\mathbf{a}_i$ are chosen so that they have the same sign.

In view of (ii) above, even when p is large so that $\widehat{\mathbf{V}}$ estimates a larger number, $p(p+1)/2$, of parameters, if many eigenvalues λ_h are small compared with the largest λ_i's, then they have small contributions to the sum in (ii), in which $\lambda_i\lambda_h/(\lambda_i - \lambda_h)^2 \doteq \lambda_h/\lambda_i$ when λ_h is much smaller than λ_i, noting that

$\text{tr}(\mathbf{a}_h\mathbf{a}_h^T) = 1$. Hence the covariance structure of the data can be approximated by a few principal components with large eigenvalues. In this approximation, only a few, say k, principal components are involved in the covariance matrix of the estimate $\widehat{\mathbf{a}}_j$ for $1 \le j \le k$.

Representation of data in terms of principal components

Let $\mathbf{x}_i = (x_{i1}, \ldots, x_{ip})^T$, $i = 1, \ldots, n$, be the multivariate sample. Let $\mathbf{X}_k = (x_{1k}, \ldots, x_{nk})^T$, $1 \le k \le p$, and define

$$\mathbf{Y}_j = \widehat{a}_{1j}\mathbf{X}_1 + \cdots + \widehat{a}_{pj}\mathbf{X}_p, \qquad 1 \le j \le p,$$

where $\widehat{\mathbf{a}}_j = (\widehat{a}_{1j}, \ldots, \widehat{a}_{pj})^T$ is the eigenvector corresponding to the jth largest eigenvalue $\widehat{\lambda}_j$ of the sample covariance matrix $\widehat{\mathbf{V}}$ with $||\widehat{\mathbf{a}}_j|| = 1$. Since the matrix $\widehat{\mathbf{A}} := (\widehat{a}_{ij})_{1 \le i,j \le p}$ is orthogonal (i.e., $\widehat{\mathbf{A}}\widehat{\mathbf{A}}^T = \mathbf{I}$), it follows that the observed data $\mathbf{X}_k$ can be expressed in terms of the "principal components" $\mathbf{Y}_j$ as

$$\mathbf{X}_k = \widehat{a}_{k1}\mathbf{Y}_1 + \cdots + \widehat{a}_{kp}\mathbf{Y}_p. \tag{2.16}$$

Moreover, the sample correlation matrix of the transformed data y_{tj} is the identity matrix.

PCA of correlation matrices

As shown above, the key ingredient of PCA is the decomposition $\mathbf{V} = \lambda_1\mathbf{a}_1\mathbf{a}_1^T + \cdots + \lambda_p\mathbf{a}_p\mathbf{a}_p^T$ of a nonnegative definite matrix $\mathbf{V}$. An alternative to $\text{Cov}(\mathbf{x})$ is the correlation matrix $\text{Corr}(\mathbf{x})$, which is also nonnegative definite, consisting of the correlation coefficients $\text{Corr}(X_i, X_j)$, $1 \le i, j \le p$. In fact, $\text{Corr}(\mathbf{x}) = \text{Cov}\big(X_1/\sigma_1, \ldots, X_p/\sigma_p\big)$ is itself a covariance matrix, where σ_i is the standard deviation of X_i. Since a primary goal of PCA is to uncover low-dimensional subspaces around which $\mathbf{x}$ tends to concentrate, scaling the components of $\mathbf{x}$ by their standard deviations may work better for this purpose, and applying PCA to sample correlation matrices may give more stable results.

2.2.3 An example: PCA of U.S. Treasury-LIBOR swap rates

PCA can be implemented by the following software packages:

```
R/Splus:     princomp, screeplot, biplot.princomp;
MATLAB:      princomp, pcacov, biplot.
```

We now apply PCA to account for the variance of daily changes in swap rates with a few principal components (or factors). Details of interest rate swap contracts, which involve exchanging the U.S. Treasury bond rate with the London Interbank Offered Rate (LIBOR), and the associated swap rates for different maturities are given in Chapter 10. The top panel of Figure 2.1 plots the daily swap rates r_{kt} for four of the eight maturities $T_k = 1, 2, 3, 4, 5, 7, 10,$ and 30 years from July 3, 2000 to July 15, 2005. The data are obtained from `www.Economagic.com`. Let $d_{kt} = r_{kt} - r_{k,t-1}$ denote the daily changes in the k-year swap rates during the period. The middle and bottom panels plot the differenced time series d_{kt} for 1-year and 30-year swap rates, respectively. Further discussion of these plots will be given in Chapter 5 (Section 5.2.3). The sample mean vector and the sample covariance and correlation matrices of the difference data $\{(d_{1t}, \ldots, d_{8t}), 1 \le t \le 1256\}$ are

$$\widehat{\boldsymbol{\mu}} = -\left(2.412\ 2.349\ 2.293\ 2.245\ 2.196\ 2.158\ 2.094\ 1.879\right)^T \times 10^{-3}.$$

$$\widehat{\boldsymbol{\Sigma}} = \begin{pmatrix} 0.233 \\ 0.300\ 0.438 \\ 0.303\ 0.454\ 0.488 \\ 0.297\ 0.453\ 0.492\ 0.508 \\ 0.292\ 0.451\ 0.494\ 0.514\ 0.543 \\ 0.268\ 0.421\ 0.466\ 0.490\ 0.513\ 0.498 \\ 0.240\ 0.384\ 0.431\ 0.458\ 0.477\ 0.472\ 0.467 \\ 0.169\ 0.278\ 0.318\ 0.344\ 0.362\ 0.365\ 0.369\ 0.322 \end{pmatrix} \times 10^{-2},$$

$$\widehat{\mathbf{R}} = \begin{pmatrix} 1.000 \\ 0.941\ 1.000 \\ 0.899\ 0.983\ 1.000 \\ 0.862\ 0.961\ 0.988\ 1.000 \\ 0.821\ 0.925\ 0.960\ 0.979\ 1.000 \\ 0.787\ 0.901\ 0.945\ 0.973\ 0.986\ 1.000 \\ 0.729\ 0.850\ 0.902\ 0.941\ 0.948\ 0.978\ 1.000 \\ 0.618\ 0.742\ 0.803\ 0.852\ 0.866\ 0.912\ 0.951\ 1.000 \end{pmatrix}.$$

Table 2.1 gives the results of PCA using both the covariance and correlation matrices. Also given there are eigenvalues, factor loadings (eigenvectors), and the proportions of overall variance explained by the principal components.

Swap rate movements indicated by PCA

Let $x_{tk} = (d_{tk} - \widehat{\mu}_k)/\widehat{\sigma}_k$, where $\widehat{\sigma}_k$ is the sample standard deviation of d_{tk} and is given by the square root of the kth diagonal element of $\widehat{\boldsymbol{\Sigma}}$. We can use (2.16) to represent $\mathbf{X}_k = (x_{1k}, \ldots, x_{nk})^T$ in terms of the principal components $\mathbf{Y}_1, \ldots, \mathbf{Y}_8$. Table 2.1 shows the PCA results using the sample covariance matrix $\widehat{\boldsymbol{\Sigma}}$ and the sample correlation matrix $\widehat{\mathbf{R}}$. From part (b) of Table 2.1,

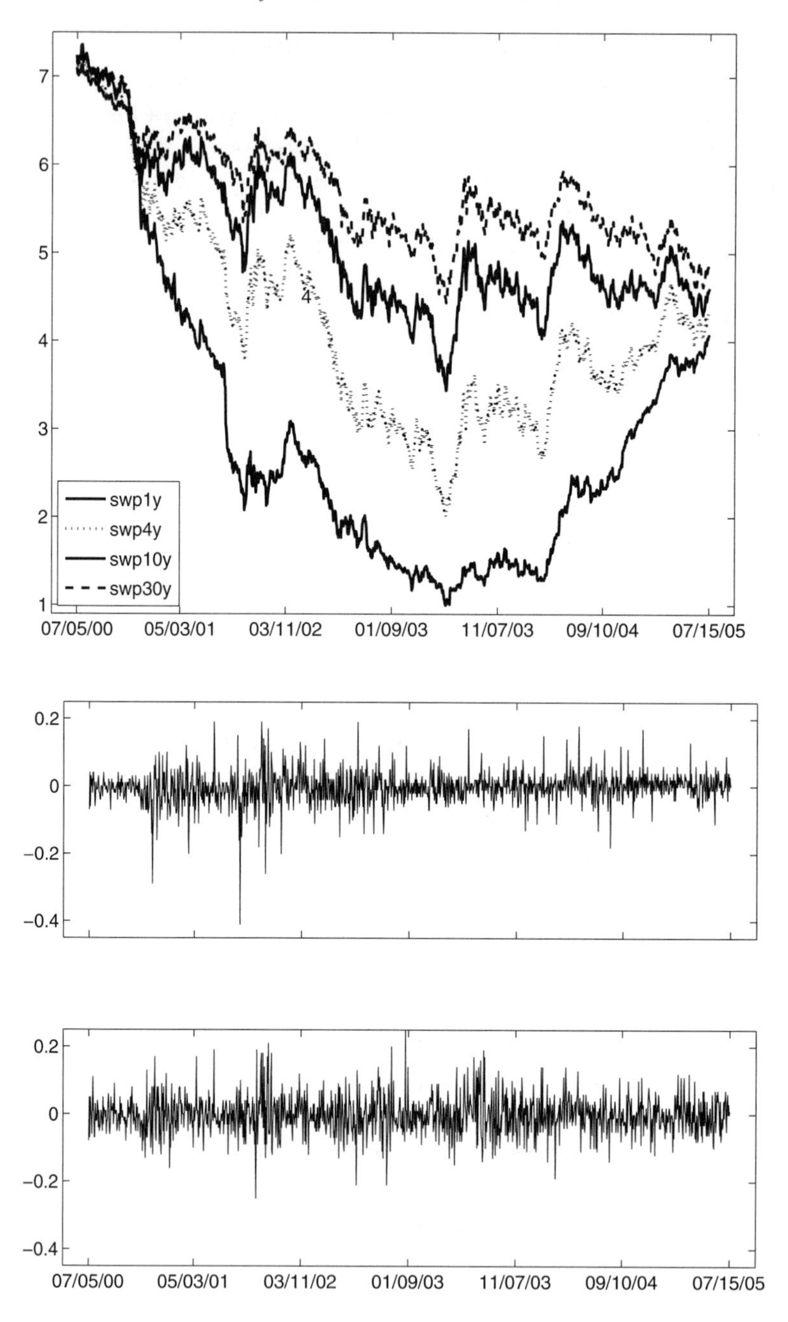

Fig. 2.1. Swap rates from July 3, 2000 to July 15, 2005. Top panel: the original time series. Middle and bottom panels: the differenced series for 1-year (middle) and 30-year (bottom) swap rates.

Table 2.1. PCA of covariance and correlation matrices of swap rate data.

	PC1	PC2	PC3	PC4	PC5	PC6	PC7	PC8
(a) Using sample covariance matrix								
Eigenvalue ($\times 10^4$)	324.1	18.44	3.486	1.652	0.984	0.473	0.292	0.253
Proportion	0.926	0.053	0.010	0.005	0.003	0.001	0.001	0.001
Factor loadings	0.231	0.491	−0.535	−0.580	−0.081	−0.275	0.006	−0.030
	0.351	0.431	−0.150	0.279	0.049	0.725	−0.050	0.246
	0.381	0.263	0.078	0.456	0.057	−0.265	−0.001	−0.706
	0.393	0.087	0.175	0.299	−0.069	−0.534	0.062	0.652
	0.404	−0.054	0.486	−0.447	0.414	0.131	0.455	−0.054
	0.387	−0.211	0.238	−0.262	−0.089	0.052	−0.818	−0.038
	0.365	−0.395	−0.127	−0.034	−0.735	0.154	0.344	−0.103
	0.279	−0.541	−0.588	0.138	0.513	−0.032	−0.003	0.027
(b) Using sample correlation matrix								
Eigenvalue	7.265	0.548	0.103	0.041	0.022	0.011	0.006	0.005
Proportion	0.908	0.067	0.013	0.005	0.003	0.001	0.001	0.001
Factor loadings	0.324	0.599	−0.586	−0.402	−0.025	−0.175	0.002	−0.013
	0.356	0.352	0.037	0.436	0.004	0.717	−0.044	0.205
	0.364	0.187	0.215	0.432	0.007	−0.349	0.007	−0.691
	0.368	0.043	0.253	0.206	−0.076	−0.533	0.045	0.681
	0.365	−0.064	0.385	−0.455	0.499	0.151	0.482	-0.056
	0.365	−0.192	0.216	−0.342	−0.014	0.075	−0.811	−0.052
	0.356	−0.348	−0.059	−0.154	−0.768	0.143	0.324	−0.098
	0.328	−0.565	−0.589	0.267	0.393	−0.025	0.002	0.020

the proportion of the overall variance explained by the first principal component is $7.265/8 = 90.8\%$. Moreover, the second principal component explains $0.548/8 = 6.7\%$ and the third principal component explains $0.103/8 = 1.3\%$ of the overall variance. Hence 98.9% of the overall variance is explained by the first three principal components, yielding the approximation

$$(\mathbf{d}_k - \widehat{\mu}_k)/\widehat{\sigma}_k \doteq \widehat{a}_{k1}\mathbf{Y}_1 + \widehat{a}_{k2}\mathbf{Y}_2 + \widehat{a}_{k3}\mathbf{Y}_3 \tag{2.17}$$

for the differenced swap rates.

In Figure 2.2, the bottom panel plots the factor loadings of the first three principal components (or the entries of the eigenvectors $\mathbf{a}_1$, $\mathbf{a}_2$, and $\mathbf{a}_3$) versus the maturities of the swap rates. The top panel shows the variances of all principal components. The graphs of the factor loadings show the following constituents of interest rate or yield curve movements documented in the literature:

(a) *Parallel shift component.* The factor loadings of the first principal component are roughly constant over different maturities. This means that a change in the swap rate for one maturity is accompanied by roughly the same change for other maturities. Indeed, if $\widehat{a}_{11}, \dots, \widehat{a}_{p1}$ (the components of $\widehat{\mathbf{a}}_1$) are equal, then the first summand on the right-hand side of (2.17) is the same for all maturities T_k.
(b) *Tilt component.* The factor loadings of the second principal component have a monotonic change with maturities. Changes in short-maturity and long-maturity swap rates in this component have opposite signs.
(c) *Curvature component.* The factor loadings of the third principal component are different for the midterm rates and the average of short- and long-term rates, revealing a curvature of the graph that resembles the convex shape of the relationship between the rates and their maturities.

2.3 Multivariate normal distribution

2.3.1 Definition and density function

(i) An $m \times 1$ random vector $\mathbf{Z} = (Z_1, \dots, Z_m)^T$ is said to have the m-variate standard normal distribution if $Z_1, \dots, Z_m$ are independent standard normal random variables. The joint density function of $\mathbf{Z}$ is therefore

$$f(\mathbf{z}) = \prod_{i=1}^{m} \left\{ \frac{1}{\sqrt{2\pi}} \exp\left(-\frac{1}{2} z_i^2 \right) \right\} = (2\pi)^{-m/2} \exp(-\mathbf{z}^T\mathbf{z}/2).$$

(ii) An $m \times 1$ random vector $\mathbf{Y}$ is said to have a multivariate normal distribution if it is of the form $\mathbf{Y} = \boldsymbol{\mu} + \mathbf{A}\mathbf{Z}$, where $\mathbf{Z}$ is standard m-variate normal and $\boldsymbol{\mu}$ and $\mathbf{A}$ are $m \times 1$ and $m \times m$ nonrandom matrices.

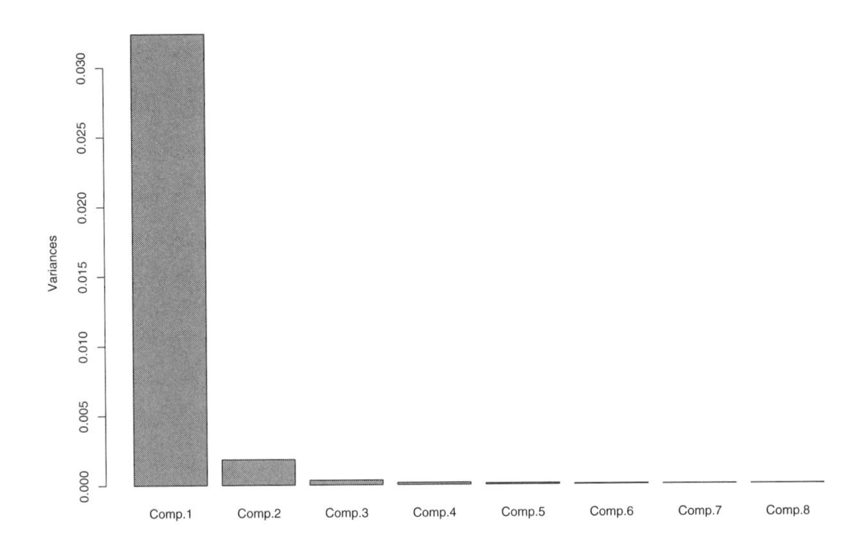

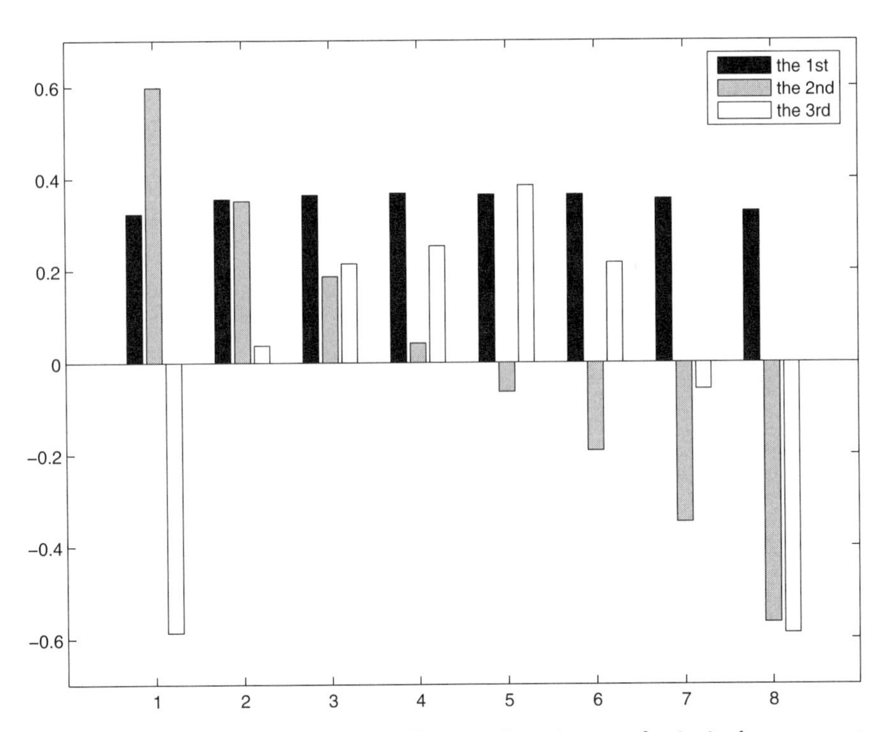

Fig. 2.2. PCA of correlation matrix. Top panel: variances of principal components. Bottom panel: eigenvectors of the first three principal components, which represent the parallel shift, tilt, and convexity components of swap rate movements.

Since $E\mathbf{Z} = \mathbf{0}$ and $\text{Cov}(\mathbf{Z}) = \mathbf{I}$, the random vector $\mathbf{Y} = \boldsymbol{\mu} + \mathbf{A}\mathbf{Z}$ has mean $\boldsymbol{\mu}$ and covariance matrix $\mathbf{V} = \mathbf{A}\mathbf{A}^T$. Analogous to a univariate normal distribution that is determined by its mean and variance, we write $\mathbf{Y} \sim N(\boldsymbol{\mu}, \mathbf{V})$. If $\mathbf{V}$ is nonsingular, then a change of variables applied to the density function of $\mathbf{Z}$ shows that the density function of $\mathbf{Y}$ is

$$f(\mathbf{y}) = \frac{1}{(2\pi)^{m/2}\sqrt{\det(\mathbf{V})}} \exp\Big\{ -\frac{1}{2}(\mathbf{y}-\boldsymbol{\mu})^T\mathbf{V}^{-1}(\mathbf{y}-\boldsymbol{\mu})\Big\}, \mathbf{y} \in \mathbb{R}^m. \quad (2.18)$$

For the case $m = 1$, (2.18) reduces to the familiar normal density function

$$f(y) = \frac{1}{\sqrt{2\pi}\sigma} \exp\{-(y-\mu)^2/2\sigma^2\}, \quad (2.19)$$

so $\mathbf{V}^{-1}$ and $\sqrt{\det(\mathbf{V})}$ in (2.18) are used to replace $1/\sigma^2$ and σ in (2.19). In the case $m = 2$, (2.18) can be written in the usual form of a bivariate normal density function:

$$f(y_1, y_2) = \frac{1}{2\pi\sigma_1\sigma_2\sqrt{1-\rho^2}} \exp\left\{ - \left[\frac{(y_1-\mu_1)^2}{\sigma_1^2} - 2\rho\frac{(y_1-\mu_1)(y_2-\mu_2)}{\sigma_1\sigma_2} + \frac{(y_2-\mu_2)^2}{\sigma_2^2}\right] \Big/ \left[2(1-\rho^2)\right] \right\}, \quad (2.20)$$

where $\sigma_i^2 = \text{Var}(Y_i)$ and ρ is the correlation coefficient between Y_1 and Y_2.

2.3.2 Marginal and conditional distributions

Suppose that $\mathbf{Y} \sim N(\boldsymbol{\mu}, \mathbf{V})$ is partitioned as

$$\mathbf{Y} = \begin{pmatrix} \mathbf{Y}_1 \\ \mathbf{Y}_2 \end{pmatrix}, \quad \boldsymbol{\mu} = \begin{pmatrix} \boldsymbol{\mu}_1 \\ \boldsymbol{\mu}_2 \end{pmatrix}, \quad \mathbf{V} = \begin{pmatrix} \mathbf{V}_{11} & \mathbf{V}_{12} \\ \mathbf{V}_{21} & \mathbf{V}_{22} \end{pmatrix},$$

where $\mathbf{Y}_1$ and $\boldsymbol{\mu}_1$ have dimension $r(< m)$ and $\mathbf{V}_{11}$ is $r \times r$. Then $\mathbf{Y}_1 \sim N(\boldsymbol{\mu}_1, \mathbf{V}_{11})$, $\mathbf{Y}_2 \sim N(\boldsymbol{\mu}_2, \mathbf{V}_{22})$, and the conditional distribution of $\mathbf{Y}_1$ given $\mathbf{Y}_2 = \mathbf{y}_2$ is $N\big(\boldsymbol{\mu}_1 + \mathbf{V}_{12}\mathbf{V}_{22}^{-1}(\mathbf{y}_2 - \boldsymbol{\mu}_2), \mathbf{V}_{11} - \mathbf{V}_{12}\mathbf{V}_{22}^{-1}\mathbf{V}_{21}\big)$, as can be shown by computing the conditional density function $f_{\mathbf{Y}_1|\mathbf{Y}_2}(\mathbf{y}_1|\mathbf{y}_2) = f(\mathbf{y})/f_{\mathbf{Y}_2}(\mathbf{y}_2)$ via (2.18); see Exercise 2.4.

2.3.3 Orthogonality and independence, with applications to regression

Let $\boldsymbol{\zeta}$ be a standard normal random vector of dimension p and let $\mathbf{A}$, $\mathbf{B}$ be $r \times p$ and $q \times p$ nonrandom matrices such that $\mathbf{A}\mathbf{B}^T = \mathbf{0}$. Then the components of $\mathbf{U}_1 := \mathbf{A}\boldsymbol{\zeta}$ are uncorrelated with those of $\mathbf{U}_2 := \mathbf{B}\boldsymbol{\zeta}$ because $E(\mathbf{U}_1\mathbf{U}_2^T) = \mathbf{A}E(\boldsymbol{\zeta}\boldsymbol{\zeta}^T)\mathbf{B}^T = \mathbf{0}$. Note that

$$\mathbf{U} = \begin{pmatrix} \mathbf{U}_1 \\ \mathbf{U}_2 \end{pmatrix} = \begin{pmatrix} \mathbf{A} \\ \mathbf{B} \end{pmatrix} \boldsymbol{\zeta}$$

has a multivariate normal distribution with mean $\mathbf{0}$ and covariance matrix

$$\mathbf{V} = \begin{pmatrix} \mathbf{A}\mathbf{A}^T & \mathbf{0} \\ \mathbf{0} & \mathbf{B}\mathbf{B}^T \end{pmatrix},$$

and therefore

$$\mathbf{V}^{-1} = \begin{pmatrix} (\mathbf{A}\mathbf{A}^T)^{-1} & \mathbf{0} \\ \mathbf{0} & (\mathbf{B}\mathbf{B}^T)^{-1} \end{pmatrix}.$$

It then follows from (2.18) that the density function of $\mathbf{U}$ is a product of the density functions of $\mathbf{U}_1$ and $\mathbf{U}_2$, so $\mathbf{U}_1$ and $\mathbf{U}_2$ are independent. Note also that $\mathbf{U}_1^T\mathbf{U}_2 = \mathbf{0}$; i.e., they are orthogonal vectors. Hence *orthogonal vectors that have a jointly normal zero-mean distribution are independent.* Since a multivariate normal vector is a linear transformation of a standard normal vector, this result implies that, for a multivariate normal vector, uncorrelated components are independent.

Let $\mathbf{e} = \mathbf{Y} - \mathbf{X}\widehat{\boldsymbol{\beta}}$ as in Section 1.4.1, where it is shown that $\mathbf{e} = (\mathbf{I} - \mathbf{H})\mathbf{Y}$. Since $(\mathbf{I} - \mathbf{H})\mathbf{H} = \mathbf{0}$ and $\mathbf{Y} = \mathbf{X}\boldsymbol{\beta} + \boldsymbol{\epsilon}$, it then follows that

$$\mathbf{e} = (\mathbf{I} - \mathbf{H})\boldsymbol{\epsilon}, \qquad \mathbf{X}(\widehat{\boldsymbol{\beta}} - \boldsymbol{\beta}) = \mathbf{H}\boldsymbol{\epsilon}, \tag{2.21}$$

are orthogonal. Assume conditions (A) and (C*) of Section 1.1.4. Since $\boldsymbol{\epsilon}/\sigma$ is standard normal, the orthogonal vectors $\mathbf{e}$ and $\mathbf{X}(\widehat{\boldsymbol{\beta}} - \boldsymbol{\beta})$ are independent $n \times 1$ normal vectors. Since $\mathbf{X}$ is a nonrandom matrix of full rank p (i.e., the $n \times p$ matrix $\mathbf{X}$ has p linearly independent rows), it then follows that $\mathbf{e}$ and $\widehat{\boldsymbol{\beta}} - \boldsymbol{\beta}$ are independent, with

$$\mathbf{e} \sim N\big(\mathbf{0}, \sigma^2(\mathbf{I} - \mathbf{H})\big), \qquad \widehat{\boldsymbol{\beta}} - \boldsymbol{\beta} \sim N\big(\mathbf{0}, \sigma^2(\mathbf{X}^T\mathbf{X})^{-1}\big).$$

This proves (1.11) and (1.13), noting that $s^2 = \sum_{i=1}^n e_i^2/(n-p)$.

Geometrically, the matrix $\mathbf{H} = \mathbf{X}(\mathbf{X}^T\mathbf{X})^{-1}\mathbf{X}^T$ is associated with projection into the linear space $\mathcal{L}$ spanned by the column vectors of $\mathbf{X}$, and $\mathbf{I} - \mathbf{H}$ is associated with projection into the linear space $\mathcal{L}^\perp$ consisting of vectors orthogonal to $\mathcal{L}$; see Section 1.1.2(b). Since $\mathbf{X}$ has full rank p, $\mathcal{L}$ has dimension p and $\mathcal{L}^\perp$ has dimension $n - p$. We can choose an orthogonal basis in $\mathcal{L}$ or $\mathcal{L}^\perp$ to perform the projection. Thus, there exists an $n \times n$ orthogonal matrix $\mathbf{Q}$ such that

$$\mathbf{Q}^T\mathbf{H}\mathbf{Q} = \begin{pmatrix} \mathbf{I}_p & \mathbf{0} \\ \mathbf{0} & \mathbf{0} \end{pmatrix}, \qquad \mathbf{Q}^T(\mathbf{I}-\mathbf{H})\mathbf{Q} = \begin{pmatrix} \mathbf{0} & \mathbf{0} \\ \mathbf{0} & \mathbf{I}_{n-p} \end{pmatrix}. \tag{2.22}$$

Let $\mathbf{Z} = \mathbf{Q}^T\boldsymbol{\epsilon}/\sigma$. Since $\mathbf{Q}^T\mathbf{Q} = \mathbf{I}$, $\mathbf{Z} \sim N(\mathbf{0}, \mathbf{I})$. Moreover,

$$\mathbf{Q}^T(\mathbf{I}-\mathbf{H})\boldsymbol{\epsilon}/\sigma = \mathbf{Q}^T(\mathbf{I}-\mathbf{H})\mathbf{Q}\mathbf{Z} = (0, \ldots, 0, Z_{p+1}, \ldots, Z_n)^T \tag{2.23}$$

by (2.22). Since $\mathbf{Q}$ is an orthogonal matrix, $||\mathbf{Q}^T\mathbf{x}||^2 = \mathbf{x}^T\mathbf{Q}\mathbf{Q}^T\mathbf{x} = ||\mathbf{x}||^2$ for all $\mathbf{x} \in \mathbb{R}^n$, and therefore

$$\sum_{t=1}^n e_t^2 = ||\mathbf{e}||^2 = ||\mathbf{Q}^T\mathbf{e}||^2 = ||\mathbf{Q}^T(\mathbf{I}-\mathbf{H})\boldsymbol{\epsilon}||^2$$

by (2.21). Hence it follows from (2.23) (in which the Z_i are i.i.d. $N(0,1)$) that

$$\sum_{t=1}^n e_t^2/\sigma^2 = \sum_{i=p+1}^n Z_i^2 \sim \chi^2_{n-p}, \tag{2.24}$$

proving (1.12).

The preceding argument can be used to prove (1.25) by showing that the orthogonal vectors $\widehat{\mathbf{Y}} - \widehat{\mathbf{Y}}_0$ and $\mathbf{Y} - \widehat{\mathbf{Y}}$ are independent normal and that $||\widehat{\mathbf{Y}} - \widehat{\mathbf{Y}}_0||^2/\sigma^2$ and $||\mathbf{Y} - \widehat{\mathbf{Y}}||^2/\sigma^2$ are independent χ^2_{p-r} and χ^2_{n-p}, respectively.

2.3.4 Sample covariance matrix and Wishart distribution

In view of (2.24), the estimate (1.10) of σ^2 in the regression model (1.1) is a χ^2 random variable divided by its degrees of freedom under assumptions (A) and (C*) of Section 1.1.4. In particular, for the special case $q = 0$ in the regression model (1.1), the estimate (1.10) reduces to the sample variance $s^2 = \sum_{i=1}^n (y_i - \bar{y})^2/(n-1)$, for which there are $n-1$ degrees of freedom due to the loss of one degree of freedom (i.e., the linear constraint $\sum_{i=1}^n (y_i - \bar{y}) = 0$) in estimating μ by $\bar{y}$. Replacing $y_i \sim N(\mu, \sigma^2)$ by $\mathbf{Y}_i \sim N(\boldsymbol{\mu}, \boldsymbol{\Sigma})$, the argument in Section 2.3.3 can be modified to prove a similar result on the independence of the sample mean vector and the sample covariance matrix.

Sample mean and covariance matrix from a multivariate normal distribution

Let $\mathbf{Y}_1, \ldots, \mathbf{Y}_n$ be independent $m \times 1$ random $N(\boldsymbol{\mu}, \boldsymbol{\Sigma})$ vectors with $n > m$ and positive definite $\boldsymbol{\Sigma}$. Define

$$\bar{\mathbf{Y}} = \frac{1}{n}\sum_{i=1}^n \mathbf{Y}_i, \quad \mathbf{W} = \sum_{i=1}^n (\mathbf{Y}_i - \bar{\mathbf{Y}})(\mathbf{Y}_i - \bar{\mathbf{Y}})^T. \tag{2.25}$$

As pointed out above, the sample mean vector $\bar{\mathbf{Y}}$ and the sample covariance matrix $\mathbf{W}/(n-1)$ are independent, generalizing the corresponding result in the case $m=1$. Moreover, since $\bar{\mathbf{Y}}$ is a linear transformation of the multivariate normal vector $(\mathbf{Y}_1^T, \ldots, \mathbf{Y}_n^T)^T$, it is also normal. Making use of (2.3) and (2.4) to derive $E(\bar{\mathbf{Y}})$ and $\mathrm{Cov}(\bar{\mathbf{Y}})$, we then obtain $\bar{\mathbf{Y}} \sim N(\boldsymbol{\mu}, \boldsymbol{\Sigma}/n)$, which generalizes the corresponding result in the case $m=1$. We next generalize $\sum_{t=1}^n (y_t - \bar{y})^2/\sigma^2 \sim \chi^2_{n-1}$ to the multivariate case.

The Wishart distribution and its properties

It is straightforward to generalize Definition 1.2(i) for the χ^2-distribution to the multivariate case, which is called the *Wishart distribution*.

Definition 2.3. Suppose $\mathbf{Y}_1, \ldots, \mathbf{Y}_n$ are independent $N(\mathbf{0}, \boldsymbol{\Sigma})$ random vectors of dimension m. Then the random matrix $\mathbf{W} = \Sigma_{i=1}^n \mathbf{Y}_i \mathbf{Y}_i^T$ is said to have a Wishart distribution, denoted by $W_m(\boldsymbol{\Sigma}, n)$.

We next use this definition to derive the density function of $\mathbf{W}$ when $\boldsymbol{\Sigma}$ is positive definite. We begin by considering the case $m=1$ and deriving the density function of χ^2_n. First note that χ^2_1 is the distribution of Z^2, where $Z \sim N(0,1)$, and a change of variables shows that $W = Z^2$ has the gamma$(1/2, 1/2)$ distribution. From this it follows that χ^2_n, which is the distribution of $Z_1^2 + \cdots + Z_n^2$ with i.i.d. standard normal Z_i, is the same as the gamma$(n/2, 1/2)$ distribution; the gamma distribution with scale parameter β and shape parameter α is denoted by gamma(α, β) and has density function

$$f(w) = \frac{\beta^\alpha}{\Gamma(\alpha)} w^{\alpha-1} e^{-\beta w}, \qquad w > 0.$$

Therefore $\sigma^2\chi^2_n$ has the density function $w^{(n-2)/2} e^{-w/(2\sigma^2)} \big/ \big[(2\sigma^2)^{n/2} \Gamma(n/2)\big]$. The density function of the Wishart distribution $W_m(\boldsymbol{\Sigma}, n)$ generalizes this to

$$f(\mathbf{W}) = \frac{\det(\mathbf{W})^{(n-m-1)/2} \exp\big\{-\frac{1}{2}\mathrm{tr}\big(\boldsymbol{\Sigma}^{-1}\mathbf{W}\big)\big\}}{[2^m \det(\boldsymbol{\Sigma})]^{n/2} \Gamma_m(n/2)}, \qquad \mathbf{W} > \mathbf{0}, \tag{2.26}$$

in which $\mathbf{W} > \mathbf{0}$ denotes that $\mathbf{W}$ is positive definite and $\Gamma_m(\cdot)$ denotes the multivariate gamma function

$$\Gamma_m(t) = \pi^{m(m-1)/4} \prod_{i=1}^m \Gamma\Big(t - \frac{i-1}{2}\Big). \tag{2.27}$$

Note that the usual gamma function corresponds to the case $m = 1$.

In Chapters 3 and 4 (Sections 3.3.3 and 4.4.1), applications of the Wishart distribution to statistical analysis of investment theories are given. The following properties of the Wishart distribution are generalizations of some well-known properties of the chi-square distribution and the gamma(α, β) distribution.

(a) If $\mathbf{W} \sim W_m(\boldsymbol{\Sigma}, n)$, then $E(\mathbf{W}) = n\boldsymbol{\Sigma}$.
(b) Let $\mathbf{W}_1, \ldots, \mathbf{W}_k$ be independently distributed with $\mathbf{W}_j \sim W_m(\boldsymbol{\Sigma}, n_j)$, $j = 1, \ldots, k$. Then $\sum_{j=1}^k \mathbf{W}_j \sim W_m(\boldsymbol{\Sigma}, \sum_{j=1}^k n_j)$.
(c) Let $\mathbf{W} \sim W_m(\boldsymbol{\Sigma}, n)$ and $\mathbf{A}$ be a nonrandom $m \times m$ nonsingular matrix. Then $\mathbf{AWA}^T \sim W_m(\mathbf{A}\boldsymbol{\Sigma}\mathbf{A}^T, n)$. In particular, $\mathbf{a}^T\mathbf{Wa} \sim (\mathbf{a}^T\boldsymbol{\Sigma}\mathbf{a})\chi_n^2$ for all nonrandom vectors $\mathbf{a} \neq \mathbf{0}$.

The multivariate t-distribution

The multivariate generalization of $\sum_{t=1}^n (y_t - \bar{y})^2 \sim \sigma^2\chi_{n-1}^2$ involves the Wishart distribution and is given by

$$\mathbf{W} := \sum_{i=1}^n (\mathbf{Y}_i - \bar{\mathbf{Y}})(\mathbf{Y}_i - \bar{\mathbf{Y}})^T \sim W_m(\boldsymbol{\Sigma}, n-1). \tag{2.28}$$

Moreover, $\mathbf{W}$ is independent of $\bar{\mathbf{Y}} \sim N(\boldsymbol{\mu}, \boldsymbol{\Sigma}/n)$, as noted above. Hence the situation is the same as in the univariate $(m = 1)$ case. It is straightforward to generalize Definition 1.2(ii) for the t-distribution to the multivariate case.

Definition 2.4. If $\mathbf{Z} \sim N(\mathbf{0}, \boldsymbol{\Sigma})$ and $\mathbf{W} \sim W_m(\boldsymbol{\Sigma}, k)$ such that the $m \times 1$ vector $\mathbf{Z}$ and the $m \times m$ matrix $\mathbf{W}$ are independent, then $(\mathbf{W}/k)^{-1/2}\mathbf{Z}$ is said to have the m-variate t-distribution with k degrees of freedom.

As will be shown in Chapter 12 (Sections 12.2.1 and 12.3.3), the univariate and multivariate t-distributions have important applications in risk management. Chapter 6 (Section 6.3) also gives applications of the t-distribution to volatility modeling. In the univariate case, we can use a change of variables (see Section 2.1.1) to find the joint density function of (T, U) in Definition 1.2(ii). We can then integrate the joint density function with respect to U to obtain the marginal density function of T. This yields an explicit formula for the density function of Student's t-distribution with k degrees of freedom:

$$f(t) = \frac{\Gamma((k+1)/2)}{\sqrt{\pi k}\,\Gamma(k/2)}\left(1 + \frac{t^2}{k}\right)^{-(k+1)/2}, \qquad -\infty < t < \infty. \tag{2.29}$$

By making use of the density function (2.26) of the Wishart distribution, the preceding argument used to derive (2.29) can be extended to show that the m-variate t-distribution with k degrees of freedom in Definition 2.4 has the density function

$$f(\mathbf{t}) = \frac{\Gamma\big((k+m)/2\big)}{(\pi k)^{m/2}\Gamma(k/2)}\left(1+\frac{\mathbf{t}^T\mathbf{t}}{k}\right)^{-(k+m)/2}, \qquad \mathbf{t}\in\mathbb{R}^m. \tag{2.30}$$

As pointed out in the remark in Section 1.2.2, the square of a t_k random variable has the $F_{1,k}$-distribution. More generally, if $\mathbf{t}$ has the m-variate t-distribution with k degrees of freedom such that $k \geq m$, then

$$\frac{k-m+1}{km}\mathbf{t}^T\mathbf{t} \text{ has the } F_{m,k-m+1}\text{-distribution.} \tag{2.31}$$

Applying (2.31) to *Hotelling's* T^2*-statistic*

$$T^2 = n(\bar{\mathbf{Y}}-\boldsymbol{\mu})^T\big(\mathbf{W}/(n-1)\big)^{-1}(\bar{\mathbf{Y}}-\boldsymbol{\mu}), \tag{2.32}$$

where $\bar{\mathbf{Y}}$ and $\mathbf{W}$ are defined in (2.25), yields

$$\frac{n-m}{m(n-1)}T^2 \sim F_{m,n-m}, \tag{2.33}$$

noting that

$$\Big[\mathbf{W}\big/(n-1)\Big]^{-1/2}\Big[\sqrt{n}(\bar{\mathbf{Y}}-\boldsymbol{\mu})\Big] = \Big[W_m(\boldsymbol{\Sigma}, n-1)\big/(n-1)\Big]^{-1/2} N(\mathbf{0},\boldsymbol{\Sigma})$$

has the multivariate t-distribution.

In Definition 2.4 of the multivariate t-distribution, it is assumed that $\mathbf{Z}$ and $\mathbf{W}$ share the same $\boldsymbol{\Sigma}$. More generally, we can consider the case where $\mathbf{Z} \sim N(\mathbf{0}, \mathbf{V})$ instead. By considering $\mathbf{V}^{-1/2}\mathbf{Z}$ instead of $\mathbf{Z}$, we can assume that $\mathbf{V} = \mathbf{I}_m$. Then the density function of $\big(\mathbf{W}/k\big)^{-1/2}\mathbf{Z}$, with independent $\mathbf{Z} \sim N(\mathbf{0}, \mathbf{I}_m)$ and $\mathbf{W} \sim W_m(\boldsymbol{\Sigma}, k)$, has the general form

$$f(\mathbf{t}) = \frac{\Gamma\big((k+m)/2\big)}{(\pi k)^{m/2}\Gamma(k/2)\sqrt{\det(\boldsymbol{\Sigma})}}\left(1+\frac{\mathbf{t}^T\boldsymbol{\Sigma}^{-1}\mathbf{t}}{k}\right)^{-(k+m)/2}, \quad \mathbf{t}\in\mathbb{R}^m. \tag{2.34}$$

This density function, which generalizes (2.30), is used in defining "multivariate t-copulas" in Section 12.3.3.

2.4 Likelihood inference

2.4.1 Method of maximum likelihood

Suppose $X_1, \ldots, X_n$ have joint density function $f_{\boldsymbol{\theta}}(x_1, \ldots, x_n)$, where $\boldsymbol{\theta}$ is an unknown parameter vector, which we shall write as a column vector of dimension p. The *likelihood function* based on the observations $X_1, \ldots, X_n$ is $L(\boldsymbol{\theta}) = f_{\boldsymbol{\theta}}(X_1, \ldots, X_n)$, and the MLE (*maximum likelihood estimate*) $\widehat{\boldsymbol{\theta}}$ of $\boldsymbol{\theta}$

is the maximizer of $L(\boldsymbol{\theta})$. Since $\log x$ is an increasing function of $x > 0$, we usually use the *log-likelihood* $l(\boldsymbol{\theta}) = \log f_{\boldsymbol{\theta}}(X_1, \ldots, X_n)$ instead.

Example: MLE in Gaussian regression models

Consider the regression model $y_t = \boldsymbol{\beta}^T \mathbf{x}_t + \epsilon_t$, in which the regressors $\mathbf{x}_t$ can depend on the past observations $y_{t-1}, \mathbf{x}_{t-1}, \ldots, y_1, \mathbf{x}_1$, and the unobservable random errors ϵ_t are independent normal with mean 0 and variance σ^2. Letting $\boldsymbol{\theta} = (\boldsymbol{\beta}, \sigma^2)$, the likelihood function is

$$L(\boldsymbol{\theta}) = \prod_{t=1}^{n} \left\{ \frac{1}{\sqrt{2\pi}\sigma} e^{-(y_t - \boldsymbol{\beta}^T \mathbf{x}_t)^2/(2\sigma^2)} \right\}. \tag{2.35}$$

It then follows that the MLE of $\boldsymbol{\beta}$ is the same as the OLS estimator $\widehat{\boldsymbol{\beta}}$ that minimizes $\sum_{t=1}^{n} (y_t - \boldsymbol{\beta}^T \mathbf{x}_t)^2$. Moreover, the MLE of σ^2 is $\widehat{\sigma}^2 = n^{-1} \sum_{t=1}^{n} (y_t - \widehat{\boldsymbol{\beta}}^T \mathbf{x}_t)^2$, which differs slightly from the s^2 in (1.10).

Jensen's inequality

Suppose X has a finite mean and $\varphi : \mathbb{R} \longrightarrow \mathbb{R}$ is convex, i.e., $\lambda\varphi(x) + (1 - \lambda)\varphi(y) \geq \varphi(\lambda x + (1-\lambda)y)$ for all $0 < \lambda < 1$ and $x, y \in \mathbb{R}$. (If the inequality above is strict, then φ is called strictly convex.) Then $E\varphi(X) \geq \varphi(EX)$, and the inequality is strict if φ is strictly convex unless X is degenerate (i.e., it takes only one value).

The classical asymptotic theory of maximum likelihood deals with i.i.d. X_t so that $l(\boldsymbol{\theta}) = \sum_{t=1}^{n} \log f_{\boldsymbol{\theta}}(X_t)$. Let $\boldsymbol{\theta}_0$ denote the true parameter value. By the law of large numbers,

$$n^{-1} l(\boldsymbol{\theta}) \to E_{\boldsymbol{\theta}_0} \{\log f_{\boldsymbol{\theta}}(X_1)\} \quad \text{as } n \to \infty, \tag{2.36}$$

with probability 1, for every fixed $\boldsymbol{\theta}$. Since $\log x$ is a (strictly) concave function, (i.e., $-\log x$ is strictly convex), it follows from Jensen's inequality that

$$E_{\boldsymbol{\theta}_0} \left(\log \frac{f_{\boldsymbol{\theta}}(X_1)}{f_{\boldsymbol{\theta}_0}(X_1)} \right) \leq \log \left(E_{\boldsymbol{\theta}_0} \frac{f_{\boldsymbol{\theta}}(X_1)}{f_{\boldsymbol{\theta}_0}(X_1)} \right) = \log 1 = 0.$$

Moreover, the inequality $\leq$ above is strict unless the random variable $f_{\boldsymbol{\theta}}(X_1) / f_{\boldsymbol{\theta}_0}(X_1)$ is degenerate under $P_{\boldsymbol{\theta}_0}$, which means $P_{\boldsymbol{\theta}_0}\{f_{\boldsymbol{\theta}}(X_1) = f_{\boldsymbol{\theta}_0}(X_1)\} = 1$ since $E_{\boldsymbol{\theta}_0}\{f_{\boldsymbol{\theta}}(X_1)/f_{\boldsymbol{\theta}_0}(X_1)\} = 1$.

Thus, except in the case where $f_{\boldsymbol{\theta}}(X_1)$ is the same as $f_{\boldsymbol{\theta}_0}(X_1)$ with probability 1, $E_{\boldsymbol{\theta}_0} \log f_{\boldsymbol{\theta}}(X_1) < E_{\boldsymbol{\theta}_0} \log f_{\boldsymbol{\theta}_0}(X_1)$ for $\boldsymbol{\theta} \neq \boldsymbol{\theta}_0$. This suggests that the MLE is *consistent* (i.e., it converges to $\boldsymbol{\theta}_0$ with probability 1 as $n \to \infty$).

A rigorous proof of consistency of the MLE involves sharpening the argument in (2.36) by considering (instead of fixed $\boldsymbol{\theta}$) $\sup_{\boldsymbol{\theta}\in U} n^{-1}l(\boldsymbol{\theta})$ over neighborhoods U of $\boldsymbol{\theta}' \neq \boldsymbol{\theta}_0$.

Suppose $f_{\boldsymbol{\theta}}(x)$ is smooth in $\boldsymbol{\theta}$. Then we can find the MLE by solving the equation $\nabla l(\boldsymbol{\theta}) = \mathbf{0}$, where ∇l is the (gradient) vector of partial derivatives $\partial l/\partial \boldsymbol{\theta}_i$. For large n, the consistent estimate $\widehat{\boldsymbol{\theta}}$ is near $\boldsymbol{\theta}_0$, and therefore Taylor's theorem yields

$$\mathbf{0} = \nabla l(\widehat{\boldsymbol{\theta}}) = \nabla l(\boldsymbol{\theta}_0) + \nabla^2 l(\boldsymbol{\theta}^*)(\widehat{\boldsymbol{\theta}} - \boldsymbol{\theta}_0), \tag{2.37}$$

where $\boldsymbol{\theta}^*$ lies between $\widehat{\boldsymbol{\theta}}$ and $\boldsymbol{\theta}_0$ (and is therefore close to $\boldsymbol{\theta}_0$) and

$$\nabla^2 l(\boldsymbol{\theta}) = \left(\frac{\partial^2 l}{\partial \boldsymbol{\theta}_i \partial \boldsymbol{\theta}_j} \right)_{1\le i,j\le p} \tag{2.38}$$

is the (Hessian) matrix of second partial derivatives. From (2.37) it follows that

$$\widehat{\boldsymbol{\theta}} - \boldsymbol{\theta}_0 = \left(-\nabla^2 l(\boldsymbol{\theta}^*)\right)^{-1} \nabla l(\boldsymbol{\theta}_0). \tag{2.39}$$

The central limit theorem can be applied to

$$\nabla l(\boldsymbol{\theta}_0) = \sum_{t=1}^{n} \nabla \log f_{\boldsymbol{\theta}}(X_t)|_{\boldsymbol{\theta}=\boldsymbol{\theta}_0}, \tag{2.40}$$

noting that the gradient vector $\nabla \log f_{\boldsymbol{\theta}}(X_t)|_{\boldsymbol{\theta}=\boldsymbol{\theta}_0}$ has mean 0 and covariance matrix

$$\mathbf{I}(\boldsymbol{\theta}_0) = -\left(E_{\boldsymbol{\theta}} \left\{ \frac{\partial^2}{\partial \boldsymbol{\theta}_i \partial \boldsymbol{\theta}_j} \log f_{\boldsymbol{\theta}}(X_t)|_{\boldsymbol{\theta}=\boldsymbol{\theta}_0} \right\} \right)_{1\le i,j\le p}. \tag{2.41}$$

The matrix $\mathbf{I}(\boldsymbol{\theta}_0)$ is called the *Fisher information matrix*. By the law of large numbers,

$$-n^{-1}\nabla^2 l(\boldsymbol{\theta}_0) = -n^{-1} \sum_{t=1}^{n} \nabla^2 \log f_{\boldsymbol{\theta}}(X_t)|_{\boldsymbol{\theta}=\boldsymbol{\theta}_0} \to \mathbf{I}(\boldsymbol{\theta}_0) \tag{2.42}$$

with probability 1. Moreover, $n^{-1}\nabla^2 l(\boldsymbol{\theta}^*) = n^{-1}\nabla^2 l(\boldsymbol{\theta}_0) + o(1)$ with probability 1 by the consistency of $\widehat{\boldsymbol{\theta}}$. From (2.39) and (2.42), it follows that

$$\sqrt{n}(\widehat{\boldsymbol{\theta}} - \boldsymbol{\theta}_0) = \left(-n^{-1}\nabla^2 l(\boldsymbol{\theta}^*)\right)^{-1} \left(\nabla l(\boldsymbol{\theta}_0)/\sqrt{n}\right) \tag{2.43}$$

converges in distribution to $N(0, \mathbf{I}^{-1}(\boldsymbol{\theta}_0))$ as $n \to \infty$, in view of the following theorem.

Slutsky's theorem. If $\mathbf{X}_n$ converges in distribution to $\mathbf{X}$ and $\mathbf{Y}_n$ converges in probability to some *nonrandom* $\mathbf{C}$, then $\mathbf{Y}_n\mathbf{X}_n$ converges in distribution to $\mathbf{CX}$.

It is often more convenient to work with the *observed Fisher information matrix* $-\nabla^2 l(\widehat{\boldsymbol{\theta}})$ than the Fisher information matrix $n\mathbf{I}(\boldsymbol{\theta}_0)$ and to use the following variant of the limiting normal distribution of (2.43):

$$(\widehat{\boldsymbol{\theta}} - \boldsymbol{\theta}_0)^T(-\nabla^2 l(\widehat{\boldsymbol{\theta}}))(\widehat{\boldsymbol{\theta}} - \boldsymbol{\theta}_0) \text{ has a limiting } \chi^2_p\text{-distribution.} \tag{2.44}$$

The preceding asymptotic theory has assumed that X_t are i.i.d. so that the law of large numbers and central limit theorem can be applied. More generally, without assuming that X_t are independent, the log-likelihood function can still be written as a sum:

$$l_n(\boldsymbol{\theta}) = \sum_{t=1}^{n} \log f_{\boldsymbol{\theta}}(X_t|X_1, \dots, X_{t-1}). \tag{2.45}$$

It can be shown that $\{\nabla l_n(\boldsymbol{\theta}_0), n \geq 1\}$ is a martingale, or equivalently,

$$E\left\{\nabla \log f_{\boldsymbol{\theta}}(X_t|X_1, \dots, X_{t-1})|_{\boldsymbol{\theta}=\boldsymbol{\theta}_0}\Big|X_1, \dots, X_{t-1}\right\} = \mathbf{0},$$

and martingale strong laws and central limit theorems can be used to establish the consistency and asymptotic normality of $\widehat{\boldsymbol{\theta}}$ under certain conditions (see Appendix A), so (2.44) still holds in this more general setting.

2.4.2 Asymptotic inference

In view of (2.44), an approximate $100(1-\alpha)\%$ confidence ellipsoid for $\boldsymbol{\theta}_0$ is

$$\left\{\boldsymbol{\theta} : (\widehat{\boldsymbol{\theta}} - \boldsymbol{\theta})^T\big(-\nabla^2 l_n(\widehat{\boldsymbol{\theta}})\big)(\widehat{\boldsymbol{\theta}} - \boldsymbol{\theta}) \leq \chi^2_{p;1-\alpha}\right\}, \tag{2.46}$$

where $\chi^2_{p;1-\alpha}$ is the $(1-\alpha)$th quantile of the χ^2_p-distribution.

Hypothesis testing

To test $H_0 : \boldsymbol{\theta} \in \Theta_0$, where Θ_0 is a q-dimensional subspace of the parameter space with $0 \leq q < p$, the *generalized likelihood ratio* (GLR) *statistic* is

$$\Lambda = 2\{l_n(\widehat{\boldsymbol{\theta}}) - \sup_{\boldsymbol{\theta}\in\Theta_0} l_n(\boldsymbol{\theta})\}, \tag{2.47}$$

where $l_n(\boldsymbol{\theta})$ is the log-likelihood function, defined in (2.45). The derivation of the limiting normal distribution of (2.43) can be modified to show that

$$\Lambda \text{ has a limiting } \chi^2_{p-q} \text{ distribution under } H_0. \tag{2.48}$$

Hence the GLR test with significance level (type I error probability) α rejects H_0 if Λ exceeds $\chi^2_{p-q;1-\alpha}$.

Delta method

Let $g : \mathbb{R}^p \to \mathbb{R}$. If $\widehat{\boldsymbol{\theta}}$ is the MLE of $\boldsymbol{\theta}$, then $g(\widehat{\boldsymbol{\theta}})$ is the MLE of $g(\boldsymbol{\theta})$. If g is continuously differentiable in some neighborhood of the true parameter value $\boldsymbol{\theta}_0$, then $g(\widehat{\boldsymbol{\theta}}) = g(\boldsymbol{\theta}_0) + (\nabla g(\boldsymbol{\theta}^*))^T (\widehat{\boldsymbol{\theta}} - \boldsymbol{\theta}_0)$, where $\boldsymbol{\theta}^*$ lies between $\widehat{\boldsymbol{\theta}}$ and $\boldsymbol{\theta}_0$. Therefore $g(\widehat{\boldsymbol{\theta}}) - g(\boldsymbol{\theta}_0)$ is asymptotically normal with mean 0 and variance

$$\left(\nabla g(\widehat{\boldsymbol{\theta}})\right)^T \left(-\nabla^2 l_n(\widehat{\boldsymbol{\theta}})\right)^{-1} \left(\nabla g(\widehat{\boldsymbol{\theta}})\right). \tag{2.49}$$

This is called the *delta method* for evaluating the asymptotic variance of $g(\widehat{\boldsymbol{\theta}})$. More precisely, the asymptotic normality of $g(\widehat{\boldsymbol{\theta}}) - g(\boldsymbol{\theta}_0)$ means that

$$\left(g(\widehat{\boldsymbol{\theta}}) - g(\boldsymbol{\theta}_0)\right) \Big/ \sqrt{\left(\nabla g(\widehat{\boldsymbol{\theta}})\right)^T \left(-\nabla^2 l_n(\widehat{\boldsymbol{\theta}})\right)^{-1} \left(\nabla g(\widehat{\boldsymbol{\theta}})\right)} \tag{2.50}$$

has a limiting standard normal distribution, which we can use to construct confidence intervals for $g(\boldsymbol{\theta}_0)$. In the time series applications in Chapters 5 and 6, instead of $g(\boldsymbol{\theta})$, one is often interested in $g(\boldsymbol{\theta}, X_n)$, and the same result applies by regarding $g(\boldsymbol{\theta}, X_n)$ as a function of $\boldsymbol{\theta}$ (with X_n fixed).

2.4.3 Parametric bootstrap

If the actual value $\boldsymbol{\theta}_0$ of $\boldsymbol{\theta}$ were known, then an alternative to relying on the preceding asymptotic theory would be to evaluate the sampling distribution of the maximum likelihood estimator (or any other estimator) by Monte Carlo simulations, drawing a large number of samples of the form $(X_1^*, \ldots, X_n^*)$ from the distribution $F_{\boldsymbol{\theta}_0}$, which has joint density function $f_{\boldsymbol{\theta}_0}(x_1, \ldots, x_n)$. Since $\boldsymbol{\theta}_0$ is actually unknown, the parametric bootstrap draws bootstrap samples $(X_{b,1}^*, \ldots, X_{b,n}^*)$, $b = 1, \ldots, B$, from $F_{\widehat{\boldsymbol{\theta}}}$, where $\widehat{\boldsymbol{\theta}}$ is the maximum likelihood estimate of $\boldsymbol{\theta}$. As in Section 1.6, we can use these bootstrap samples to estimate the bias and standard error of $\widehat{\boldsymbol{\theta}}$. To construct a bootstrap confidence region for $g(\boldsymbol{\theta}_0)$, we can use (2.50) as an approximate *pivot* whose (asymptotic) distribution does not depend on the unknown $\boldsymbol{\theta}_0$ and use the quantiles of its bootstrap distribution in place of standard normal quantiles.

Exercises

2.1. The file d_swap.txt contains the daily swap rates r_{kt} for eight maturities $T_k = 1, 2, 3, 4, 5, 7, 10$, and 30 years from July 3, 2000 to July 15, 2007. In the PCA of Section 2.2.3, (2.17) is an approximation to

$$(\mathbf{d}_k - \widehat{\mu}_k)/\widehat{\sigma}_k = \sum_{j=1}^{8} \widehat{a}_{kj} \mathbf{Y}_j.$$

The ratio $a_{kj}^2 / \left(\sum_{i=1}^{8} a_{ki}^2 \right)$ represents the proportion that the jth principal component contributes to the variance of the daily changes in the swap rate with maturity T_k. Compute this ratio for the first three principal components and each swap rate. Compare the results with those in Section 2.2.3, where we consider the overall variance instead of the individual variances for different swap rates. Discuss your finding.

2.2. The file m_swap.txt contains the monthly swap rates r_{kt} for eight maturities $T_k = 1, 2, 3, 4, 5, 7, 10$, and 30 years from July 2000 to June 2007.
   (a) Perform a principal component analysis (PCA) of the data using the sample covariance matrix.
   (b) Perform a PCA of the data using the sample correlation matrix.
   (c) Compare your results with those in Section 2.2.3. Discuss the influence of sampling frequency on the result.

2.3. The file d_logret_12stocks.txt contains the daily log returns of 12 stocks from January 3, 2001 to December 30, 2005. The 12 stocks include Apple Computer, Adobe Systems, Automatic Data Processing, Advanced Micro Devices, Dell, Gateway, Hewlett-Packard Company, International Business Machines Corp., Microsoft Corp., Orcale Corp., Sun Microsystems, and Yahoo!.
   (a) Perform a principal component analysis (PCA) of the data using the sample covariance matrix.
   (b) Perform a PCA of the data using the sample correlation matrix.

2.4. Consider an $n \times n$ nonsingular matrix

$$\mathbf{A} = \begin{pmatrix} \mathbf{A}_{11} & \mathbf{A}_{12} \\ \mathbf{A}_{21} & \mathbf{A}_{22} \end{pmatrix},$$

where $\mathbf{A}_{11}$ is $m \times m$ ($m < n$). Assume that $\mathbf{A}_{22}$ and $\widetilde{\mathbf{A}}_{11} := \mathbf{A}_{11} - \mathbf{A}_{12}\mathbf{A}_{22}^{-1}\mathbf{A}_{21}$ are nonsingular.
   (a) Show that $\mathbf{A}^{-1}$ is given by

$$\begin{pmatrix} \widetilde{\mathbf{A}}_{11}^{-1} & -\widetilde{\mathbf{A}}_{11}^{-1}\mathbf{A}_{12}\mathbf{A}_{22}^{-1} \\ -\mathbf{A}_{22}^{-1}\mathbf{A}_{21}\widetilde{\mathbf{A}}_{11}^{-1} & \mathbf{A}_{22}^{-1} + \mathbf{A}_{22}^{-1}\mathbf{A}_{21}\widetilde{\mathbf{A}}_{11}^{-1}\mathbf{A}_{12}\mathbf{A}_{22}^{-1} \end{pmatrix}.$$

(b) Show that $\det(\mathbf{A}) = \det(\mathbf{A}_{22}) \cdot \det(\mathbf{A}_{11} - \mathbf{A}_{12}\mathbf{A}_{22}^{-1}\mathbf{A}_{21})$.

(c) Use (a) and (b) to derive the conditional distribution of $\mathbf{Y}_1$ given $\mathbf{Y}_2 = \mathbf{y}_2$ in Section 2.3.2.

2.5. Suppose (X_1, X_2) has the bivariate normal distribution with density function

$$f(x_1, x_2) = \frac{1}{2\pi\sqrt{(1-\rho^2)}} \exp\left\{ -\frac{x_1^2 - 2\rho x_1 x_2 + x_2^2}{2(1-\rho^2)} \right\}.$$

Show that $\mathrm{Corr}(X_1^2, X_2^2) = \rho^2$.

2.6. Let (U_i, V_i) be i.i.d. bivariate normal random vectors. The sample correlation coefficient is given by $\widehat{\rho} = s_{UV}/(s_U s_V)$, where $s_U^2 = n^{-1}\sum_{i=1}^n (U_i - \overline{U})^2$, $s_V^2 = n^{-1}\sum_{i=1}^n (V_i - \overline{V})^2$, $s_{UV} = n^{-1}\sum_{i=1}^n (U_i - \overline{U})(V_i - \overline{V})$, $\overline{U} = n^{-1}\sum_{i=1}^n U_i$, and $\overline{V} = n^{-1}\sum_{i=1}^n V_i$.

(a) Let $\mathbf{x}_i = (U_i, V_i, U_i^2, V_i^2, U_i V_i)^T$, $1 \le i \le n$, and $\overline{\mathbf{x}} = n^{-1}\sum_{i=1}^n \mathbf{x}_i$. Show that $\widehat{\rho} = g(\overline{\mathbf{x}})$, where

$$g((x_1, x_2, x_3, x_4, x_5)^T) = \frac{x_5 - x_1 x_2}{(x_3 - x_1^2)^{1/2}(x_4 - x_2^2)^{1/2}}.$$

(b) Use (a) and the delta method to prove

$$\sqrt{n}(\widehat{\rho} - \rho) \to N(0, (1-\rho^2)^2).$$

2.7. The file `w_logret_3automanu.txt` contains the weekly log returns of three auto manufacturers, General Motors Corp., Ford Motor Corp., and Toyota Motor Corp., from the week of January 3, 1994 to the week of June 25, 2007.

(a) Write down the sample covariance matrix $\widehat{\mathbf{V}}$ of these returns.

(b) For each pair of these three auto manufacturers, use the result in Exercise 2.6(b) to construct a 95% confidence interval for the correlation coefficient of their returns.

(c) What assumptions have you made for (b)? Perform some graphical checks, such as data plots and Q-Q plots, on these assumptions.

2.8. Consider the linear regression model $y_t = \boldsymbol{\beta}^T \mathbf{x}_t + \epsilon_t$ satisfying assumptions (A) and (C*) in Section 1.1.4. Show that minimizing Akaike's information criterion (1.32) is equivalent to minimizing (1.33).

2.9. Let $X_1, \ldots, X_n$ be n observations for which the joint density function $f_{\boldsymbol{\theta}}(x_1, \ldots, x_n)$ depends on an unknown parameter vector $\boldsymbol{\theta}$. Assuming that f is a smooth function of $\boldsymbol{\theta}$, show that:

(a) $E\big(\boldsymbol{\nabla} \log f_{\boldsymbol{\theta}}(X_1, \ldots, X_n)\big) = \mathbf{0}$;

(b) $E\big(-\boldsymbol{\nabla}^2 l(\boldsymbol{\theta})\big) = \mathrm{Cov}\big(\boldsymbol{\nabla} \log f_{\boldsymbol{\theta}}(X_1, \ldots, X_n)\big)$, where $l(\boldsymbol{\theta})$ denotes the log-likelihood function.

3

Basic Investment Models and Their Statistical Analysis

Three cornerstones of quantitative finance are asset returns, interest rates, and volatilities. They appear in many fundamental formulas in finance. In this chapter, we consider their interplay and the underlying statistical issues in a classical topic in quantitative finance, namely portfolio theory. Sections 3.2–3.4 give an overview of Markowitz's mean-variance theory of optimal portfolios, Sharpe's CAPM (capital asset pricing model), and the arbitrage pricing theory (APT) developed by Ross. These theories all involve the means and standard deviations (volatilities) of returns on assets and their portfolios, and CAPM and APT also involve interest rates. Section 3.1 introduces the concept of asset returns and associated statistical models. This chapter uses the simplest statistical model for returns data, namely i.i.d. random variables for daily, weekly, or yearly returns. More refined time series models for asset returns are treated in Chapters 6, 9, and 11 (in which Section 11.2 deals with high-frequency data).

Sections 3.2–3.4 also consider statistical methods and issues in the implementation of these investment theories. Although these single-period theories and the underlying statistical model assuming i.i.d. asset returns seem to suggest that the statistical methods needed would be relatively simple, it turns out that the statistical problems encountered in the implementation of these theories require much deeper and more powerful methods. Whereas CAPM is the simplest to implement because it only involves simple linear regression, its ease of implementation has led to extensive empirical testing of the model. Section 3.3.4 gives a summary of these empirical studies and statistical issues such as sample selection bias. Empirical evidence has indicated that CAPM does not completely explain the cross section of expected asset returns and that additional factors are needed, as in the multifactor pricing models suggested by APT. However, this leaves open the question concerning how the factors should be chosen, which is discussed in Sections 3.4.2 and 3.4.3. In this connection, factor analysis, which is a statistical method widely used in psychometrics, is also introduced. The parameters in Markowitz's portfolio theory are the mean and covariance matrix of the vector of asset returns in the portfolio under consideration, and it is straightforward to estimate them from empirical data of asset returns. However, the uncertainties in these estimates and their impact on implementing Markowitz's efficient portfolios have led to two major approaches in the past decade to address the statistical issues. Section 3.5 reviews one approach, which involves the bootstrap method described in Section 1.6. The other approach uses multivariate analysis and Bayesian methods from Chapters 2 and 4 and will be described in Chapter 4.

3.1 Asset returns

3.1.1 Definitions

One-period net returns and gross returns

Let P_t denote the asset price at time t. Suppose the asset does not have dividends over the period from time $t-1$ to time t. Then the *one-period net return* on this asset is

$$R_t = \frac{P_t - P_{t-1}}{P_{t-1}}, \tag{3.1}$$

which is the profit rate of holding the asset during the period. Another concept is the *gross return* P_t/P_{t-1}, which is equal to $1 + R_t$.

Multiperiod returns

One-period returns can be extended to the multiperiod case as follows. The *gross return over k periods* is then defined as

$$1 + R_t(k) = \frac{P_t}{P_{t-k}} = \prod_{j=0}^{k-1}(1 + R_{t-j}), \tag{3.2}$$

and the *net return over these periods* is $R_t(k)$. In practice, we usually use years as the time unit. The *annualized gross return* for holding an asset over k years is $\left(1+R_t(k)\right)^{1/k}$, and the *annualized net return* is $\left(1+R_t(k)\right)^{1/k} - 1$.

Continuously compounded return (log return)

Let $p_t = \log P_t$. The *logarithmic return* or *continuously compounded return* on an asset is defined as

$$r_t = \log(P_t/P_{t-1}) = p_t - p_{t-1}. \tag{3.3}$$

One property of log returns is that, as the time step Δt of a period approaches 0, the log return r_t is approximately equal to the net return:

$$r_t = \log(P_t/P_{t-1}) = \log(1 + R_t) \approx R_t.$$

Another property is the additivity of multiperiod returns: A k-period log return is the sum of k simple single-period log returns:

$$r_t[k] = \log \frac{P_t}{P_{t-k}} = \sum_{j=0}^{k-1} \log(1 + R_{t-j}) = \sum_{j=0}^{k-1} r_{t-j}.$$

Adjustment for dividends

Many assets pay dividends periodically. In this case, the definition of asset returns has to be modified to incorporate dividends. Let D_t be the *dividend* payment between times $t-1$ and t. The net return and the continuously compounded return are modified as

$$R_t = \frac{P_t + D_t}{P_{t-1}} - 1, \quad r_t = \log(P_t + D_t) - \log P_{t-1}.$$

Multiperiod returns can be similarly modified. In particular, a k-period log return now becomes

$$r_t[k] = \log \left[\prod_{j=0}^{k-1} \frac{P_{t-j} + D_{t-j}}{P_{t-j-1}} \right] = \sum_{j=0}^{k-1} \log \left(\frac{P_{t-j} + D_{t-j}}{P_{t-j-1}} \right).$$

Excess returns

In many applications, it is convenient to use the *excess return*, which refers to the difference $r_t - r_t^*$ between the asset's log return r_t and the log return r_t^* on some reference asset, which is usually taken to be a riskless asset such as a short-term U.S. Treasury bill.

Portfolio returns

Suppose one has a portfolio consisting of p different assets. Let w_i be the weight, often expressed as a percentage, of the portfolio's value invested in asset i. Thus, the value of asset i is $w_i P_t$ when the total value of the portfolio is P_t. Suppose R_{it} and r_{it} are the net return and log return of asset i at time t, respectively. The value of the portfolio provided by the asset at time t is $w_i P_{t-1}(1 + R_{it})$, so the total value of the portfolio is $(1 + \sum_{i=1}^{p} w_i R_{it}) P_{t-1}$. Therefore the overall net return R_t and a corresponding formula for the log return r_t of the portfolio are

$$R_t = \sum_{i=1}^{p} w_i R_{it}, \quad r_t = \log\left(1 + \sum_{i=1}^{p} w_i R_{it}\right) \approx \sum_{i=1}^{p} w_i r_{it}. \tag{3.4}$$

3.1.2 Statistical models for asset prices and returns

In the case of a "risk-free" asset (e.g. a Treasury bond), the rate of return is called an *interest rate.* If the interest rate is a constant R and is compounded once per unit time, then the value of the risk-free asset at time t is $P_t = P_0(1+R)^t$, which is a special case of (3.2) with $R_{t-j} = R$. If the interest rate is continuously compounded at rate r, then $P_t = P_0 e^{rt}$, so $\log(P_t/P_{t-1}) = r$ as in (3.3), and $dP_t/P_t = r dt$. A commonly used model for risky asset prices simply generalizes this ordinary differential equation to a stochastic differential equation of the form

$$\frac{dP_t}{P_t} = \theta dt + \sigma dw_t, \tag{3.5}$$

where $\{w_t, t \geq 0\}$ is Brownian motion; see Appendix A. The price process P_t is called *geometric Brownian motion* (GBM), with *volatility* σ and instantaneous rate of return θ, and has the explicit representation $P_t = P_0 \exp\{(\theta - \frac{\sigma^2}{2})t + \sigma w_t\}$.

The discrete-time analog of this price process has returns $r_t = \log(P_t/P_{t-1})$ that are i.i.d. $N(\mu, \sigma^2)$ with $\mu = \theta - \sigma^2/2$. Note that $P_t = P_0 \exp\left(\sum_{i=1}^t r_i\right)$ has the *lognormal* distribution (i.e., $\log P_t$ is normal); see Figure 3.1 for the density function of a lognormal distribution. Note that $N(\mu, \sigma^2)$ is the sum of m i.i.d. $N(\mu/m, \sigma^2/m)$ random variables, which is the "infinitely divisible" property of the normal distribution. Hence, the mean and standard deviation (SD, also called volatility) of the annual log return r_{year} are related to those of the monthly log return r_{month} by

$$E(r_{\text{year}}) = 12E(r_{\text{month}}), \qquad \text{SD}(r_{\text{year}}) = \sqrt{12}\,\text{SD}(r_{\text{month}}). \tag{3.6}$$

For daily returns, we consider only the number of trading days in the year (often taken to be 252).

The convention above is for relating the annual mean return and its volatility to their monthly or daily counterparts. This convention is based on the i.i.d. assumption of daily returns. Daily prices of securities quoted in the financial press are usually closing prices. Market data are actually much more complicated and voluminous than those summarized in the financial press. Transaction databases consist of historical prices, traded quantities, and bid-ask prices and sizes, transaction by transaction. These "high-frequency" data provide information on the "market microstructure"; see Chapter 11 (Section 11.2) for further details. In Chapter 6, we consider time series models (in

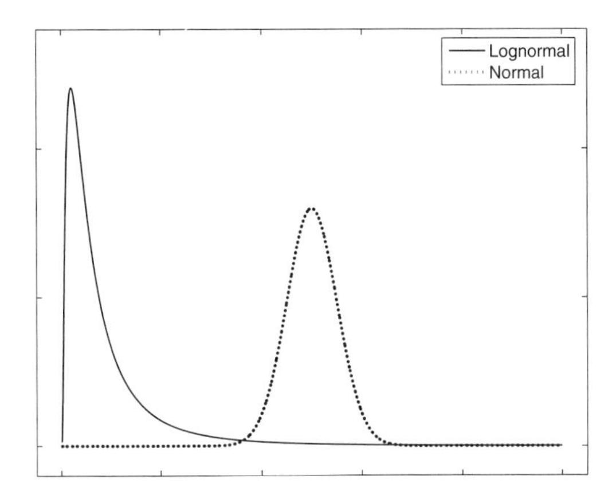

Fig. 3.1. Density functions of normal and lognormal random variables.

contrast to the i.i.d. returns considered here) of daily, weekly, and monthly asset returns.

3.2 Markowitz's portfolio theory

3.2.1 Portfolio weights

For a single-period portfolio of p assets with weights w_i, the return of the portfolio over the period can be represented by $R = \sum_{i=1}^{p} w_i R_i$; see (3.4), in which we suppress the index t since we consider here a fixed time t. The mean μ and variance σ^2 of the portfolio return R are given by

$$\mu = \sum_{i=1}^{p} w_i E(R_i), \qquad \sigma^2 = \sum_{1 \le i,j \le p} w_i w_j \mathrm{Cov}(R_i, R_j). \tag{3.7}$$

With the weights w_i satisfying the constraints

(i) $\sum_{i=1}^{p} w_i = 1$,
(ii) $1 \ge w_i \ge 0$,

diversification via a portfolio tends to reduce the "risk," as measured by the return's standard deviation, of the risky investment. For example, when the R_i are uncorrelated or negatively correlated and the weights w_i satisfy (i) and (ii),

$$\sigma^2 \le \sum_{i=1}^{p} w_i^2 \mathrm{Var}(R_i) \le \sum_{i=1}^{p} w_i \mathrm{Var}(R_i).$$

In particular, in the case $w_i = 1/p$ and $\mathrm{Var}(R_i) = v$ for all i, $\sigma^2 \le v/p$, which is only a fraction, $1/p$, of the individual asset return's variance. To reduce risk through diversification, an investor usually has to settle for lower expected returns than those of the riskier assets with high mean returns. Markowitz's theory of optimal portfolios relates to the optimal trade-off between the portfolio's mean return and its standard deviation (volatility).

Sometimes it is possible to sell an asset that one does not own by *short selling* (or *shorting*) the asset. Short selling involves borrowing the asset from a lender (e.g., a brokerage firm) and then selling the asset to a buyer. At a later date, one has to purchase the asset and return it to the lender. Assumption (ii) on the weights can be dropped if short selling is allowed. In this case, the theory becomes simpler.

3.2.2 Geometry of efficient sets

Feasible region

To begin, consider the case of $p = 2$ risky assets whose returns have means μ_1, μ_2, standard deviations σ_1, σ_2, and correlation coefficient ρ. Let $w_1 = \alpha$ and $w_2 = 1 - \alpha$ $(0 \le \alpha \le 1)$. Then the mean return of the portfolio is $\mu(\alpha) = \alpha\mu_1 + (1-\alpha)\mu_2$, and its volatility $\sigma(\alpha)$ is given by $\sigma^2(\alpha) = \alpha^2\sigma_1^2 + 2\rho\alpha(1-\alpha)\sigma_1\sigma_2 + (1-\alpha)^2\sigma_2^2$. Figure 3.2 plots the curve $\{(\sigma(\alpha), \mu(\alpha)) : 0 \le \alpha \le 1\}$ for different values of ρ. The points P_1 and P_2 correspond to $\alpha = 1$ and $\alpha = 0$, respectively. The dotted lines denote the triangular boundaries of the (σ, μ) region for portfolio returns when short selling is allowed in the case $\rho = -1$, and the point A corresponds to one such portfolio.

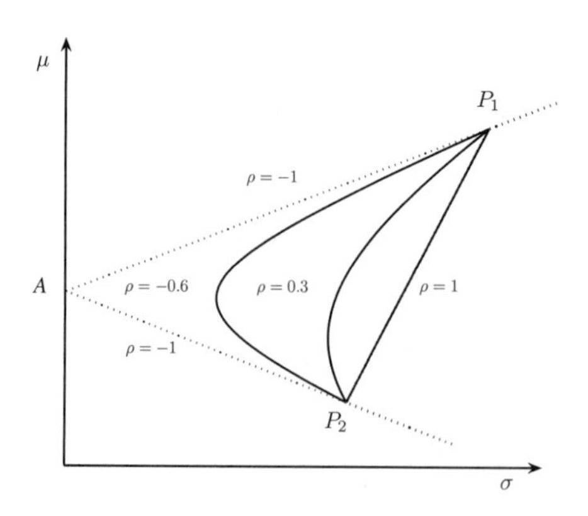

Fig. 3.2. Feasible region for two assets.

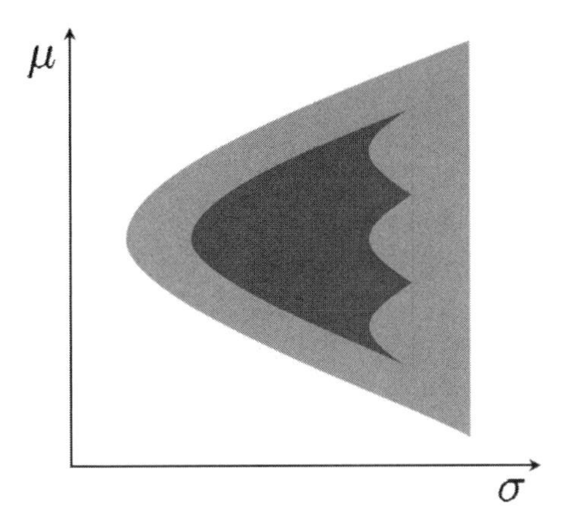

Fig. 3.3. Feasible region for $p \geq 3$ assets.

The set of points in the (σ, μ) plane that correspond to the returns of portfolios of the p assets is called a *feasible region*. For $p \geq 3$, the feasible region is a connected two-dimensional set. It is also convex to the left in the sense that given any two points in the region, the line segment joining them does not cross the left boundary of the region. Figure 3.3 gives a schematic plot of the feasible region without short selling (the darkened region) and with short selling (the shaded larger region).

Minimum-variance set and the efficient frontier

The left boundary of the feasible region is called the *minimum-variance set*. For a given value μ of the mean return, the feasible point with the smallest σ lies on this left boundary; it corresponds to the *minimum-variance portfolio* (MVP). For a given value σ of volatility, investors prefer the portfolio with the largest mean return, which is achieved at an upper left boundary point of the feasible region. The upper portion of the minimum-variance set is called the *efficient frontier*; see Figure 3.4.

3.2.3 Computation of efficient portfolios

Let $\mathbf{r} = (R_1, \ldots, R_p)^T$ denote the vector of returns of p assets, $\mathbf{1} = (1, \ldots, 1)^T$, $\mathbf{w} = (w_1, \ldots, w_p)^T$,

$$\begin{aligned} \boldsymbol{\mu} &= (\mu_1, \ldots, \mu_p)^T = (E(R_1), \ldots, E(R_p))^T, \\ \boldsymbol{\Sigma} &= (\mathrm{Cov}(R_i, R_j))_{1 \leq i \leq j \leq p}. \end{aligned} \tag{3.8}$$

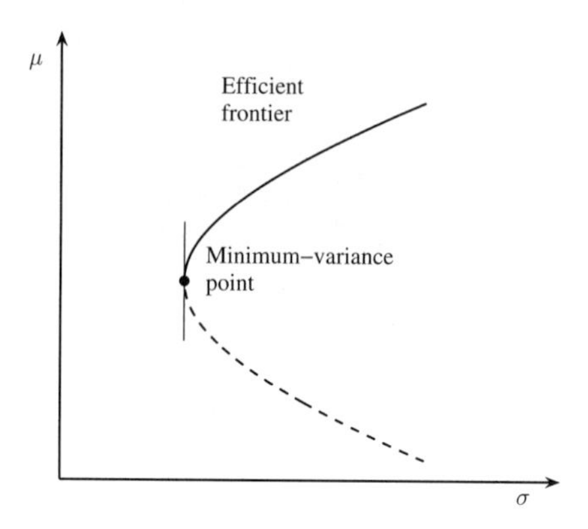

Fig. 3.4. Efficient frontier and minimum-variance point.

We first consider the case where short selling is allowed (so that w_i can be negative); Lagrange multipliers can be used to come up with explicit formulas for the efficient portfolio weights. We then impose the constraint $0 \le w_i \le 1$ and show how quadratic programming can be used to solve for efficient portfolios.

Lagrange multipliers under short selling

Given a target value μ_* for the mean return of the portfolio, the weight vector $\mathbf{w}$ of an efficient portfolio can be characterized by

$$\mathbf{w}_{\text{eff}} = \arg\min_{\mathbf{w}} \mathbf{w}^T \boldsymbol{\Sigma} \mathbf{w} \quad \text{subject to } \mathbf{w}^T \boldsymbol{\mu} = \mu_*, \ \mathbf{w}^T \mathbf{1} = 1, \tag{3.9}$$

when short selling is allowed. The method of Lagrange multipliers leads to the optimization problem $\min_{\mathbf{w},\lambda,\gamma} \left\{ \mathbf{w}^T \boldsymbol{\Sigma} \mathbf{w} - 2\lambda(\mathbf{w}^T \boldsymbol{\mu} - \mu_*) - 2\gamma(\mathbf{w}^T \mathbf{1} - 1) \right\}$. Differentiating the objective function with respect to $\mathbf{w}$, λ, and γ and setting the derivatives equal to 0 leads to the system of linear equations

$$\boldsymbol{\Sigma}\mathbf{w} = \lambda \boldsymbol{\mu} + \gamma \mathbf{1}, \quad \mathbf{w}^T \boldsymbol{\mu} = \mu_*, \quad \mathbf{w}^T \mathbf{1} = 1, \tag{3.10}$$

which has the explicit solution

$$\mathbf{w}_{\text{eff}} = \left\{ B\boldsymbol{\Sigma}^{-1}\mathbf{1} - A\boldsymbol{\Sigma}^{-1}\boldsymbol{\mu} + \mu_*\left(C\boldsymbol{\Sigma}^{-1}\boldsymbol{\mu} - A\boldsymbol{\Sigma}^{-1}\mathbf{1}\right) \right\} \Big/ D \tag{3.11}$$

when $\boldsymbol{\Sigma}$ is nonsingular, where

$$A = \boldsymbol{\mu}^T \boldsymbol{\Sigma}^{-1} \mathbf{1} = \mathbf{1}^T \boldsymbol{\Sigma}^{-1} \boldsymbol{\mu}, \quad B = \boldsymbol{\mu}^T \boldsymbol{\Sigma}^{-1} \boldsymbol{\mu}, \quad C = \mathbf{1}^T \boldsymbol{\Sigma}^{-1} \mathbf{1}, \quad D = BC - A^2.$$

The variance of the return on this efficient portfolio is

$$\sigma_{\text{eff}}^2 = \left(B - 2\mu_* A + \mu_*^2 C\right)/D. \tag{3.12}$$

The μ_* that minimizes σ_{eff}^2 is given by

$$\mu_{\text{minvar}} = \frac{A}{C}, \tag{3.13}$$

which corresponds to the *global MVP* with variance $\sigma_{\text{minvar}}^2 = 1/C$ and weight vector

$$\mathbf{w}_{\text{minvar}} = \boldsymbol{\Sigma}^{-1}\mathbf{1}/C. \tag{3.14}$$

For two MVPs with mean returns μ_p and μ_q, their weight vectors are given by (3.11) with $\mu_* = \mu_p, \mu_q$, respectively. From this it follows that the covariance of the returns r_p and r_q is given by

$$\text{Cov}(r_p, r_q) = \frac{C}{D}\left(\mu_p - \frac{A}{C}\right)\left(\mu_q - \frac{A}{C}\right) + \frac{1}{C}. \tag{3.15}$$

Combining (3.15) and (3.13) shows that $\text{Cov}(r_{\text{minvar}}, r_q) = 1/C$ for any MVP q whose return is denoted by r_q.

Quadratic programming under nonnegativity constraints on w

When short selling is not allowed, we need to add the constraint $w_i \geq 0$ for all i (denoted by $\mathbf{w} \geq \mathbf{0}$). Hence the optimization problem (3.9) has to be modified as

$$\mathbf{w}_{\text{eff}} = \arg\min_{\mathbf{w}} \mathbf{w}^T \boldsymbol{\Sigma} \mathbf{w} \quad \text{subject to } \mathbf{w}^T \boldsymbol{\mu} = \mu_*, \ \mathbf{w}^T \mathbf{1} = 1, \ \mathbf{w} \geq \mathbf{0}. \tag{3.16}$$

Such problems do not have explicit solutions by transforming them to a system of equations via Lagrange multipliers. Instead, we can use quadratic programming to minimize the quadratic objective function $\mathbf{w}^T \boldsymbol{\Sigma} \mathbf{w}$ under linear equality constraints and nonnegativity constraints.

The implementation of quadratic programming can use the `MATLAB` optimization toolbox; in particular, the function `quadprog(H, f, A, b, Aeq, beq, LB, UB)` performs quadratic programming to minimize $\frac{1}{2}\mathbf{x}^T \mathbf{H} \mathbf{x} + \mathbf{f}^T \mathbf{x}$, in which $\mathbf{H}$ is an $N \times N$ matrix and $\mathbf{f}$ is an $N \times 1$ vector, subject to the inequality constraint $\mathbf{A}\mathbf{x} \leq \mathbf{b}$, equality constraint $\mathbf{A}_{\text{eq}}\mathbf{x} = \mathbf{b}_{\text{eq}}$, and additional inequality constraints of the form $LB_i \leq x_i \leq UB_i$ for $i = 1, \ldots, N$.

3.2.4 Estimation of $\boldsymbol{\mu}$ and $\boldsymbol{\Sigma}$ and an example

Table 3.1 gives the means and covariances of the monthly log returns of six stocks, estimated from 63 monthly observations during the period August

2000 to October 2005. The stocks cover six sectors in the Dow Jones Industrial Average: American Express (AXP, consumer finance), Citigroup Inc. (CITI, banking), Exxon Mobil Corp. (XOM, integrated oil and gas), General Motors (GM, automobiles), Intel Corp. (INTEL, semiconductors), and Pfizer Inc. (PFE, pharmaceuticals).

Table 3.1. Estimated mean (in parentheses, multiplied by 10^2) and covariance matrix (multiplied by 10^4) of monthly log returns.

		AXP	CITI	XOM	GM	INTEL	PFE
AXP	(0.033)	9.01					
CITI	(0.034)	5.69	9.64				
XOM	(0.317)	2.39	1.89	5.25			
GM	(−0.338)	5.97	4.41	2.40	20.2		
INTEL	(−0.701)	10.1	12.1	0.59	12.4	46.0	
PFE	(−0.414)	1.46	2.34	9.85	1.86	1.06	5.33

Figure 3.5 shows the "plug-in" efficient frontier for these six stocks allowing short selling and using the approximation in (3.4). By "plug-in" we mean that the mean $\boldsymbol{\mu}$ and covariance $\boldsymbol{\Sigma}$ in (3.11) are substituted by the estimated values $\widehat{\boldsymbol{\mu}}$ and $\widehat{\boldsymbol{\Sigma}}$ given in Table 3.1. To adjust for sampling variability of the estimates $\widehat{\boldsymbol{\mu}}$ and $\widehat{\boldsymbol{\Sigma}}$, Michaud (1989) advocates using a "resampled" efficient frontier instead of the plug-in frontier; details are given in Section 3.5.1.

3.3 Capital asset pricing model (CAPM)

The capital asset pricing model, introduced by Sharpe (1964) and Lintner (1965), builds on Markowitz's portfolio theory to develop economy-wide implications of the trade-off between return and risk, assuming that there is a risk-free asset and that all investors have homogeneous expectations and hold mean-variance-efficient portfolios.

3.3.1 The model

The one-fund theorem

Suppose the market has a risk-free asset with return r_f (interest rate) besides n risky assets. If both lending and borrowing of the risk-free asset at rate r_f are allowed, the feasible region is an infinite triangular region. The efficient

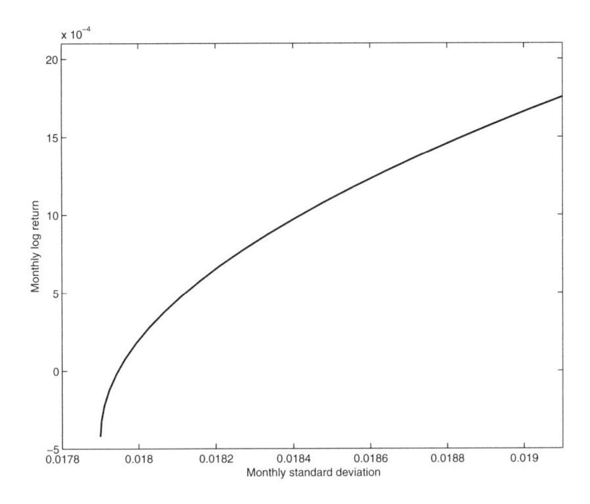

Fig. 3.5. Estimated efficient frontier of portfolios that consist of six assets.

frontier is a straight line that is tangent to the original feasible region of the n risky assets at a point M, called the *tangent point*; see Figure 3.6. This tangent point M can be thought of as an index fund or market portfolio.

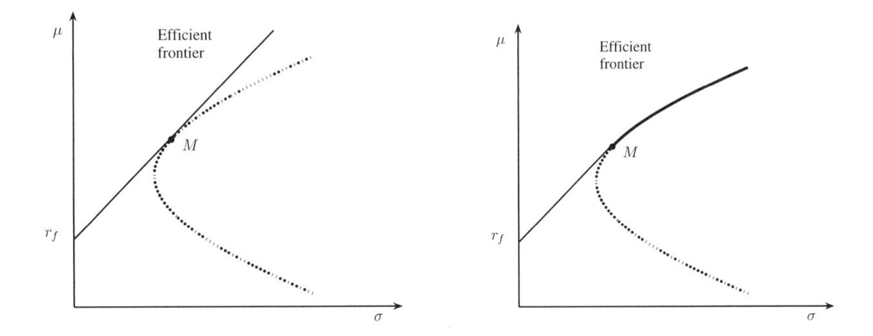

Fig. 3.6. Minimum-variance portfolios of risky assets and a risk-free asset. Left panel: short selling is allowed. Right panel: short selling is not allowed.

One-fund theorem. There is a single fund M of risky assets such that any efficient portfolio can be constructed as a linear combination of the fund M and the risk-free asset.

When short selling is allowed, the minimum-variance portfolio (MVP) with the expected return μ_* can be computed by solving the optimization problem

$$\min_{\mathbf{w}} \mathbf{w}^T \boldsymbol{\Sigma} \mathbf{w} \text{ subject to } \mathbf{w}^T \boldsymbol{\mu} + (1 - \mathbf{w}^T \mathbf{1}) r_f = \mu_*, \tag{3.17}$$

in which $\boldsymbol{\mu}$ and $\boldsymbol{\Sigma}$ are defined in (3.8).

Note that the constraint $\mathbf{w}^T \mathbf{1} = 1$ in the Markowitz theory is no longer necessary since $(1 - \mathbf{w}^T \mathbf{1})$ can be invested in the risk-free asset. The Lagrangian is the function $L = \mathbf{w}^T \boldsymbol{\Sigma} \mathbf{w} + 2\lambda\{\mathbf{w}^T \boldsymbol{\mu} + (1 - \mathbf{w}^T \mathbf{1}) r_f - \mu_*\}$. Differentiating L with respect to $\mathbf{w}$ and λ yields

$$\begin{cases} \boldsymbol{\Sigma} \mathbf{w} + \lambda(\boldsymbol{\mu} - r_f \mathbf{1}) = \mathbf{0}, \\ \mathbf{w}^T \boldsymbol{\mu} + (1 - \mathbf{w}^T \mathbf{1}) r_f = \mu_*, \end{cases} \tag{3.18}$$

which has an explicit solution for $\mathbf{w}$ when $\boldsymbol{\Sigma}$ is nonsingular:

$$\mathbf{w}_{\text{eff}} = \frac{(\mu_* - r_f)}{(\boldsymbol{\mu} - r_f \mathbf{1})^T \boldsymbol{\Sigma}^{-1} (\boldsymbol{\mu} - r_f \mathbf{1})} \boldsymbol{\Sigma}^{-1} (\boldsymbol{\mu} - r_f \mathbf{1}). \tag{3.19}$$

Note that $\mathbf{w}_{\text{eff}}$ is a scalar multiple (depending on μ_*) of the vector $\boldsymbol{\Sigma}^{-1}(\boldsymbol{\mu} - r_f \mathbf{1})$ that does not depend on μ_*. Hence all MVPs are linear combinations of the risk-free asset and a particular portfolio with weight vector $\boldsymbol{\Sigma}^{-1}(\boldsymbol{\mu} - r_f \mathbf{1}) / \{\mathbf{1}^T \boldsymbol{\Sigma}^{-1} (\boldsymbol{\mu} - r_f \mathbf{1})\}$, which is the "fund" (or market portfolio M) in the one-fund theorem. The market portfolio is a portfolio consisting of weighted assets in a universe of investments. A widely used proxy for the market portfolio is the Standard and Poor's (S&P) 500 index, which is considered a leading index of the U.S. equity market. It includes 500 stocks chosen for their market capitalization, liquidity, and industry sector, as illustrated by Figures 3.7–3.9.

Sharpe ratio and the capital market line

For a portfolio whose return has mean μ and variance σ^2, its *Sharpe ratio* is $(\mu - r_f)/\sigma$, which is the expected excess return per unit of risk. The straight line joining $(0, r_f)$ and the tangent point M in Figure 3.6, which is the efficient frontier in the presence of a risk-free asset, is called the *capital market line.* Note that this line is defined by the linear equation

$$\mu = r_f + \frac{\mu_M - r_f}{\sigma_M} \sigma; \tag{3.20}$$

i.e., the Sharpe ratio of any efficient portfolio is the same as that of the market portfolio.

Beta and the security market line

The *beta*, denoted by β_i, of risky asset i that has return r_i is defined by $\beta_i = \text{Cov}(r_i, r_M)/\sigma_M^2$. The capital asset pricing model (CAPM) relates the

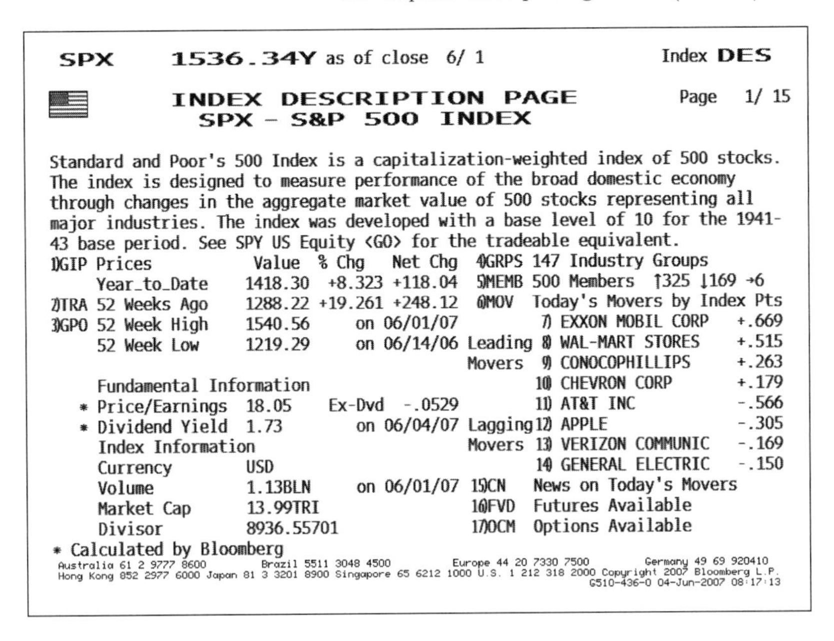

SPX 1536.34Y as of close 6/ 1 Index DES

INDEX DESCRIPTION PAGE Page 1/ 15
SPX - S&P 500 INDEX

Standard and Poor's 500 Index is a capitalization-weighted index of 500 stocks. The index is designed to measure performance of the broad domestic economy through changes in the aggregate market value of 500 stocks representing all major industries. The index was developed with a base level of 10 for the 1941-43 base period. See SPY US Equity <GO> for the tradeable equivalent.

		Value	% Chg	Net Chg
1)GIP	Prices			
	Year_to_Date	1418.30	+8.323	+118.04
2)TRA	52 Weeks Ago	1288.22	+19.261	+248.12
3)GPO	52 Week High	1540.56	on 06/01/07	
	52 Week Low	1219.29	on 06/14/06	
	Fundamental Information			
	* Price/Earnings	18.05	Ex-Dvd	-.0529
	* Dividend Yield	1.73	on 06/04/07	
	Index Information			
	Currency	USD		
	Volume	1.13BLN	on 06/01/07	
	Market Cap	13.99TRI		
	Divisor	8936.55701		

4)GRPS 147 Industry Groups
5)MEMB 500 Members ↑325 ↓169 →6
6)MOV Today's Movers by Index Pts

Leading Movers	7)	EXXON MOBIL CORP	+.669
	8)	WAL-MART STORES	+.515
	9)	CONOCOPHILLIPS	+.263
	10)	CHEVRON CORP	+.179
Lagging Movers	11)	AT&T INC	-.566
	12)	APPLE	-.305
	13)	VERIZON COMMUNIC	-.169
	14)	GENERAL ELECTRIC	-.150

15)CN News on Today's Movers
16)FVD Futures Available
17)OCM Options Available

* Calculated by Bloomberg

Australia 61 2 9777 8600 Brazil 5511 3048 4500 Europe 44 20 7330 7500 Germany 49 69 920410
Hong Kong 852 2977 6000 Japan 81 3 3201 8900 Singapore 65 6212 1000 U.S. 1 212 318 2000 Copyright 2007 Bloomberg L.P.
G510-436-0 04-Jun-2007 08:17:13

Fig. 3.7. S&P 500 index description. Used with permission of Bloomberg LP.

Page P216 Index DES

S&P 500 INDEX Page 2/ 15

Historical Return Analysis

	Value	Prc Appr	Annual
1 Day	1530.62	.374	40.547
5 Days	1507.51	1.912	99.660
MTD	1530.62	.374	40.547
QTD	1420.86	8.127	54.055
Ytd	1418.30	8.323	20.425
1 Month	1505.62	2.040	26.848
3 Months	1387.17	10.754	48.676
6 Months	1409.12	9.028	18.928

Historical Return Analysis

	Value	Prc Appr	Annual
1 Year	1288.22	19.261	19.146
2 Years	1196.02	28.454	13.318
5 Years	1040.69	47.627	8.097
Qtr 1:06	1248.29	3.728	16.004
Qtr 2:06	1294.83	-1.902	-7.493
Qtr 3:06	1270.20	5.168	22.676
Qtr 4:06	1335.85	6.172	27.493
Qtr 1:07	1418.30	.180	.734

Fundamental Analysis

P/E	18.05
Est P/E	16.41
Price/Book	2.998
Price/Sales	1.683
Diluted P/E Cont Ops	17.117
EBITDA	185.939

Fundamental Analysis

EPS	85.135
Est EPS	93.623
Book/Share	512.496
Sales/Share	913.118
Diluted EPS Cont Ops	89.755
Cashflow/share	129.004

*Calculated by Bloomberg Daily

Australia 61 2 9777 8600 Brazil 5511 3048 4500 Europe 44 20 7330 7500 Germany 49 69 920410
Hong Kong 852 2977 6000 Japan 81 3 3201 8900 Singapore 65 6212 1000 U.S. 1 212 318 2000 Copyright 2007 Bloomberg L.P.
G510-436-0 04-Jun-2007 08:18:38

Fig. 3.8. Historical return analysis of S&P 500 index on June 4, 2007. Used with permission of Bloomberg LP.

Page P216 Index DES

S&P 500 INDEX Page 3/ 15

Index Members			Index Members		
1) 3M CO	(MMM	UN)	21) ALTERA CORP	(ALTR	UQ)
2) ABBOTT LABS	(ABT	UN)	22) ALTRIA GROUP INC	(MO	UN)
3) ABERCROMBIE & FI	(ANF	UN)	23) AMAZON.COM INC	(AMZN	UW)
4) ACE LTD	(ACE	UN)	24) AMBAC FINL GROUP	(ABK	UN)
5) ADC TELECOM INC	(ADCT	UW)	25) AMER INTL GROUP	(AIG	UN)
6) ADOBE SYS INC	(ADBE	UW)	26) AMEREN CORP	(AEE	UN)
7) ADV MICRO DEVICE	(AMD	UN)	27) AMERICAN ELECTRI	(AEP	UN)
8) AES CORP	(AES	UN)	28) AMERICAN EXPRESS	(AXP	UN)
9) AETNA INC	(AET	UN)	29) AMERICAN STANDAR	(ASD	UN)
10) AFFIL COMPUTER-A	(ACS	UN)	30) AMERIPRISE FINAN	(AMP	UN)
11) AFLAC INC	(AFL	UN)	31) AMERISOURCEBERGE	(ABC	UN)
12) AGILENT TECH INC	(A	UN)	32) AMGEN INC	(AMGN	UW)
13) AIR PRODS & CHEM	(APD	UN)	33) ANADARKO PETROLE	(APC	UN)
14) ALCOA INC	(AA	UN)	34) ANALOG DEVICES	(ADI	UN)
15) ALLEGHENY ENERGY	(AYE	UN)	35) ANHEUSER BUSCH	(BUD	UN)
16) ALLEGHENY TECH	(ATI	UN)	36) AON CORP	(AOC	UN)
17) ALLERGAN INC	(AGN	UN)	37) APACHE CORP	(APA	UN)
18) ALLIED WASTE IND	(AW	UN)	38) APARTMENT INVEST	(AIV	UN)
19) ALLSTATE CORP	(ALL	UN)	39) APOLLO GROUP-A	(APOL	UW)
20) ALLTEL CORP	(AT	UN)	40) APPLE	(AAPL	UW)

Australia 61 2 9777 8600 Brazil 5511 3048 4500 Europe 44 20 7330 7500 Germany 49 69 920410
Hong Kong 852 2977 6000 Japan 81 3 3201 8900 Singapore 65 6212 1000 U.S. 1 212 318 2000 Copyright 2007 Bloomberg L.P.
G510-436-0 04-Jun-2007 08:19:25

Fig. 3.9. Index members of S&P 500. Used with permission of Bloomberg LP.

expected excess return (also called the *risk premium*) $\mu_i - r_f$ of asset i to its beta via

$$\mu_i - r_f = \beta_i(\mu_M - r_f). \tag{3.21}$$

The *security market line* refers to the linear relationship (3.21) between the expected excess return $\mu_i - r_f$ and $\mu_M - r_f$. To derive (3.21), consider the mean return $\mu(\alpha)$ of the portfolio consisting of a portion α invested in the asset i and $1 - \alpha$ invested in the market portfolio, so $\mu(\alpha) = \alpha\mu_i + (1-\alpha)\mu_M$. The variance $\sigma^2(\alpha)$ of the portfolio's return satisfies $\sigma^2(\alpha) = \alpha^2\sigma_i^2 + (1-\alpha)^2\sigma_M^2 + 2\alpha(1-\alpha)\mathrm{Cov}(r_i, r_M)$. The tangency condition at M (corresponding to $\alpha = 0$) can be translated into the condition that the slope of the curve $(\sigma(\alpha), \mu(\alpha))$ at $\alpha = 0$ is equal to the slope of the capital market line. Since

$$\frac{d}{d\alpha}\mu(\alpha) = \mu_i - \mu_M, \qquad \frac{d}{d\alpha}\sigma(\alpha) = \frac{\mathrm{Cov}(r_i, r_M) - \sigma_M^2}{\sigma_M} = (\beta_i - 1)\sigma_M,$$

and since $d\mu(\alpha)/d\sigma(\alpha) = (d\mu(\alpha)/d\alpha)/(d\sigma(\alpha)/d\alpha)$, it then follows that

$$\frac{\mu_i - \mu_M}{(\beta_i - 1)\sigma_M} = \frac{\mu_M - r_f}{\sigma_M},$$

the right-hand side of which is the slope of the capital market line, proving (3.21).

Since $\beta_i = \text{Cov}(r_i, r_M)/\sigma_M^2$ and since $Er_i = \mu_i$ and $Er_M = \mu_M$, we can rewrite (3.21) as a regression model:

$$r_i - r_f = \beta_i(r_M - r_f) + \epsilon_i, \tag{3.22}$$

in which $E(\epsilon_i) = 0$ and $\text{Cov}(\epsilon_i, r_M) = 0$; see Section 1.5. From (3.22), it follows that

$$\sigma_i^2 = \beta_i^2 \sigma_M^2 + \text{Var}(\epsilon_i), \tag{3.23}$$

decomposing the variance σ_i^2 of the ith asset return as a sum of the *systematic risk* $\beta_i^2\sigma_M^2$, which that is associated with the market, and the *idiosyncratic risk*, which is unique to the asset and uncorrelated with market movements.

In view of the definition $\beta_i = \text{Cov}(r_i, r_M)/\sigma_M^2$, beta can be used as a measure of the sensitivity of the asset return to market movements. The asset is said to be *aggressive* if its beta exceeds 1, *defensive* if its beta falls below 1, and *neutral* if its beta is 1. High-tech companies, whose activities are sensitive to changes in the market, usually have high betas. On the other hand, food manufacturers, which depend more on demand than on economic cycles, often have low betas.

3.3.2 Investment implications

Besides the Sharpe ratio $(\mu - r_f)/\sigma$, the *Treynor index* and the *Jensen index* are commonly used by practitioners to evaluate the performance of a portfolio. The higher the Sharpe ratio, the better an investment performed. Using β instead of σ as a measure of risk, the Treynor index is defined by $(\mu - r_f)/\beta$.

The Jensen index is the α in the generalization of CAPM to $\mu - r_f = \alpha + \beta(\mu_M - r_f)$. Whereas CAPM postulates $\alpha = 0$ (see (3.22)), Jensen (1968) considered a more general regression model $r - r_f = \alpha + \beta(\mu_M - r_f) + \epsilon$ in his empirical study of 115 mutual funds during the period 1945–1965. An investment with a positive α is considered to perform better than the market. Jensen's findings were that 76 funds had negative alphas and 39 funds had positive alphas, and that the average alpha value was -0.011. These findings support the *efficient market hypothesis*, according to which it is not possible to outperform the market portfolio in an efficient market; see Chapter 11.

3.3.3 Estimation and testing

Let $\mathbf{y}_t$ be a $q \times 1$ vector of excess returns on q assets and let x_t be the excess return on the market portfolio (or, more precisely, its proxy) at time t. The capital asset pricing model can be associated with the null hypothesis $H_0 : \boldsymbol{\alpha} = \mathbf{0}$ in the regression model

$$\mathbf{y}_t = \boldsymbol{\alpha} + x_t\boldsymbol{\beta} + \boldsymbol{\epsilon}_t, \qquad 1 \le t \le n, \tag{3.24}$$

where $E(\boldsymbol{\epsilon}_t) = \mathbf{0}$, $\mathrm{Cov}(\boldsymbol{\epsilon}_t) = \mathbf{V}$, and $E(x_t\boldsymbol{\epsilon}_t) = 0$. Note that (3.24) is a multivariate representation of q regression models of the form $y_{tj} = \alpha_j + x_t\beta_j + \epsilon_{tj}$, $j = 1, \ldots q$. Letting $\bar{x} = n^{-1}\sum_{t=1}^n x_t$ and $\bar{\mathbf{y}} = n^{-1}\sum_{t=1}^n \mathbf{y}_t$, the OLS estimates of $\boldsymbol{\alpha}$ and $\boldsymbol{\beta}$ are given by

$$\widehat{\boldsymbol{\beta}} = \frac{\sum_{t=1}^n (x_t - \bar{x})\mathbf{y}_t}{\sum_{t=1}^n (x_t - \bar{x})^2}, \qquad \widehat{\boldsymbol{\alpha}} = \bar{\mathbf{y}} - \bar{x}\widehat{\boldsymbol{\beta}}. \tag{3.25}$$

The maximum likelihood estimate of $\mathbf{V}$ is

$$\widehat{\mathbf{V}} = n^{-1}\sum_{t=1}^n (\mathbf{y}_t - \widehat{\boldsymbol{\alpha}} - \widehat{\boldsymbol{\beta}}x_t)(\mathbf{y}_t - \widehat{\boldsymbol{\alpha}} - \widehat{\boldsymbol{\beta}}x_t)^T. \tag{3.26}$$

The properties of OLS in Sections 1.1.4 and 1.5.3 can be used to establish the asymptotic normality of $\widehat{\boldsymbol{\beta}}$ and $\widehat{\boldsymbol{\alpha}}$, yielding the approximations

$$\widehat{\boldsymbol{\beta}} \approx N\left(\boldsymbol{\beta}, \frac{\widehat{\mathbf{V}}}{\sum_{t=1}^n (x_t - \bar{x})^2}\right), \qquad \widehat{\boldsymbol{\alpha}} \approx N\left(\boldsymbol{\alpha}, \left(\frac{1}{n} + \frac{\bar{x}^2}{\sum_{t=1}^n (x_t - \bar{x})^2}\right)\widehat{\mathbf{V}}\right),$$

from which approximate confidence regions for $\boldsymbol{\beta}$ and $\boldsymbol{\alpha}$ can be constructed.

In the case where $\boldsymbol{\epsilon}_t$ are i.i.d. normal and are independent of the market excess returns x_t, $\widehat{\boldsymbol{\alpha}}$, $\widehat{\boldsymbol{\beta}}$, and $\widehat{\mathbf{V}}$ are the maximum likelihood estimates of $\boldsymbol{\alpha}$, $\boldsymbol{\beta}$, and $\mathbf{V}$. Moreover, the results of Sections 2.3.3 and 2.3.4 can be used to obtain the conditional distribution of $(\widehat{\boldsymbol{\alpha}}, \widehat{\boldsymbol{\beta}}, \widehat{\mathbf{V}})$ given $(x_1, \ldots, x_n)$:

$$\widehat{\boldsymbol{\alpha}} \sim N\left(\boldsymbol{\alpha}, \left(\frac{1}{n} + \frac{\bar{x}^2}{\sum_{t=1}^n (x_t - \bar{x})^2}\right)\mathbf{V}\right), \quad \widehat{\boldsymbol{\beta}} \sim N\left(\boldsymbol{\beta}, \frac{\mathbf{V}}{\sum_{t=1}^n (x_t - \bar{x})^2}\right),$$
$$n\widehat{\mathbf{V}} \sim W_q(\mathbf{V}, n-2), \tag{3.27}$$

with $\widehat{\mathbf{V}}$ independent of $(\widehat{\boldsymbol{\alpha}}, \widehat{\boldsymbol{\beta}})$. Moreover, arguments similar to those in (2.32) and (2.33) can be used to show that

$$\left(\frac{n-q-1}{q}\right)\widehat{\boldsymbol{\alpha}}^T\widehat{\mathbf{V}}^{-1}\widehat{\boldsymbol{\alpha}} \Big/ \left\{1 + \frac{\bar{x}^2}{n^{-1}\sum_{t=1}^n (x_t - \bar{x})^2}\right\} \sim F_{q,n-q-1} \text{ under } H_0; \tag{3.28}$$

see Exercise 3.2. Although the F-test of CAPM assumes that ϵ_t are i.i.d. normal, arguments similar to those in Section 1.5.3 can be used to show that (3.28) still holds approximately without the normality assumption when $n - q - 1$ is moderate or large.

3.3.4 Empirical studies of CAPM

An illustrative example

We illustrate the statistical analysis of CAPM with the monthly returns data of the six stocks in Section 3.2.4 using the Dow Jones Industrial Average index as the market portfolio M and the 3-month U.S. Treasury bill as the risk-free asset. These data are used to estimate $\mu_i - r_f$, $\mu_M - r_f$, β_i, the Jensen index α_i, and the associated p-value of the t-test of H_0: $\alpha_i = 0$, the Sharpe ratio, and the Treynor index of stock i. The results are given in Table 3.2. All Jensen indices in the first row of Table 3.2 are very small and are not significantly different from 0 ($p > 0.5$, t-test). The third row of the table gives the beta of each stock, which measures its market risk. Note that the beta of Intel is markedly larger than the other betas, which are smallest for Exxon-Mobil and Pfizer. The fourth and fifth rows of Table 3.2 show the systematic and idiosyncratic risks, respectively. The last two rows of the table give the Sharpe ratio and Treynor index of each stock. Exxon-Mobil has the highest Sharpe ratio and Treynor index, which may be a consequence of the surging oil and gas prices since 2000. The Sharpe ratio of the Dow Jones Industrial Average index during the period is −0.106. The F-statistic in (3.28) has the value 1.07, so the CAPM hypothesis $H_0 : \boldsymbol{\alpha} = \mathbf{0}$ is not rejected at the .10 level ($F_{6,56;.90} = 1.88$). Figure 3.10 plots the security market line fitted to the data (which are also shown in the figure) for each stock.

Table 3.2. Performance of six stocks from August 2000 to October 2005.

	AXP	CITI	XOM	GM	INTEL	PFE
$\alpha \times 10^3$	0.87	0.81	2.23	−2.41	−4.31	−5.21
p-value	0.72	0.76	0.40	0.59	0.52	0.06
β	1.23	1.20	0.52	1.44	2.28	0.46
$\beta^2\sigma_M^2 \times 10^4$	5.77	5.48	1.04	7.91	19.8	0.80
$\sigma_\epsilon^2 \times 10^4$	3.50	4.18	4.22	12.0	26.7	4.64
Sharpe$\times 10^2$	−5.49	−5.36	5.05	−12.1	−13.2	−26.4
Treynor$\times 10^3$	−1.35	−1.38	2.22	−3.74	−3.96	−13.4

Summary of empirical literature

Since the development of CAPM in the 1960s, a large body of literature has evolved on empirical evidence for or against the model. The early evidence

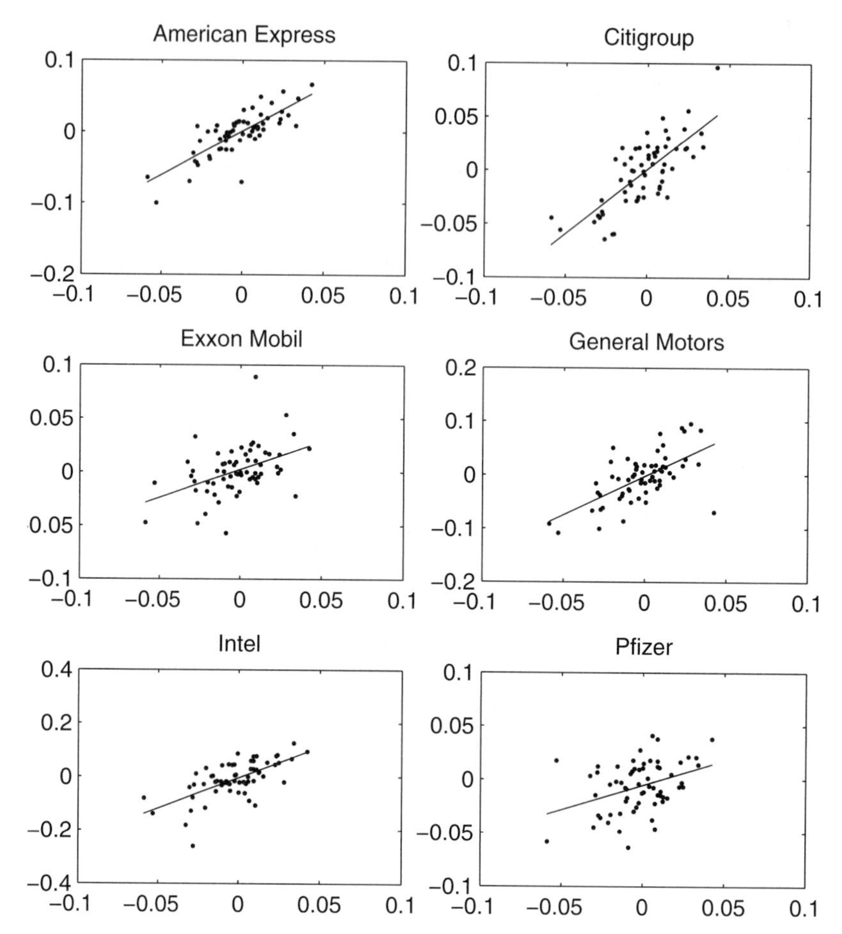

Fig. 3.10. Security market lines for six stocks. Horizontal axis: monthly excess return of S&P 500. Vertical axis: monthly excess return of listed stock.

was largely positive, but in the late 1970s, less favorable evidence began to appear. Basu (1977) reported the "price–earnings-ratio effect": Firms with low price–earnings ratios have higher average returns, and firms with high price–earnings ratios have lower average returns than the values implied by CAPM. Banz (1981) noted the "size effect," that firms with low market capitalizations have higher average returns than under CAPM. More recently, Fama and French (1992, 1993) have found that beta cannot explain the difference in returns between portfolios formed on the basis of the ratio of book value to market value of equity. Jegadesh and Titman (1995) have noted that a portfolio formed by buying stocks whose values have declined and selling stocks whose values have risen has a higher average return than predicted by CAPM.

There has been controversy concerning the issues of *data snooping*, *selection bias*, and *proxy bias* for a market portfolio in the empirical evidence for or against CAPM. Data snooping refers to using information from data to guide subsequent research that uses the same or similar data due to the non-experimental nature of the empirical studies; it is impossible to run a prospective experiment to generate a new dataset. Lo and MacKinlay (1990) have analyzed the potential magnitude of data-snooping biases in tests of CAPM. Chapter 11 (Section 11.1.4) will describe methods to adjust data-snooping biases for valid inference. Selection bias arises when data availability causes certain stocks to be excluded from the analysis. For example, Kothari, Shanken, and Sloan (1995) have noted that studies involving book–market ratios exclude failing stocks, but failing stocks are expected to have low returns and high book–market ratios, resulting in an upward bias of the average return of the high book–market-ratio stocks that are included in the study. The market portfolio in CAPM is a theoretical investment that is unobserved, and proxies are actually used in empirical studies of CAPM. Roll (1977) argues that tests of CAPM only reject the mean-variance efficiency of the proxy but not the efficiency of the true market portfolio. Stambaugh (1982) examines the sensitivity of tests of CAPM to the exclusion of assets by using broader proxies for the market portfolio and shows that the inferences are similar whether one uses a stock-based proxy or a stock-and-bond-based proxy, suggesting that Roll's concern may not be a serious problem.

3.4 Multifactor pricing models

Multifactor pricing models generalize CAPM by embedding it in a regression model of the form

$$r_i = \alpha_i + \boldsymbol{\beta}_i^T \mathbf{f} + \epsilon_i, \qquad i = 1, \cdots, p, \tag{3.29}$$

in which the r_i is the return on the ith asset, α_i and $\boldsymbol{\beta}_i$ are unknown regression parameters, $\mathbf{f} = (f_1, \ldots, f_k)^T$ is a regression vector of *factors*, and ϵ_i is an unobserved random disturbance that has mean 0 and is uncorrelated with $\mathbf{f}$. The case $k = 1$ is called a *single-factor* (also called *single-index*) model, and CAPM is a single-factor model with $f = r_M - r_f$ and $\alpha_i = r_f$; see (3.22).

3.4.1 Arbitrage pricing theory

Whereas the mean-variance efficiency theory in Section 3.3.1 provides the theoretical background of (3.22) for the single-factor CAPM and decomposes risk into a systematic component and an idiosyncratic component in (3.23), the arbitrage pricing theory (APT) introduced by Ross (1976) allows for multiple risk factors. Unlike (3.21), APT does not require identification of the market portfolio and relates the expected return μ_i of the ith asset to the risk-free return, or to a more general parameter λ_0 without assuming the existence of a risk-free asset, and to a $k \times 1$ vector $\boldsymbol{\lambda}$ of risk premiums:

$$\mu_i \approx \lambda_0 + \boldsymbol{\beta}_i^T \boldsymbol{\lambda}, \qquad i = 1, \ldots, p, \tag{3.30}$$

in which the approximation is a consequence of the absence of arbitrage in a large economy that has a "well-diversified" portfolio whose variance approaches 0 as $p \to \infty$. Since (3.30) is only an approximation, it does not produce directly testable restrictions on the asset returns. While APT provides an economic theory underlying multifactor models of asset returns, the theory does not identify the factors. Approaches to the choice of factors can be broadly classified as economic and statistical. The economic approach specifies (i) macroeconomic and financial market variables that are thought to capture the systematic risks of the economy or (ii) characteristics of firms that are likely to explain differential sensitivity to the systematic risks, forming factors from portfolios of stocks based on the characteristics. For example, the factors used by Chen, Roll, and Ross (1986) include the yield spread between long- and short-term interest rates for U.S. government bonds, inflation, industrial production growth, and the yield spread between corporate high- and low-grade bonds. Fama and French's (1993) three-factor model, described in Section 3.4.4, uses firm characteristics to form factor portfolios, including a proxy of the market portfolio as one factor. The statistical approach uses factor analysis or PCA (principal component analysis) to estimate the parameters of model (3.29) from a set of observed asset returns.

3.4.2 Factor analysis

Letting $\mathbf{r} = (r_1, \ldots, r_p)^T$, $\boldsymbol{\alpha} = (\alpha_1, \ldots, \alpha_p)^T$, $\boldsymbol{\epsilon} = (\epsilon_1, \ldots, \epsilon_p)^T$, and $\mathbf{B}$ to be the $p \times k$ matrix whose ith row vector is $\boldsymbol{\beta}_i^T$, we can write (3.29) as $\mathbf{r} = \boldsymbol{\alpha} + \mathbf{B}\mathbf{f} + \boldsymbol{\epsilon}$ with $E\boldsymbol{\epsilon} = E\mathbf{f} = \mathbf{0}$ and $E(\mathbf{f}\boldsymbol{\epsilon}^T) = \mathbf{0}$. Although this looks like a regression model, the fact that the regressor $\mathbf{f}$ is unobservable means that least squares regression cannot be applied to estimate $\mathbf{B}$. The model, however, is the same as that in a multivariate statistical method called *factor analysis*. Let $\mathbf{r}_t$, $t = 1, \ldots, n$, be independent observations from the model so that $\mathbf{r}_t = \boldsymbol{\alpha} + \mathbf{B}\mathbf{f}_t + \boldsymbol{\epsilon}_t$ and $E\boldsymbol{\epsilon}_t = E\mathbf{f}_t = \mathbf{0}$, $E(\mathbf{f}_t\boldsymbol{\epsilon}_t^T) = \mathbf{0}$, $\mathrm{Cov}(\mathbf{f}_t) = \boldsymbol{\Omega}$, and $\mathrm{Cov}(\boldsymbol{\epsilon}_t) = \mathbf{V}$. Then

$$E(\mathbf{r}_t) = \boldsymbol{\alpha}, \qquad \mathrm{Cov}(\mathbf{r}_t) = \mathbf{B}\boldsymbol{\Omega}\mathbf{B}^T + \mathbf{V}. \tag{3.31}$$

The decomposition of the covariance matrix $\boldsymbol{\Sigma}$ of $\mathbf{r}_t$ in (3.31) is the essence of factor analysis, which separates variability into a *systematic* part due to the variability of certain unobserved factors, represented by $\mathbf{B}\boldsymbol{\Omega}\mathbf{B}^T$, and an error (*idiosyncratic*) part, represented by $\mathbf{V}$. In psychometrics, for a battery of mental tests given to a group of individuals, it is often observed that an individual's test score consists of a part that is peculiar to this particular test ("error") and a part that is a function of more fundamental quantities ("factors").

Identifiability and orthogonal factor model

Standard factor analysis assumes a strict factor structure in which the factors account for all the pairwise covariances of the asset returns, so $\mathbf{V}$ is assumed to be diagonal; i.e., $\mathbf{V} = \text{diag}(v_1, \dots, v_p)$. Since $\mathbf{B}$ and $\boldsymbol{\Omega}$ are not uniquely determined by $\boldsymbol{\Sigma} = \mathbf{B}\boldsymbol{\Omega}\mathbf{B}^T + \mathbf{V}$, the *orthogonal factor model* assumes that $\boldsymbol{\Omega} = \mathbf{I}$ so that $\mathbf{B}$ is unique up to an orthogonal transformation and $\mathbf{r} = \boldsymbol{\alpha} + \mathbf{B}\boldsymbol{\beta} + \boldsymbol{\epsilon}$ with $\text{Cov}(\mathbf{f}) = \boldsymbol{\Omega}$ yields

$$\text{Cov}(\mathbf{r}, \mathbf{f}) = E\{(\mathbf{r} - \boldsymbol{\alpha})\mathbf{f}^T\} = \mathbf{B}\boldsymbol{\Omega} = \mathbf{B}, \tag{3.32}$$

$$\text{Var}(r_i) = \sum_{j=1}^{k} b_{ij}^2 + \text{Var}(\epsilon_i), \qquad 1 \le i \le p, \tag{3.33}$$

$$\text{Cov}(r_i, r_j) = \sum_{l=1}^{k} b_{il} b_{jl}, \qquad 1 \le i, j \le p. \tag{3.34}$$

Maximum likelihood estimation

Assuming the observed $\mathbf{r}_t$ to be independent $N(\boldsymbol{\alpha}, \boldsymbol{\Sigma})$, the likelihood function is

$$L(\boldsymbol{\alpha}, \boldsymbol{\Sigma}) = (2\pi)^{-pn/2} (\det \boldsymbol{\Sigma})^{-n/2} \exp\left\{ -\frac{1}{2} \sum_{t=1}^{n} (\mathbf{r}_t - \boldsymbol{\alpha})^T \boldsymbol{\Sigma}^{-1} (\mathbf{r}_t - \boldsymbol{\alpha}) \right\},$$

with $\boldsymbol{\Sigma}$ constrained to be of the form $\boldsymbol{\Sigma} = \mathbf{B}\mathbf{B}^T + \text{diag}(v_1, \dots, v_p)$, in which $\mathbf{B}$ is $p \times k$. The MLE $\widehat{\boldsymbol{\alpha}}$ of $\boldsymbol{\alpha}$ is $\bar{\mathbf{r}} := n^{-1} \sum_{t=1}^{n} \mathbf{r}_t$, and we can maximize $-\frac{1}{2} n \log\det(\boldsymbol{\Sigma}) - \frac{1}{2} \text{tr}(\mathbf{W}\boldsymbol{\Sigma}^{-1})$ over $\boldsymbol{\Sigma}$ of the form above, where $\mathbf{W} = \sum_{t=1}^{n} (\mathbf{r}_t - \bar{\mathbf{r}})(\mathbf{r}_t - \bar{\mathbf{r}})^T$. Iterative algorithms can be used to find the maximizer $\widehat{\boldsymbol{\Sigma}}$ and therefore also $\widehat{\mathbf{B}}$ and $\widehat{v}_1, \dots, \widehat{v}_p$.

Factor rotation and factor scores

In factor analysis, the entries of the matrix $\widehat{\mathbf{B}}$ are called *factor loadings*. Since $\widehat{\mathbf{B}}$ is unique only up to orthogonal transformations, the usual practice is to multiply $\widehat{\mathbf{B}}$ by a suitably chosen orthogonal matrix $\mathbf{Q}$, called a *factor rotation*, so that the factor loadings have a simple interpretable structure. Letting $\widehat{\mathbf{B}}^* = \widehat{\mathbf{B}}\mathbf{Q}$, a popular choice of $\mathbf{Q}$ is that which maximizes the *varimax criterion*

$$C = p^{-1} \sum_{j=1}^{k} \left[\sum_{i=1}^{p} \widehat{b}_{ij}^{*4} - \left(\sum_{i=1}^{p} \widehat{b}_{ij}^{*2} \right)^2 \Big/ p \right]$$
$$\propto \sum_{j=1}^{k} \operatorname{Var}\big(\text{squared loadings of the } j\text{th factor}\big).$$

Intuitively, maximizing C corresponds to spreading out the squares of the loadings on each factor as much as possible. Varimax rotations are given in popular factor analysis computer programs; see the illustrative example below.

Since the values of the factors $\mathbf{f}_t$, $1 \le t \le n$, are unobserved, it is often of interest to impute these values, called *factor scores*, for model diagnostics. From the model $\mathbf{r} - \boldsymbol{\alpha} = \mathbf{B}\mathbf{f} + \boldsymbol{\epsilon}$ with $\operatorname{Cov}(\boldsymbol{\epsilon}) = \mathbf{V}$, the generalized least squares estimate of $\mathbf{f}$ when $\mathbf{B}$, $\mathbf{V}$, and $\boldsymbol{\alpha}$ are known is

$$\widehat{\mathbf{f}} = (\mathbf{B}^T\mathbf{V}^{-1}\mathbf{B})^{-1}\mathbf{B}^T\mathbf{V}^{-1}(\mathbf{r}_t - \boldsymbol{\alpha});$$

see (1.53). Bartlett (1937) therefore suggested estimating $\mathbf{f}_t$ by

$$\widehat{\mathbf{f}}_t = (\widehat{\mathbf{B}}^T\widehat{\mathbf{V}}^{-1}\widehat{\mathbf{B}})^{-1}\widehat{\mathbf{B}}^T\widehat{\mathbf{V}}^{-1}(\mathbf{r}_t - \bar{\mathbf{r}}). \tag{3.35}$$

Number of factors

The theory underlying multifactor pricing models and factor analysis assumes that the number k of factors has been specified and does not indicate how to specify it. For the theory to be useful, however, k has to be reasonably small. The latitude in the choice of k has led to two approaches in empirical work. One approach is to repeat the estimation for a variety of choices of k to see if the results are sensitive to increasing the number of factors. For example, Connor and Korajczyk (1988) consider five and ten factors and find not much difference in the results, therefore suggesting that five is an adequate number of factors.

The second approach involves more formal hypothesis testing that the k-factor model indeed holds when the $\mathbf{r}_t$ are independent $N(\boldsymbol{\alpha}, \boldsymbol{\Sigma})$. The null hypothesis H_0 is that $\boldsymbol{\Sigma}$ is of the form $\mathbf{B}\mathbf{B}^T + \mathbf{V}$ with $\mathbf{V}$ diagonal and $\mathbf{B}$ being $p \times k$. The generalized likelihood ratio (GLR) statistic (see Section 2.4) that tests this null hypothesis against unconstrained $\boldsymbol{\Sigma}$ is of the form

$$\Lambda = n\Big\{ \log\det\big(\widehat{\mathbf{B}}\widehat{\mathbf{B}}^T + \widehat{\mathbf{V}}\big) - \log\det\big(\widehat{\boldsymbol{\Sigma}}\big)\Big\}, \tag{3.36}$$

where $\widehat{\boldsymbol{\Sigma}} = n^{-1}\sum_{t=1}^{n}(\mathbf{r}_t - \bar{\mathbf{r}})(\mathbf{r}_t - \bar{\mathbf{r}})^T$ is the unconstrained MLE of $\boldsymbol{\Sigma}$. Under H_0, Λ is approximately χ^2 with

$$\frac{1}{2}p(p+1) - \left\{p(k+1) - \frac{1}{2}k(k-1)\right\} = \frac{1}{2}\{(p-k)^2 - p - k\}$$

degrees of freedom. Bartlett (1954) has shown that the χ^2 approximation to the distribution of (3.36) can be improved by replacing n in (3.36) by $n - 1 - (2p + 4k + 5)/6$, which is often used in empirical studies. For example, Roll and Ross (1980) use this approach and conclude that three is an adequate number of factors in their empirical investigation of APT.

An illustrative example

Factor analysis can be implemented by the `R` functions `factanal` and `varimax` or the function `factoran` in the `MATLAB Statistics` toolbox. The varimax rotations can also be specified in these functions. We illustrate this with the monthly excess returns data in Section 3.3.4 on six stocks. To choose the number of factors, we start with a one-factor model for which the GLR statistic is 18.54 and has p-value 0.0294 (based on the χ^2_9 approximation to GLR under the null hypothesis, which is therefore rejected). We then consider a two-factor model for which the GLR statistic is 5.88 and has p-value 0.209 (based on the χ^2_4 approximation), leading to the choice of two factors. Table 3.3 shows the factor loadings estimated by maximum likelihood and the rotated factor loadings (by varimax rotations) for $k = 2$ factors. The last two rows of the table give the sample variance of each factor and the proportion it contributes to the overall variance. The varimax rotation matrix is

$$\begin{pmatrix} 0.936 & 0.353 \\ -0.353 & 0.936 \end{pmatrix}.$$

3.4.3 The PCA approach

The fundamental decomposition $\boldsymbol{\Sigma} = \lambda_1 \mathbf{a}_1 \mathbf{a}_1^T + \cdots + \lambda_p \mathbf{a}_p \mathbf{a}_p^T$ in PCA (see Section 2.2.2) can be used to decompose $\boldsymbol{\Sigma}$ into $\boldsymbol{\Sigma} = \mathbf{B}\mathbf{B}^T + \mathbf{V}$. Here $\lambda_1 \geq \cdots \geq \lambda_p$ are the ordered eigenvalues of $\boldsymbol{\Sigma}$, $\mathbf{a}_i$ is the unit eigenvector associated with λ_i, and

$$\mathbf{B} = (\sqrt{\lambda_1}\mathbf{a}_1, \ldots, \sqrt{\lambda_k}\mathbf{a}_k), \qquad \mathbf{V} = \sum_{l=k+1}^{p} \lambda_l \mathbf{a}_l \mathbf{a}_l^T. \tag{3.37}$$

As pointed out in Section 2.2.2, PCA is particularly useful when most eigenvalues of $\boldsymbol{\Sigma}$ are small in comparison with the k largest ones, so that k

Table 3.3. Factor analysis of the monthly excess returns of six stocks.

	Estimates of Factor Loadings		Rotated Factor Loadings	
	f_1	f_2	f_1^*	f_2^*
AXP	0.755	0.152	0.653	0.409
CITI	0.769	0	0.705	0.311
XOM	0.341	0.619	0.101	0.699
GM	0.574	0	0.521	0.246
INTEL	0.758	−0.371	0.840	0
PFE	0.297	0.392	0.140	0.472
Variance	2.271	0.701	1.930	1.042
Proportion	0.379	0.117	0.322	0.174

principal components of $\mathbf{r}_t - \bar{\mathbf{r}}$ account for a large proportion of the overall variance. In this case, we can use PCA to determine k. Instead of working with $\widehat{\boldsymbol{\Sigma}} = (\widehat{\sigma}_{ij})_{1\le i,j\le p}$, for which the units of the variables may not be commensurate, it is often better to work with the correlation matrix using principal components of the standardized variables $\big((r_{t1}-\bar{r}_1)/\sqrt{\widehat{\sigma}_{11}}, \dots, (r_{tp}-\bar{r}_p)/\sqrt{\widehat{\sigma}_{pp}}\big)^T$. Standardization avoids having the variable with largest variance unduly influence the determination of the factor loadings. If we use the sample version of (3.37) for the correlation matrix $\mathbf{R}$ to estimate $\mathbf{B}$ and $\mathbf{V}$, the resultant $\widehat{\mathbf{V}}$ is not diagonal. An often-used procedure is simply setting $\widehat{\mathbf{V}} = \operatorname{diag}\big(1 - \sum_{j=1}^k \widehat{b}_{1j}^2, \dots, 1 - \sum_{j=1}^k \widehat{b}_{kj}^2\big)$ as an approximation, since most of the overall variability is already accounted for by the first k principal components of the correlation matrix $\widehat{\mathbf{R}}$. Instead of (3.35), the factor scores using this PCA approximation are given by the usual OLS estimates

$$\widehat{\mathbf{f}}_t = \big(\widehat{\mathbf{B}}^T\widehat{\mathbf{B}}\big)^{-1}\widehat{\mathbf{B}}^T\Big((\bar{r}_{t1} - \bar{r}_1)/\sqrt{\widehat{\sigma}_{11}}, \dots, (\bar{r}_{tp} - \bar{r}_p)/\sqrt{\widehat{\sigma}_{pp}}\Big)^T.$$

3.4.4 The Fama-French three-factor model

Fama and French (1993, 1996) propose a model with three factors in which the expected return of a portfolio in excess of the risk-free rate, $E(r_i) - r_f$, can be explained by a linear regression of its return on the three factors. The first factor is the excess return on a market portfolio, $r_M - r_f$, which is the only factor in the CAPM. The second factor is the difference between the return on a portfolio S of small stocks and the return on a portfolio L of large stocks, $r_S - r_L$, which captures the risk factor in returns related to size. Here "small" and "large" refer to the market value of equity, which is the price of the stock multiplied by the number of shares outstanding. The third factor is the difference between the return on a portfolio H of high book-to-market stocks and the return on a portfolio L of low book-to-market stocks, $r_H - r_L$, which captures the risk factor in returns related to the book-to-market equity. Therefore the Fama-French three-factor model is of the form (3.29) with $\mathbf{f} = (r_M - r_f, r_S - r_L, r_H - r_L)^T$. Fama and French (1992, 1993) argue that their three-factor model removes most of the pricing anomalies with CAPM. Because the factors in the Fama-French model are specified (instead of being determined from data by factor analysis or PCA as in Sections 3.4.2 and 3.4.3), one can use standard regression analysis to test the model and estimate its parameters.

3.5 Applications of resampling to portfolio management

Before considering the *resampled efficient frontier* to implement Markowitz's theory, we illustrate how bootstrap resampling can be applied to assess the biases and standard errors of the estimates of the CAPM parameters and related performance indices in Table 3.2 based on the monthly excess log returns of six stocks from August 2000 to October 2005. The α and β in the table are estimated by applying OLS to the regression model $r_i - r_f = \alpha_i + \beta_i(r_M - r_f) + \epsilon_i$, in which the market portfolio M is taken to be the Dow Jones Industrial Average index and r_f is the annualized rate of the 3-month U.S. Treasury bill. Let x_t denote the excess log return $r_{M,t} - r_{f,t}$ of the market portfolio M at time t, and let $y_{i,t} = r_{i,t} - r_{f,t}$ denote the corresponding excess log return for the ith stock. We draw $B = 500$ bootstrap samples $\{(x_t^*, y_{i,t}^*), 1 \le t \le n = 63\}$ from the observed sample $\{(x_t, y_{i,t}), 1 \le t \le n = 63\}$ and compute the OLS estimates $\widehat{\alpha}_i^*$ and $\widehat{\beta}_i^*$ for the regression model $y_{i,t}^* = \alpha_i^* + \beta_i^* x_t^* + \epsilon_t^*$. The average values of $\widehat{\alpha}_i^*$, $\widehat{\beta}_i^*$ and corresponding Sharpe index and Treynor index of the B bootstrap samples are given in Table 3.4. The standard deviations of $\widehat{\alpha}_i^*$ and $\widehat{\beta}_i^*$ in the B bootstrap samples provide estimated standard errors of $\widehat{\alpha}_i$ and $\widehat{\beta}_i$ and are given in parentheses; see Section 1.6.2. Table 3.4 also gives the biases and standard errors of the estimated Sharpe and Treynor indices.

3.5.1 Michaud's resampled efficient frontier

As pointed out in Section 3.2.4, the estimated ("plug-in") efficient frontier replaces $\boldsymbol{\mu}$ and $\boldsymbol{\Sigma}$ in the optimal portfolio weight vector $\mathbf{w}_{\text{eff}}$ defined in (3.11) by the sample mean $\widehat{\boldsymbol{\mu}}$ and covariance matrix $\widehat{\boldsymbol{\Sigma}}$, leading to the sample version $\widehat{\mathbf{w}}$. Figure 3.5 plots $\sqrt{\widehat{\mathbf{w}}^T \widehat{\boldsymbol{\Sigma}} \widehat{\mathbf{w}}}$ versus $\widehat{\mathbf{w}}^T \widehat{\boldsymbol{\mu}} = \mu_*$ for a fine grid of μ_* values,

Table 3.4. Bootstrapping CAPM.

	$\alpha \times 10^3$	β	Sharpe $\times 10^2$	Treynor $\times 10^3$
AXP	1.00 (0.28)	1.23 (0.02)	−4.51 (1.61)	−1.14 (0.40)
CITI	0.84 (0.32)	1.20 (0.02)	−5.16 (1.60)	−1.36 (0.41)
XOM	2.24 (0.33)	0.53 (0.02)	4.69 (1.59)	2.27 (0.75)
GM	−1.99 (0.58)	1.44 (0.04)	−12.2 (1.65)	−4.02 (0.57)
Intel	−4.33 (0.74)	2.29 (0.05)	−13.0 (1.44)	−3.95 (0.44)
Pfizer	−5.23 (0.33)	0.45 (0.02)	−26.2 (1.62)	−13.5 (4.05)

thereby displaying the estimated efficient frontier. Since $\widehat{\boldsymbol{\mu}}$ and $\widehat{\boldsymbol{\Sigma}}$ differ from $\boldsymbol{\mu}$ and $\boldsymbol{\Sigma}$, this "plug-in" frontier is in fact suboptimal. Frankfurter, Phillips, and Seagle (1971) and Jobson and Korkie (1980) have found that portfolios thus constructed may perform worse than the equally weighted portfolio. Michaud (1989) proposes to use instead of $\widehat{\mathbf{w}}$ the average of bootstrap weights

$$\bar{\mathbf{w}} = B^{-1} \sum_{b=1}^{B} \widehat{\mathbf{w}}_b^*, \tag{3.38}$$

where $\widehat{\mathbf{w}}_b^*$ is the estimated optimal portfolio weight vector based on the bth bootstrap sample $\{\mathbf{r}_{b1}^*, \ldots, \mathbf{r}_{bn}^*\}$ drawn with replacement from the observed sample $\{\mathbf{r}_1, \ldots, \mathbf{r}_n\}$; see Section 1.6.2. Specifically, the bth bootstrap sample has sample mean vector $\widehat{\boldsymbol{\mu}}_b^*$ and covariance matrix $\widehat{\boldsymbol{\Sigma}}_b^*$, which can be used to replace $\boldsymbol{\mu}$ and $\boldsymbol{\Sigma}$ in (3.9) or (3.16) with given μ_*, thereby yielding $\widehat{\mathbf{w}}_b^*$. Thus, Michaud's resampled efficient frontier corresponds to the plot $\sqrt{\bar{\mathbf{w}}^T \widehat{\boldsymbol{\Sigma}} \bar{\mathbf{w}}}$ versus $\bar{\mathbf{w}}^T \bar{\mathbf{r}} = \mu_*$ for a fine grid of μ_* values, as shown in Figure 3.11 (in which we have used $B = 1000$) for the six stocks considered in Figure 3.5.

Although Michaud claims that (3.38) provides an improvement over $\widehat{\mathbf{w}}$, there have been no convincing theoretical developments and simulation studies to support the claim. In Chapter 4 (Section 4.4), we describe recent developments that provide alternative approaches to this problem based on Bayesian methods, which are introduced in Section 4.3.

3.5.2 Bootstrap estimates of performance

Whereas simulation studies of performance require specific distributional assumptions on $\mathbf{r}_t$, it is desirable to be able to assess performance nonparametrically and the bootstrap method of Section 1.6.1 provides a practical way

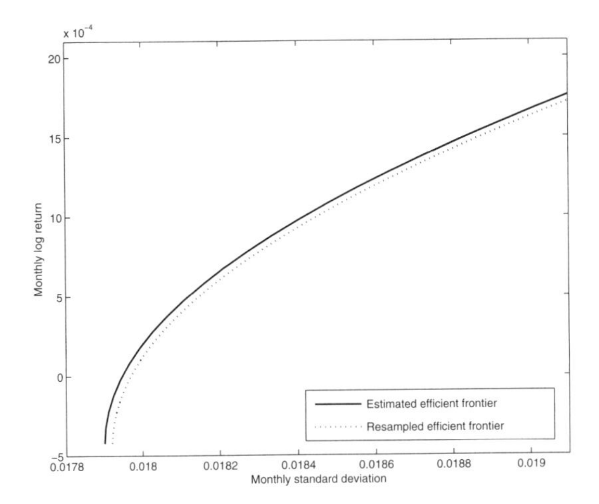

Fig. 3.11. The estimated efficient frontier (solid curve) and the resampled efficient frontier (dotted curve) of six U.S. stocks.

to do so. The bootstrap samples $\{\mathbf{r}_{b1}^*, \ldots, \mathbf{r}_{bn}^*; \mathbf{r}_b^*\}$, $1 \leq b \leq B$, can be used to estimate the means $E(\mathbf{w}_P^T \mathbf{r})$ and variances $\text{Var}(\mathbf{w}_P^T \mathbf{r})$ of various portfolios P whose weight vectors $\mathbf{w}_P$ may depend on the observed data (for which $E(\mathbf{w}_P^T \mathbf{r})$ can no longer be written as $\mathbf{w}_P^T E(\mathbf{r})$ since $\mathbf{w}_P$ is random). Details and illustrative examples are given in Lai, Xing, and Chen (2007).

Exercises

3.1. Prove (3.19).

3.2. Prove (3.27) and (3.28).

3.3. Let r_{0t} be the return of a stock index at time t. Sharpe's *single-index model* assumes that the log returns of the n stocks in the index are generated by $r_{it} = \alpha_i + \beta_i r_{0t} + \varepsilon_{it}$, $1 \leq i \leq p$, where ϵ_{it} is uncorrelated with r_{0t} and $\text{Cov}(\epsilon_{it}, \epsilon_{jt}) = \sigma^2 \mathbf{1}_{\{i=j\}}$. The model also assumes that $(r_{0t}, \ldots, r_{pt})$, $1 \leq t \leq n$, are i.i.d. vectors.

(a) Suppose $\text{Var}(r_{0t}) = \sigma_0^2$. Show that the covariance matrix $\mathbf{F} = (f_{ij})$ of the log return of the n stocks under the single-index model is given by $\mathbf{F} = \sigma_0^2 \boldsymbol{\beta}\boldsymbol{\beta}^T + \sigma^2 \mathbf{I}$, where $\boldsymbol{\beta} = (\beta_1, \ldots, \beta_p)^T$.

(b) Let $\sigma_{ij} = \text{Cov}(r_{it}, r_{jt})$ and $\boldsymbol{\Sigma} = (\sigma_{ij})_{1 \leq i,j \leq p}$. Let $\mathbf{S} = (s_{ij})$ be the sample covariance matrix based on $(r_{it}, \ldots, r_{pt})^T$, $1 \leq t \leq n$. Let $\mathbf{R}(\alpha) = \alpha \mathbf{F} + (1 - \alpha)\mathbf{S}$. Consider the quadratic loss function $L(\alpha) = ||\cdot||^2$, where $||\mathbf{A}||$ is the Frobenius norm of a square matrix $\mathbf{A}$ defined by $||\mathbf{A}||^2 = \text{tr}(\mathbf{A}^T \mathbf{A})$. Show that the minimizer α^* of $E[L(\alpha)]$ is given by

$$\alpha^* = \frac{\sum_{i=1}^p \sum_{j=1}^p [\text{Var}(s_{ij}) - \text{Cov}(f_{ij}, s_{ij})]}{\sum_{i=1}^p \sum_{j=1}^p \left[\text{Var}(f_{ij} - s_{ij}) + (E(f_{ij}) - \sigma_{ij})^2\right]}.$$

3.4. Let $\widehat{\boldsymbol{\mu}}$ and $\widehat{\boldsymbol{\Sigma}}$ denote the sample mean and the sample covariance matrix of the observed sample $\{\mathbf{r}_1, \ldots, \mathbf{r}_n\}$, and let $\widehat{\mathbf{w}}_b^*$ be the estimated optimal portfolio weight vector based on the bth bootstrap sample $\{\mathbf{r}_{b1}^*, \ldots, \mathbf{r}_{bn}^*\}$ drawn with replacement from the observed sample $\{\mathbf{r}_1, \ldots, \mathbf{r}_n\}$. The file `d_logret_6stocks.txt` contains the log returns of six stocks in Section 3.2.4.

(a) Use one bootstrap sample to plot the curve $(\sqrt{(\widehat{\mathbf{w}}_b^*)^T \widehat{\boldsymbol{\Sigma}} \widehat{\mathbf{w}}_b^*}, (\widehat{\mathbf{w}}_b^*)^T \widehat{\boldsymbol{\mu}})$ for different target mean returns $\boldsymbol{\mu}_*$.

(b) Repeat (a) for $B = 10$ bootstrap samples. Explain why all the plotted curves are below the estimated efficient frontier based on $\widehat{\boldsymbol{\mu}}$ and $\widehat{\boldsymbol{\Sigma}}$.

(c) Compute Michaud's resampled efficient frontier by using $B = 500$ bootstrap samples.

3.5. The file `m_ret_10stocks.txt` contains the monthly returns of ten stocks from January 1994 to December 2006. The ten stocks include Apple Computer, Adobe Systems, Automatic Data Processing, Advanced Micro Devices, Dell, Gateway, Hewlett-Packard Company, International Business Machines Corp., Microsoft Corp., and Oracle Corp. Consider portfolios that consist of these ten stocks.

(a) Compute the sample mean $\widehat{\boldsymbol{\mu}}$ and the sample covariance matrix $\widehat{\boldsymbol{\Sigma}}$ of the log returns.

(b) Assume that the monthly target return is 0.3% and that short selling is allowed. Estimate the optimal portfolio weights by replacing $(\boldsymbol{\mu}, \boldsymbol{\Sigma})$ in Markowitz's theory by $\widehat{\boldsymbol{\mu}}, \widehat{\boldsymbol{\Sigma}}$.

(c) Do the same as in (b) for Michaud's resampled weights (3.38) using $B = 500$ bootstrap samples.

(d) Plot the estimated efficient frontier (by varying μ_* over a grid) that uses $(\widehat{\boldsymbol{\mu}}, \widehat{\boldsymbol{\Sigma}})$ to replace $(\boldsymbol{\mu}, \boldsymbol{\Sigma})$ in Markowitz's efficient frontier.

(e) Plot Michaud's resampled efficient frontier using $B = 500$ bootstrap samples. Compare with the plot in (d).

3.6. The file `m_sp500ret_3mtcm.txt` contains three columns. The second column gives the monthly returns of the S&P 500 index from January 1994 to December 2006. The third column gives the monthly rates of the 3-month U. S. Treasury bill in the secondary market, which is obtained from the Federal Reserve Bank of St. Louis and used as the risk-free asset here. Consider the ten monthly returns in the file `m_ret_10stocks.txt`.

(a) Fit CAPM to the ten stocks. Give point estimates and 95% confidence intervals of α, β, the Sharpe index, and the Treynor index. (*Hint*: Use the delta method for the Sharpe and Treynor indices.)

(b) Use the bootstrap procedure in Section 3.5 to estimate the standard errors of the point estimates of α, β, and the Sharpe and Treynor indices.
(c) Test for each stock the null hypothesis $\alpha = 0$.
(d) Use the regression model (3.24) to test for the ten stocks the null hypothesis $\boldsymbol{\alpha} = \mathbf{0}$.
(e) Perform a factor analysis on the excess returns of the ten stocks. Show the factor loadings and rotated factor loadings. Explain your choice of the number of factors.
(f) Consider the model

$$r_t^e = \beta_1 \mathbf{1}_{\{t<t_0\}} r_M^e + \beta_2 \mathbf{1}_{\{t \geq t_0\}} r_M^e + \epsilon_t,$$

in which $r_t^e = r_t - r_f$ and $r_M^e = r_M - r_f$ are the excess returns of the stock and the S&P 500 index. The model suggests that the β in the CAPM might not be a constant (i.e., $\beta_1 \neq \beta_2$). Taking February 2001 as the month t_0, test for each stock the null hypothesis that $\beta_1 = \beta_2$.
(g) Estimate t_0 in (f) by the least squares criterion that minimizes the residual sum of squares over (β_1, β_2, t_0).

4

Parametric Models and Bayesian Methods

Parametric statistical models relate the observed data to the postulated stochastic mechanisms that generate them and that completely specified except for certain parameters. These parameters are assumed to be unknown and to be inferred from the data. A powerful and widely used approach to parametric inference is based on the likelihood function introduced in Section 2.4. We consider in Section 4.1 computational issues and apply maximum likelihood to regression problems. As pointed out in Section 2.4, the least squares method is equivalent to maximum likelihood when the random errors ϵ_t are assumed to be i.i.d. normal with mean 0. There are many applications in which this assumption is clearly violated. In particular, when the response variable y_t in a regression model is a binary variable that only takes the values 0 and 1, a natural extension of the linear regression function $\boldsymbol{\theta}^T\mathbf{x}_t$ is to relate how the parameter of the Bernoulli distribution of y_t depends on $\boldsymbol{\theta}^T\mathbf{x}_t$. *Logistic regression* provides such an extension and is described in Section 4.1.2, which also extends the basic ideas of logistic regression to more general settings, called *generalized linear models.*

Besides generalized linear models, another extension of the linear regression models in Chapter 1 is introduced in this chapter. Section 4.2 considers *nonlinear regression models*, which involve nonlinear functions $f(\boldsymbol{\theta}, \mathbf{x}_t)$ of the regression parameters, and shows how the method of least squares can be used to estimate the regression parameters. The least squares estimate is the same as the MLE when $y_t - f(\boldsymbol{\theta}, \mathbf{x}_t)$, $1 \le t \le n$, are i.i.d. normal. The sum of squared residuals now becomes

$$S(\boldsymbol{\theta}) := \sum_{t=1}^{n} (y_t - f(\boldsymbol{\theta}, \mathbf{x}_t))^2, \tag{4.1}$$

which generalizes (1.2) but does not have a closed-form expression for its minimizer when f is nonlinear in $\boldsymbol{\theta}$. However, one can successively approximate

$f(\boldsymbol{\theta}, \mathbf{x}_t)$ by linear functions of $\boldsymbol{\theta}$, thereby obtaining an explicit OLS expression for $\widehat{\boldsymbol{\theta}}^{(i)}$ at the ith iteration. This is the *Gauss-Newton* method described in Section 4.2.1, where some refinements of the method are also given. Section 4.2.2 considers statistical inference in nonlinear regression models and provides a financial application as an illustration.

Section 4.3 describes the Bayesian approach to parametric inference, which is based on the *posterior distribution* of the unknown parameter vector $\boldsymbol{\theta}$. It begins with a brief introduction to statistical decision theory in which a prior distribution π is put on the parameter space and the objective is to choose a statistical decision rule to minimize $\int E_{\boldsymbol{\theta}} L(\boldsymbol{\theta}, d(\mathbf{X})) d\pi(\boldsymbol{\theta})$, where $L(\boldsymbol{\theta}, a)$ is the loss function and $\mathbf{X}$ represents the observed data. The optimal decision rule is called the *Bayes rule.* Section 4.4 gives concrete examples of Bayes estimators and applies the Bayesian approach to the problem of mean-variance portfolio optimization when the means and covariances of the asset returns are unknown and have to be estimated from historical data.

4.1 Maximum likelihood and generalized linear models

4.1.1 Numerical methods for computing MLE

Except for simple special cases, MLEs do not have closed-form expressions. Numerical methods for function maximization are often needed to maximize the log-likelihood function $l(\boldsymbol{\theta})$ introduced in Section 2.4. When θ is one-dimensional, one can use a grid search that involves computation of the likelihood function on the grid. For multivariate $\boldsymbol{\theta}$, a grid search becomes expensive and one can use a gradient-type search instead. The steepest ascent method involves the gradient $\nabla l(\boldsymbol{\theta})$, which can be computed by numerical differentiation (difference quotients) if l is difficult to differentiate analytically. The Newton-Raphson method involves the inverse of the Hessian matrix $\left(\nabla^2 l(\boldsymbol{\theta})\right)^{-1}$, and Davidon (1959) and Fletcher and Powell (1963) proposed modifications of the Newton-Raphson method using efficient approximations to $\left(\nabla^2 l(\boldsymbol{\theta})\right)^{-1}$ in the iterative search for the maximum of l.

Often one has some constraints on the parameter values. For example, the covariance matrix $\boldsymbol{\Sigma}$ of a k-variate normal distribution has to be nonnegative definite. To satisfy this nonnegativity constraint when we maximize the likelihood function over $\boldsymbol{\Sigma}$, a useful trick is to use the Cholesky decomposition $\boldsymbol{\Sigma} = \mathbf{P}\mathbf{P}^T$, where

$$\mathbf{P} = \begin{pmatrix} p_{11} & 0 & \dots & \dots & 0 \\ p_{21} & p_{22} & 0 & \dots & 0 \\ \dots & \dots & \dots & \dots & \dots \\ p_{k1} & p_{k2} & p_{k3} & \cdots & p_{kk} \end{pmatrix}$$

is a lower-triangular matrix. Regarding the likelihood function as a function of $\mathbf{P}$ (rather than $\boldsymbol{\Sigma}$), we can circumvent the nonnegativity constraint since $\mathbf{P}\mathbf{P}^T$ is always symmetric nonnegative definite. Algorithms for numerical optimization can be found in Press et al. (1992, Chapter 10), and some general-purpose software packages are

```
R:          nlm, optim,
MATLAB:     fminsearch, fminunc, fmincon, fminimax.
```

For specific models, there are also built-in optimization routines to compute the MLE and perform likelihood inference. We shall refer to these software packages when we introduce the specific models.

4.1.2 Generalized linear models

A *generalized linear model* relates the response y_t to the p-dimensional regressor $\mathbf{x}_t$ by the following assumptions that generalize the classical linear regression model with normal errors in Section 2.4.1. For notational simplicity, we drop the subscript t when we describe the first two assumptions, in which the univariate parameters θ and ϕ may vary with t through $\mathbf{x}_t$.

(G1) y has density function

$$f(y;\theta,\phi) = \exp\Big\{\big[y\theta - b(\theta)\big]/\phi + c(y,\phi)\Big\}. \tag{4.2}$$

(G2) $g(\theta) = \boldsymbol{\beta}^T\mathbf{x}$ for some given smooth increasing function g and unknown parameter $\boldsymbol{\beta}$.

(G3) $(\mathbf{x}_t, y_t)$, $1 \le t \le n$, are independent.

The parametric family of density functions (4.2) is called an *exponential family* with *canonical parameter* θ and *dispersion parameter* $\phi > 0$. It includes the normal $N(\theta,\sigma^2)$ family with $b(\theta) = \theta^2/2$, $\phi = \sigma^2$, and $c(y,\phi) = -\{y^2/\sigma^2 + \log(2\pi\sigma^2)\}/2$ and the Poisson density $f(y;\lambda) = e^{-\lambda}\lambda^y/y!$ $(y = 0,1,2,\dots)$ with $\theta = \log\lambda$, $b(\theta) = e^\theta$, and $\phi = 1$. The mean and variance of the distribution in the exponential family (4.2) are given by

$$\text{mean} = b'(\theta), \qquad \text{variance} = \phi b''(\theta); \tag{4.3}$$

see Exercise 4.2. From the second formula in (4.3), it follows that b is convex and therefore b' is increasing. The case $g(\theta) = \theta$ in (G2) is called the *canonical link*.

An important special case of (4.2) is the Bernoulli density $f(y) = p^y(1-p)^{1-y}$ $(y = 0,1)$. Here $p = e^\theta/(1+e^\theta)$, or equivalently $\theta = \text{logit}(p)$, where

$$\text{logit}(p) = \log\big[p/(1-p)\big], \qquad 0 < p < 1. \tag{4.4}$$

Moreover, $\phi = 1$ and $b(\theta) = -\log(1-p)$, so $b'(\theta) = (1-p)^{-1}dp/d\theta = p$. In this case, using the canonical link function $g(\theta) = \theta$ in (G2) yields the *logistic regression model* $\text{logit}\big(P\{y_t = 1|\mathbf{x}_t\}\big) = \boldsymbol{\beta}^T\mathbf{x}_t$ or equivalently

$$P\{y_t = 1|\mathbf{x}_t\} = \frac{\exp(\boldsymbol{\beta}^T\mathbf{x}_t)}{1+\exp(\boldsymbol{\beta}^T\mathbf{x}_t)}, \quad P\{y_t = 0|\mathbf{x}_t\} = \frac{1}{1+\exp(\boldsymbol{\beta}^T\mathbf{x}_t)}. \tag{4.5}$$

Logistic regression is widely used to model how the probability of the occurrence of an event (e.g., default of a firm or delinquent payment of a loan in credit risk analysis) varies with explanatory variables (predictors). In Chapter 11 (Section 11.2.4), it is used to model the probability of a price change, and the conditional probability that the price change is positive given its occurrence, in high-frequency trading in security markets.

An alternative choice of g in (G2) that is used in some applications is $g(\theta) = \Phi^{-1}(p)$, where Φ is the standard normal distribution function, yielding the *probit model*

$$P\{y_t = 1|\mathbf{x}_t\} = \Phi(\boldsymbol{\beta}^T\mathbf{x}_t). \tag{4.6}$$

An application of the probit model is given in Section 11.2.4 on modeling the price-generating process from transaction price quotes.

Likelihood functions for generalized linear models and iteratively reweighted least squares

Given the observed $\{(\mathbf{x}_i, y_i) : 1 \le i \le n\}$, the log-likelihood associated with (G1)–(G3) is

$$l(\boldsymbol{\beta}, \phi) = \sum_{i=1}^{n} \Big\{ \big[y_i g^{-1}(\boldsymbol{\beta}^T\mathbf{x}_i) - b(g^{-1}(\boldsymbol{\beta}^T\mathbf{x}_i))\big]/\phi + c(y_i, \phi) \Big\}. \tag{4.7}$$

In particular, it follows from (4.5) that the log-likelihood function in logistic regression is

$$l(\boldsymbol{\beta}) = \sum_{i=1}^{n} \Big\{ y_i\boldsymbol{\beta}^T\mathbf{x}_i + \log\Big[1+\exp(\boldsymbol{\beta}^T\mathbf{x}_i)\Big]\Big\}. \tag{4.8}$$

We next describe an algorithm called *iteratively reweighted least squares* to compute the MLE of $\boldsymbol{\beta}$ in the case where $g(\theta) = \theta$ and $\phi = 1$. Let $\mathbf{Y} = (y_1, \dots, y_n)^T, \mathbf{X} = (\mathbf{x}_1, \dots, \mathbf{x}_n)^T, \boldsymbol{\mu}(\boldsymbol{\beta}) = (b'(\boldsymbol{\beta}^T\mathbf{x}_1), \dots, b'(\boldsymbol{\beta}^T\mathbf{x}_n))^T, \mathbf{W}(\boldsymbol{\beta}) = \text{diag}\big(b''(\boldsymbol{\beta}^T\mathbf{x}_1), \dots, b''(\boldsymbol{\beta}^T\mathbf{x}_n)\big)$. From (4.7) with $\phi = 1$ and $g^{-1}(\theta) = \theta$, it follows that

$$\boldsymbol{\nabla} l(\boldsymbol{\beta}) = \mathbf{X}^T\big[\mathbf{Y} - \boldsymbol{\mu}(\boldsymbol{\beta})\big], \quad \boldsymbol{\nabla}^2 l(\boldsymbol{\beta}) = -\mathbf{X}^T\mathbf{W}(\boldsymbol{\beta})\mathbf{X}. \tag{4.9}$$

Hence a single Newton-Raphson update that starts with $\boldsymbol{\beta}_{\text{old}}$ is

$$\begin{aligned}\boldsymbol{\beta}_{\text{new}} &= \boldsymbol{\beta}_{\text{old}} - \left[\nabla^2 l(\boldsymbol{\beta})\right]^{-1}\left(\nabla l(\boldsymbol{\beta})\right)\Big|_{\boldsymbol{\beta}=\boldsymbol{\beta}_{\text{old}}} \\ &= \boldsymbol{\beta}_{\text{old}} + \left[\mathbf{X}^T\mathbf{W}(\boldsymbol{\beta}_{\text{old}})\mathbf{X}\right]^{-1}\mathbf{X}^T\left[\mathbf{Y} - \boldsymbol{\mu}(\boldsymbol{\beta}_{\text{old}})\right]. \end{aligned} \tag{4.10}$$

The following software packages can be used to fit generalized linear models:

```
R:        glm(formula, family, data, weights, control),
MATLAB:   glmfit(X, y, distribution).
```

Details can be found in Venables and Ripley (2002, Chapter 7) and the help function in MATLAB.

Analysis of deviance

The analysis of variance for linear regression models in Section 1.2.2 can be extended to generalized linear models by using the generalized likelihood ratio (GLR) statistics of Section 2.4.2. Let $(\widehat{\boldsymbol{\beta}}, \widehat{\phi})$ be the unconstrained MLE of $(\boldsymbol{\beta}, \phi)$ and $(\widetilde{\boldsymbol{\beta}}, \widetilde{\phi})$ be the MLE when $\boldsymbol{\beta}$ is constrained to belong to an r-dimensional linear subspace $\mathcal{L}_0$ with $r < p$. The GLR statistic (2.47) then reduces to $2\{l(\widehat{\boldsymbol{\beta}}, \widehat{\phi}) - l(\widetilde{\boldsymbol{\beta}}, \widetilde{\phi})\}$, which is called the *deviance* between the full model and the submodel $\boldsymbol{\beta} \in \mathcal{L}_0$. The F-statistic (1.25) for ANOVA in linear regression is a scalar multiple of the deviance in this case; see Exercise 4.3, which also provides an analog of ANOVA for logistic regression via analysis of deviance. Therefore the partial F-tests for model selection can be extended to generalized linear models by using the corresponding GLR tests.

4.2 Nonlinear regression models

A nonlinear regression model is of the form

$$y_t = f(\boldsymbol{\theta}, \mathbf{x}_t) + \epsilon_t \quad (t = 1, \ldots, n), \tag{4.11}$$

where $\boldsymbol{\theta}$ is a p-dimensional unknown parameter vector, f is a nonlinear function of $\boldsymbol{\theta}$, and ϵ_t are uncorrelated random errors with $E\epsilon_t = 0$ and $\text{Var}(\epsilon_t) = \sigma^2$. In empirical modeling for which one has the freedom to choose approximations of the underlying relationship between y_t and $\mathbf{x}_t$, it is not clear why one should complicate matters by choosing a functional form that is nonlinear in the parameters. In fact, in Chapter 7, on nonparametric regression, we recommend approximating the regression function by $\boldsymbol{\theta}^T\mathbf{g}(\mathbf{x}_t)$, where $\mathbf{g} = (g_1, \ldots, g_p)^T$ is a vector of basis functions. However, when one deals with subject-matter models, domain knowledge may prescribe regression models that are nonlinear in physically meaningful parameters. A classic example is the Michaelis-Menten model in enzyme kinetics, for which $f(\boldsymbol{\theta}, x) = \theta_1 x/(\theta_2 + x)$; other important examples are given in Chapter 10 and related to interest rate models.

Although the Michaelis-Menten model is nonlinear in $\boldsymbol{\theta}$, it is actually linear in θ_1 when θ_2 is fixed. Such models are called *partially linear*. If $f(\boldsymbol{\theta}, \mathbf{x}_t)$ is linear in q parameters, which we assume to form a subvector $\boldsymbol{\theta}_1$, and is nonlinear in the other parameters, forming a subvector $\boldsymbol{\theta}_2$, then we can minimize (4.1) by minimizing $S(\widehat{\boldsymbol{\theta}}_1(\boldsymbol{\theta}_2), \boldsymbol{\theta}_2)$ over $\boldsymbol{\theta}_2$, where $\widehat{\boldsymbol{\theta}}_1(\boldsymbol{\theta}_2)$ is the OLS estimate (given in Section 1.1.1) of $\boldsymbol{\theta}_1$ for a fixed value of $\boldsymbol{\theta}_2$. This reduces the dimensionality of the minimization problem from p to $p-q$. The Michaelis-Menten model is in fact also *transformably linear*, as we can rewrite f in the form

$$\frac{1}{f} = \frac{1}{\theta_1} + \frac{\theta_2}{\theta_1 x} = \beta_1 + \beta_2 u,$$

where $\beta_1 = 1/\theta_1$, $\beta_2 = \theta_2/\theta_1$, and $u = 1/x$. This suggests fitting the linear regression model $\beta_1 + \beta_2 u_t$ to $1/y_t$.

For empirical modeling, Box and Cox (1964) proposed to include a parametric transformation of the response variable that is positive so that the regression model can be written in the linear form $y_t^{(\lambda)} = \boldsymbol{\theta}^T \mathbf{x}_t + \epsilon_t$, where

$$y_t^{(\lambda)} = \begin{cases} (y_t^\lambda - 1)/\lambda, & \lambda \neq 0, \\ \log y_t, & \lambda = 0. \end{cases} \tag{4.12}$$

Note that the *Box-Cox transformation* (4.12) is smooth in λ since $y = e^{\log y}$. The parameters λ, $\boldsymbol{\theta}$, and $\sigma^2 := \mathrm{Var}(\epsilon_t)$ can be estimated jointly by maximum likelihood, assuming the ϵ_t to be normal. The maximization in fact only needs to be carried out over λ because for a given λ the MLE of $\boldsymbol{\theta}$ is simply the OLS estimate and the MLE of σ^2 is the residual sum of squares divided by the sample size.

4.2.1 The Gauss-Newton algorithm

To compute the minimizer $\widehat{\boldsymbol{\theta}}$ of (4.1), we initialize with $\widehat{\boldsymbol{\theta}}^{(0)}$ and approximate $f(\boldsymbol{\theta}, \mathbf{x}_t)$ after the jth iteration, which yields $\widehat{\boldsymbol{\theta}}^{(j)}$ by

$$f(\boldsymbol{\theta}, \mathbf{x}_t) \approx f(\widehat{\boldsymbol{\theta}}^{(j)}, \mathbf{x}_t) + (\boldsymbol{\theta} - \widehat{\boldsymbol{\theta}}^{(j)})^T \nabla f(\widehat{\boldsymbol{\theta}}^{(j)}, \mathbf{x}_t)$$

so that (4.11) can be approximated by the linear regression model

$$y_t - f(\widehat{\boldsymbol{\theta}}^{(j)}, \mathbf{x}_t) = (\boldsymbol{\theta} - \widehat{\boldsymbol{\theta}}^{(j)})^T \nabla f(\widehat{\boldsymbol{\theta}}^{(j)}, \mathbf{x}_t) + \epsilon_t. \tag{4.13}$$

The OLS estimate $\widehat{\boldsymbol{\theta}}^{(j+1)}$ of $\boldsymbol{\theta}$ in (4.13) is given explicitly by

$$\widehat{\boldsymbol{\theta}}^{(j+1)} - \widehat{\boldsymbol{\theta}}^{(j)} = (\mathbf{X}_{(j)}^T \mathbf{X}_{(j)})^{-1} \mathbf{X}_{(j)}^T \left(\mathbf{Y} - \boldsymbol{\eta}(\widehat{\boldsymbol{\theta}}^{(j)})\right), \tag{4.14}$$

where

$$\mathbf{Y} = \begin{pmatrix} y_1 \\ \vdots \\ y_n \end{pmatrix}, \quad \boldsymbol{\eta}(\boldsymbol{\theta}) = \begin{pmatrix} f(\boldsymbol{\theta}, \mathbf{x}_1) \\ \vdots \\ f(\boldsymbol{\theta}, \mathbf{x}_n) \end{pmatrix},$$

$$\mathbf{X}_{(j)} = \begin{pmatrix} \frac{\partial f}{\partial \theta_1}(\widehat{\boldsymbol{\theta}}^{(j)}, \mathbf{x}_1) & \cdots & \frac{\partial f}{\partial \theta_p}(\widehat{\boldsymbol{\theta}}^{(j)}, \mathbf{x}_1) \\ \cdots & \cdots & \cdots \\ \frac{\partial f}{\partial \theta_1}(\widehat{\boldsymbol{\theta}}^{(j)}, \mathbf{x}_n) & \cdots & \frac{\partial f}{\partial \theta_p}(\widehat{\boldsymbol{\theta}}^{(j)}, \mathbf{x}_n) \end{pmatrix}. \tag{4.15}$$

The iterative scheme (4.14) is called the *Gauss-Newton algorithm*, and $\boldsymbol{\delta}_{j+1} := \boldsymbol{\theta}^{(j+1)} - \boldsymbol{\theta}^{(j)}$ is called the *Gauss increment*.

Matrix inversion

For numerical stability, care has to be taken in taking the inverse in (4.14). The QR decomposition of $\mathbf{X}_{(j)}$ in Section 1.1.3 is recommended. The Gauss-Newton algorithm is aborted at the jth step when one gets a singular (or nearly singular) $\mathbf{X}_{(j)}^T \mathbf{X}_{(j)}$. It may also stop after reaching a prescribed upper bound on the number of iterations without convergence. When one does not get an answer from the Gauss-Newton algorithm, one should choose another starting value and repeat the algorithm.

Step factor

The Gauss increment $\boldsymbol{\delta}_{j+1}$ may produce an increase in $S(\boldsymbol{\theta})$ when it is outside the region where the linear approximation holds. To ensure a decrease in $S(\boldsymbol{\theta})$, use a step factor $0 < \lambda \leq 1$ so that $S(\boldsymbol{\theta}^{(j)} + \lambda \boldsymbol{\delta}^{(j)}) < S(\boldsymbol{\theta}^{(j)})$. A commonly used method is to start with $\lambda = 1$ and halve it until we have $S(\boldsymbol{\theta}^{(j+1)}) < S(\boldsymbol{\theta}^{(j)})$.

Convergence criterion

A commonly used criterion for numerical convergence is the size of the parameter increment relative to the parameter value. Another criterion is that the relative change in $S(\boldsymbol{\theta})$ be small. A third criterion is that $\mathbf{Y} - \boldsymbol{\eta}(\boldsymbol{\theta}^{(j)})$ be nearly orthogonal to the tangent space of $\boldsymbol{\eta}(\boldsymbol{\theta}) := \left(f(\boldsymbol{\theta}, \mathbf{x}_1), \ldots, f(\boldsymbol{\theta}, \mathbf{x}_n)\right)^T$ at $\boldsymbol{\theta}^{(j)}$. Using the QR decomposition of $\mathbf{X}_{(j)}$ and writing $\mathbf{Q} = (\mathbf{Q}_1, \mathbf{Q}_2)$, this criterion requires sufficiently small values of

$$\frac{\|\mathbf{Q}_1^T(\mathbf{Y}-\eta(\boldsymbol{\theta}^{(j)}))\|/\sqrt{p}}{\|\mathbf{Q}_2^T(\mathbf{Y}-\eta(\boldsymbol{\theta}^{(j)}))\|/\sqrt{n-p}}.$$

The Levenberg-Marquardt modification

Since the Gauss-Newton iterative scheme is aborted whenever $\mathbf{X}_j^T\mathbf{X}_j$ is singular or nearly singular, it is desirable to avoid such difficulties in matrix inversion. This has led to the modification of (4.14) by

$$\boldsymbol{\delta}_{j+1} = (\mathbf{X}_{(j)}^T\mathbf{X}_{(j)} + \kappa\mathbf{D})^{-1}\mathbf{X}_{(j)}^T(\mathbf{Y} - \boldsymbol{\eta}(\widehat{\boldsymbol{\theta}}^{(j)})), \tag{4.16}$$

where $\mathbf{D}$ is a diagonal matrix whose diagonal elements are the same as those of $\mathbf{X}_{(j)}^T\mathbf{X}_{(j)}$, proposed by Marquardt as a refinement of an earlier proposal $\mathbf{D} = \mathbf{I}$ by Levenberg. In `MATLAB`, the function `lsqnonlin` has the option of using the Levenberg-Marquardt algrithm instead of the Gauss-Newton algorithm. In `R`, the function `nls.lm` uses the Levenberg-Marquardt algorithm to compute the nonlinear least squares estimate.

4.2.2 Statistical inference

Let $\widehat{\boldsymbol{\theta}}$ be the least squares estimate of $\boldsymbol{\theta}$ in the nonlinear regression model (4.1). Let $\boldsymbol{\theta}_0$ denote the true value of $\boldsymbol{\theta}$. First assume that the $\mathbf{x}_t$ are nonrandom as in Section 1.1.4. Then, by (4.1) and (4.11),

$$E[S(\boldsymbol{\theta})] = \sum_{i=1}^{n} [f(\boldsymbol{\theta}, \mathbf{x}_t) - f(\boldsymbol{\theta}_0, \mathbf{x}_t)]^2 + n\sigma^2,$$

recalling that $E(\epsilon_t) = 0$ and $\mathrm{Var}(\epsilon_t) = \sigma^2$. Therefore,

$$E[S(\boldsymbol{\theta})] - n\sigma^2 \begin{cases} = 0 & \text{if } \boldsymbol{\theta} = \boldsymbol{\theta}_0, \\ \to \infty & \text{if } \boldsymbol{\theta} \neq \boldsymbol{\theta}_0, \end{cases} \tag{4.17}$$

under the assumption

$$\sum_{t=1}^{\infty} [f_t(\boldsymbol{\theta}) - f_t(\boldsymbol{\theta}_0)]^2 = \infty \text{ for } \boldsymbol{\theta} \neq \boldsymbol{\theta}_0, \tag{4.18}$$

where $f_t(\boldsymbol{\theta}) = f(\boldsymbol{\theta}, \mathbf{x}_t)$. In the linear case $f_t(\boldsymbol{\theta}) = \boldsymbol{\theta}^T\mathbf{x}_t$, (4.18) is equivalent to $\lim_{n\to 0} \lambda_{\min}\left(\sum_{t=1}^n \mathbf{x}_t\mathbf{x}_t^T\right) = \infty$, which is equivalent to $(\mathbf{X}^T\mathbf{X})^{-1} \to \mathbf{0}$ in the notation of Section 1.1.4 and using the notation $\lambda_{\min}(\cdot)$ to denote the minimum eigenvalue of a nonnegative definite matrix. Since $\widehat{\boldsymbol{\theta}}$ is the minimizer of $S(\boldsymbol{\theta})$, (4.17) suggests that $\widehat{\boldsymbol{\theta}}$ is consistent. A rigorous proof involves considering $S(\boldsymbol{\theta})$ as a random function of $\boldsymbol{\theta}$ (instead of fixed $\boldsymbol{\theta}$ in (4.16)) and requires additional assumptions as in the consistency of the MLE in Section 2.4.1.

Consistency of $\widehat{\boldsymbol{\theta}}$ leads easily to its asymptotic normality since we can approximate $f_t(\widehat{\boldsymbol{\theta}})$ by $f_t(\boldsymbol{\theta}_0)+(\widehat{\boldsymbol{\theta}}-\boldsymbol{\theta}_0)^T\nabla f_t(\boldsymbol{\theta}_0)$ when $\widehat{\boldsymbol{\theta}}$ is near $\boldsymbol{\theta}_0$. The asymptotic properties of $\widehat{\boldsymbol{\theta}}$ are therefore the same as those of OLS in Section 1.1.4 or Section 1.5.3. Specifically, analogous to (1.11), we now have

$$\widehat{\boldsymbol{\theta}} \approx N\big(\boldsymbol{\theta}_0, \sigma^2(\mathbf{X}^T\mathbf{X})^{-1}\big), \tag{4.19}$$

where $\mathbf{X}$ is the same as $\mathbf{X}_{(j)}$ defined in (4.15) but with $\widehat{\boldsymbol{\theta}}^{(j)}$ replaced by $\boldsymbol{\theta}_0$ or by $\widehat{\boldsymbol{\theta}}$, using Slutsky's theorem as in (2.43). Moreover, σ^2 can be consistently estimated by

$$\widehat{\sigma}^2 = \sum_{t=1}^{n} \big(y_t - f_t(\widehat{\boldsymbol{\theta}})\big)^2 \Big/ n. \tag{4.20}$$

For nonlinear functions $g(\boldsymbol{\theta}_0)$, we can apply the delta method (see Section 2.4.2), which uses the Taylor expansion

$$g(\widehat{\boldsymbol{\theta}}) - g(\boldsymbol{\theta}_0) \doteq \Big(\nabla g(\boldsymbol{\theta}_0)\Big)^T (\widehat{\boldsymbol{\theta}} - \boldsymbol{\theta}_0) \tag{4.21}$$

to approximate $g(\widehat{\boldsymbol{\theta}})-g(\boldsymbol{\theta}_0)$ by a linear function. The delta method yields the asymptotic normality of $g(\widehat{\boldsymbol{\theta}})$ with mean $g(\boldsymbol{\theta}_0)$ and covariance matrix

$$\sigma^2\Big(\nabla g(\boldsymbol{\theta}_0)\Big)^T\Big(\mathbf{X}^T\mathbf{X}\Big)^{-1}\Big(\nabla g(\boldsymbol{\theta}_0)\Big). \tag{4.22}$$

Therefore we can construct confidence intervals for $g(\boldsymbol{\theta}_0)$ as in Section 1.5(ii), with $\mathbf{x}$ replaced by $\nabla g(\boldsymbol{\theta})$. Note that the square root of (4.22) also gives the estimated standard error for $g(\widehat{\boldsymbol{\theta}})$ if we replace the unknown σ and $\boldsymbol{\theta}_0$ in (4.22) by $\widehat{\sigma}$ and $\widehat{\boldsymbol{\theta}}$.

The adequacy of this normal expansion is questionable for highly nonlinear g, as the one-term Taylor expansion in the delta method can be quite poor. An alternative to the delta method that involves asymptotic approximations is the *bootstrap method*, which uses Monte Carlo simulations to obtain standard errors and confidence intervals. Details are similar to those in Section 1.6.

4.2.3 Implementation and an example

The nonlinear least squares procedure is implemented by many numerical software packages. The following are functions in `MATLAB` and `R`:

```
R:        nls.lm, nls,
MATLAB:   nlinfit, nlintool, nlparci, nlpredci, lsqnonlin.
```

In `MATLAB`, the function `nlinfit` (with the incorporation of a Levenberg-Marquardt step) estimates the coefficients of a nonlinear function using least squares, `nlparci` computes 95% confidence intervals for the regression parameters, and `nlpredci` gives predicted responses and the associated confidence intervals.

As an illustration, we apply the nonlinear least squares procedure to estimate the volatility of yields in the Hull-White model of interest rates. Under the Hull-White model, which is described in Chapter 10 (Section 10.4.1), the variance of the yield with maturity τ (in years) is

$$v_\tau = \left(\frac{1 - e^{-\kappa\tau}}{\kappa\tau}\right)^2 \frac{\sigma^2}{2\kappa}, \tag{4.23}$$

where κ and σ are the parameters of the Hull-White model. Table 4.1 gives the variances v_τ of the yields of U.S. τ-year zero-coupon bonds for $\tau = 1, 2, 3, 5, 7, 10$, and 20, calculated from the weekly Treasury Constant Maturity Rates from October 1, 1993 to March 23, 2007 obtained from the Federal Reserve Bank of St. Louis.

Table 4.1. Variances of historical zero-coupon bond yields with different maturities.

Maturity τ (years)	1	2	3	5	7	10	20
Variance ($\times 10^4$)	2.742	2.503	2.154	1.599	1.331	1.090	0.856

The parameter $\boldsymbol{\theta} = (\kappa, \sigma)$ is estimated from the variances v_τ in Table 4.1 by minimizing

$$\sum_{\tau \in \{1,2,3,5,7,10,20\}} \left\{ \left(\frac{1 - e^{-\kappa\tau}}{\kappa\tau}\right)^2 \frac{\sigma^2}{2\kappa} - v_\tau \right\}^2.$$

The `MATLAB` function `nlinfit` is used and gives the estimated parameters $\widehat{\kappa} = 0.1081$ and $\widehat{\sigma} = 0.0080$, with respective 95% confidence intervals $[0.0631, 0.1531]$ and $[0.0058, 0.0103]$.

4.3 Bayesian inference

4.3.1 Prior and posterior distributions

Whereas the likelihood approach to inference in Section 2.4 is based on a function (the likelihood function) of the unknown parameter $\boldsymbol{\theta}$, Bayesian inference is based on a conditional distribution, called the *posterior distribution*, of $\boldsymbol{\theta}$ given the data, treating the unknown $\boldsymbol{\theta}$ as a random variable that has a *prior distribution* with density function π. The joint density function of $(\boldsymbol{\theta}, X_1, \dots, X_n)$ is $\pi(\boldsymbol{\theta}) f_{\boldsymbol{\theta}}(x_1, \dots, x_n)$. Hence the posterior density function $\pi(\boldsymbol{\theta}|X_1, \dots, X_n)$ of $\boldsymbol{\theta}$ given $X_1, \dots, X_n$ is proportional to $\pi(\boldsymbol{\theta}) f_{\boldsymbol{\theta}}(X_1, \dots, X_n)$, with the constant of proportionality chosen so that the posterior density function integrates to 1:

$$\pi(\boldsymbol{\theta}|X_1, \dots, X_n) = \pi(\boldsymbol{\theta}) f_{\boldsymbol{\theta}}(X_1, \dots, X_n)/g(X_1, \dots, X_n), \tag{4.24}$$

where $g(X_1, \dots, X_n) = \int \pi(\boldsymbol{\theta}) f_{\boldsymbol{\theta}}(X_1, \dots, X_n) d\boldsymbol{\theta}$ (with the integral replaced by a sum when the X_i are discrete) is the marginal density of $X_1, \dots, X_n$ in the Bayesian model. Often one can ignore the constant of proportionality $1/g(X_1, \dots, X_n)$ in identifying the posterior distribution with density function (4.24), as illustrated by the following example.

Example 4.1. Let $X_1, \dots, X_n$ be i.i.d. $N(\mu, \sigma^2)$, where σ is assumed to be known and μ is the unknown parameter. Suppose μ has a prior $N(\mu_0, v_0)$ distribution. Letting $\bar{X} = n^{-1} \sum_{i=1}^n X_i$, the posterior density function of μ given $X_1, \dots, X_n$ is proportional to

$$\begin{aligned} & e^{-(\mu-\mu_0)^2/(2v_0)} \prod_{i=1}^n e^{-(X_i-\mu)^2/(2\sigma^2)} \\ & \propto \exp\left\{ -\frac{\mu^2}{2}\left[\frac{1}{v_0} + \frac{n}{\sigma^2}\right] + \mu\left[\frac{\mu_0}{v_0} + \frac{n\bar{X}}{\sigma^2}\right]\right\} \\ & \propto \exp\left\{ -\frac{1}{2}\left(\frac{1}{v_0} + \frac{n}{\sigma^2}\right)\left[\mu - \left(\frac{\mu_0}{v_0} + \frac{n\bar{X}}{\sigma^2}\right)\Big/\left(\frac{1}{v_0} + \frac{n}{\sigma^2}\right)\right]^2\right\} \end{aligned}$$

which, as a function of μ, is proportional to a normal density function. Hence the posterior distribution of μ given $(X_1, \dots, X_n)$ is

$$N\left(\left(\frac{\mu_0}{v_0} + \frac{n\bar{X}}{\sigma^2}\right)\Big/\left(\frac{1}{v_0} + \frac{n}{\sigma^2}\right), \left(\frac{1}{v_0} + \frac{n}{\sigma^2}\right)^{-1}\right). \tag{4.25}$$

Note that we have avoided the integration to determine the constant of proportionality in the preceding argument. From (4.25), the mean of the posterior distribution is a weighted average of the sample mean $\bar{X}$ and the prior mean μ_0. Moreover, the reciprocal of the posterior variance is the sum of the reciprocal of the prior variance and n/σ^2, which is the reciprocal of the variance of $\bar{X}$.

4.3.2 Bayes procedures

Bayes procedures are defined by minimizing certain functionals of the posterior distribution of $\boldsymbol{\theta}$ given the data. To describe these functionals, we first give a brief introduction to statistical decision theory. In this theory, statistical decision problems such as estimation and hypothesis testing have the following ingredients:

- a parameter space Θ and a family of distribution $\{P_\theta, \theta \in \Theta\}$;
- data $(X_1, \dots, X_n) \in \mathcal{X}$ sampled from the distribution P_θ when θ is the true parameter, where $\mathcal{X}$ is called the "sample space";
- an action space $\mathcal{A}$ consisting of all available actions to be chosen; and
- a loss function $L : \Theta \times \mathcal{A} \to [0, \infty)$ representing the loss $L(\theta, a)$ when θ is the parameter value and action a is taken.

A *statistical decision rule* is a function $d : \mathcal{X} \to \mathcal{A}$ that takes action $d(\mathbf{X})$ when $\mathbf{X} = (X_1, \dots, X_n)$ is observed. Its performance is evaluated by the *risk function*

$$R(\theta, d) = E_\theta L\big(\theta, d(\mathbf{X})\big), \qquad \theta \in \Theta.$$

A statistical decision rule d is as good as d^* if $R(\theta, d) \le R(\theta, d^*)$ for all $\theta \in \Theta$. In this case, d is better than d^* if strict inequality also holds at some θ. A statistical decision rule is said to be *inadmissible* if there exists another rule that is better; otherwise, it is called *admissible.*

Given a prior distribution π on Θ, the *Bayes risk* of a statistical decision rule d is

$$B(d) = \int R(\theta, d) d\pi(\theta). \tag{4.26}$$

A *Bayes rule* is a statistical decision rule that minimizes the Bayes risk. It can be determined by

$$d(\mathbf{X}) = \arg\min_{a \in \mathcal{A}} E\big[L(\theta, a) | \mathbf{X}\big]. \tag{4.27}$$

This follows from the key observation that (4.26) can be expressed as

$$B(d) = E\big\{E\big[L(\theta, d(\mathbf{X})\big|\theta\big]\big\} = EL(\theta, d(\mathbf{X})) = E\big\{E\big[L(\theta, d(\mathbf{X}))|\mathbf{X}\big]\big\}, \tag{4.28}$$

where we have used the *tower property* $E(Z) = E[E(Z|W)]$ of conditional expectations for the second and third equalities. Hence, to minimize $B(\theta, d)$ over statistical decision functions $d : \mathcal{X} \to \mathcal{A}$, it suffices to choose the action

$a \in \mathcal{A}$ that minimizes for the observed data $\mathbf{X}$ the *posterior loss* $E\big[L(\theta, a)\big|\mathbf{X}\big]$, which is the expected value $\int L(\theta, a) d\pi(\theta|\mathbf{X})$ with respect to the posterior distribution $\pi(\cdot|\mathbf{X})$ of θ. This shows the central role of the posterior distribution in Bayes rules. Under some weak conditions, Bayes rules are admissible, and admissible statistical decision rules are Bayes rules or limits of Bayes rules.

Example 4.2. Estimation of $\boldsymbol{\theta} \in \mathbb{R}^d$ is a statistical decision problem with $d(\mathbf{X})$ being the estimator. The usual squared error loss in estimation theory corresponds to the loss function $L(\boldsymbol{\theta}, \mathbf{a}) = ||\boldsymbol{\theta} - \mathbf{a}||^2$. Hence $\Theta = \mathcal{A} = \mathbb{R}^d$ and $E\big(||\boldsymbol{\theta} - \mathbf{a}||^2\big|\mathbf{X}\big)$ is minimized by the posterior mean $\mathbf{a} = E(\boldsymbol{\theta}|\mathbf{X})$. In particular, by (4.25), the Bayes estimator of μ (with respect to squared error loss) in Example 4.1 is

$$E(\mu|X_1, \dots, X_n) = \left(\frac{\mu_0}{v_0} + \frac{n\bar{X}}{\sigma^2}\right) \Big/ \left(\frac{1}{v_0} + \frac{n}{\sigma^2}\right). \tag{4.29}$$

4.3.3 Bayes estimators of multivariate normal mean and covariance matrix

Let $\mathbf{X}_1, \dots, \mathbf{X}_n$ be i.i.d. observations from an m-variate normal distribution $N(\boldsymbol{\mu}, \boldsymbol{\Sigma})$. Let $\bar{\mathbf{X}} = n^{-1} \sum_{i=1}^n \mathbf{X}_i$ be the sample mean vector.

Bayes estimator of $\boldsymbol{\mu}$ when $\boldsymbol{\Sigma}$ is known

Suppose $\boldsymbol{\mu}$ has a prior $N(\boldsymbol{\mu}_0, \mathbf{V}_0)$ distribution. The method used to derive (4.29) for the univariate case can be easily extended to obtain its multivariate analog: The posterior distribution of $\boldsymbol{\mu}$ given $\mathbf{X}_1, \dots, \mathbf{X}_n$ is normal with mean

$$E(\boldsymbol{\mu}|\mathbf{X}) = \mathbf{V}_0(\mathbf{V}_0 + n^{-1}\boldsymbol{\Sigma})^{-1}\bar{\mathbf{X}} + n^{-1}\boldsymbol{\Sigma}(\mathbf{V}_0 + n^{-1}\boldsymbol{\Sigma})^{-1}\boldsymbol{\mu}_0 \tag{4.30}$$

and covariance matrix $\mathbf{V}_0 - \mathbf{V}_0(\mathbf{V}_0 + n^{-1}\boldsymbol{\Sigma})^{-1}\mathbf{V}_0$. Note that the Bayes estimator $\widehat{\mu} = E(\boldsymbol{\mu}|\mathbf{X})$ shrinks $\bar{\mathbf{X}}$ (the MLE) toward the prior mean $\boldsymbol{\mu}_0$. Hence, if the prior distribution is normal, then so is the posterior distribution.

The inverted Wishart distribution

A parametric family of distributions is called a *conjugate family* for a Bayesian statistical decision problem if both the prior and posterior distributions belong to the family. The preceding paragraph shows that the normal family is a conjugate family for the problem of estimating the mean $\boldsymbol{\mu}$ of a multivariate normal distribution whose covariance matrix $\boldsymbol{\Sigma}$ is known. When $\boldsymbol{\Sigma}$ is also unknown, the Bayesian approach also needs to put a prior distribution on $\boldsymbol{\Sigma}$. A conjugate family for the problem of estimating $\boldsymbol{\Sigma}$ is the *inverted*

Wishart family $IW_m(\boldsymbol{\Psi}, k)$, which we describe below. Recall the definition of the Wishart distribution in Section 2.3.4. If $\mathbf{W}$ has the Wishart distribution $W_m(\boldsymbol{\Phi}, k)$, then $\mathbf{V} = \mathbf{W}^{-1}$ is said to have the inverted Wishart distribution, which is denoted by $IW_m(\boldsymbol{\Psi}, k)$ with $\boldsymbol{\Psi} = \boldsymbol{\Phi}^{-1}$. Making use of the density function (2.26) of the Wishart distribution and the formula (2.2) for change of variables, it can be shown that the density function of $IW_m(\boldsymbol{\Psi}, k)$ is

$$f(\mathbf{V}) = \frac{(\det \boldsymbol{\Psi})^{m/2} (\det \mathbf{V})^{-(k+m+1)/2} \exp\{-\frac{1}{2}\mathrm{tr}(\boldsymbol{\Psi}\mathbf{V}^{-1})\}}{2^{mk/2}\Gamma_m(k/2)}, \quad \mathbf{V} > \mathbf{0}, \tag{4.31}$$

in which $\mathbf{V} > \mathbf{0}$ denotes that $\mathbf{V}$ is positive definite. Moreover,

$$E(\mathbf{V}) = \boldsymbol{\Psi}/(k - m - 1). \tag{4.32}$$

An important property of the Wishart distribution is that if $\mathbf{W} \sim W_m(\boldsymbol{\Sigma}, n)$ and $\boldsymbol{\Sigma}$ has the prior distribution $IW_m(\boldsymbol{\Psi}, n_0)$, then the posterior distribution of $\boldsymbol{\Sigma}$ given $\mathbf{W}$ is $IW_m(\mathbf{W} + \boldsymbol{\Psi}, n + n_0)$; see Exercise 4.4.

Bayes estimator of $(\boldsymbol{\mu}, \boldsymbol{\Sigma})$

Let $\mathbf{X}_1, \dots, \mathbf{X}_n$ be i.i.d. observations from an m-variate normal distribution $N(\boldsymbol{\mu}, \boldsymbol{\Sigma})$ and let $W := \sum_{i=1}^n (\mathbf{X}_i - \bar{\mathbf{X}})(\mathbf{X}_i - \bar{\mathbf{X}})^T$. Recall that the inverted Wishart family is a conjugate family for $\mathbf{W}$ and that the normal family is a conjugate family for $\mathbf{X}$ when $\boldsymbol{\Sigma}$ is known. This suggests using the following prior distribution of $(\boldsymbol{\mu}, \boldsymbol{\Sigma})$:

$$\boldsymbol{\mu}|\boldsymbol{\Sigma} \sim N(\boldsymbol{\nu}, \boldsymbol{\Sigma}/\kappa), \qquad \boldsymbol{\Sigma} \sim IW_m(\boldsymbol{\Psi}, n_0). \tag{4.33}$$

Indeed (4.33) is a conjugate family since the posterior distribution of $(\boldsymbol{\mu}, \boldsymbol{\Sigma})$ given $\mathbf{X}_1, \dots, \mathbf{X}_n$ is given by

$$\begin{aligned} &\boldsymbol{\mu}|\boldsymbol{\Sigma} \sim N\left(\frac{n\bar{\mathbf{X}} + \kappa\boldsymbol{\nu}}{n+\kappa}, \frac{\boldsymbol{\Sigma}}{n+\kappa}\right), \\ &\boldsymbol{\Sigma} \sim IW_m\left(\boldsymbol{\Psi} + \sum_{i=1}^n (\mathbf{X}_i - \bar{\mathbf{X}})(\mathbf{X}_i - \bar{\mathbf{X}})^T + \frac{n\kappa}{n+\kappa}(\bar{\mathbf{X}} - \boldsymbol{\nu})(\bar{\mathbf{X}} - \boldsymbol{\nu})^T, n + n_0\right); \end{aligned} \tag{4.34}$$

see Anderson (2003, pp. 274-275). Hence it follows from (4.32) that the Bayes estimators of $\boldsymbol{\mu}$ and $\boldsymbol{\Sigma}$, which are the posterior means, are given by

$$\widehat{\boldsymbol{\mu}} = \frac{n\bar{\mathbf{X}} + \kappa\boldsymbol{\nu}}{n+\kappa}$$

and

$$\widehat{\boldsymbol{\Sigma}} = \frac{1}{n + n_0 - m - 1} \Bigg\{ \boldsymbol{\Psi} + \sum_{i=1}^{n} (\mathbf{X}_i - \bar{\mathbf{X}})(\mathbf{X}_i - \bar{\mathbf{X}})^T + \frac{n\kappa}{n+\kappa} (\bar{\mathbf{X}} - \boldsymbol{\nu})(\bar{\mathbf{X}} - \boldsymbol{\nu})^T \Bigg\}. \tag{4.35}$$

4.3.4 Bayes estimators in Gaussian regression models

Consider the regression model $y_t = \boldsymbol{\beta}^T \mathbf{x}_t + \epsilon_t$, in which the regressors $\mathbf{x}_t$ can depend on the past observations $y_{t-1}, \mathbf{x}_{t-1}, \ldots, y_1, \mathbf{x}_1$, and the unobservable random errors ϵ_t are independent normal with mean 0 and variance σ^2. The likelihood function is given by (2.35).

First suppose that σ^2 is known and that $\boldsymbol{\beta}$ follows a prior distribution $N(\mathbf{z}, \sigma^2 \mathbf{V})$. Let $\mathbf{Y} = (y_1, \cdots, y_n)^T$ and $\mathbf{X}$ be the $n \times p$ matrix whose ith row is $\mathbf{x}_i^T$, as in Section 1.1.1. Then the posterior distribution of $\boldsymbol{\beta}$ given $(\mathbf{Y}, \mathbf{X})$ is $N(\boldsymbol{\beta}_n, \sigma^2 \mathbf{V}_n)$, where

$$\mathbf{V}_n = (\mathbf{V}^{-1} + \mathbf{X}^T \mathbf{X})^{-1}, \qquad \boldsymbol{\beta}_n = \mathbf{V}_n (\mathbf{V}^{-1}\mathbf{z} + \mathbf{X}^T \mathbf{Y}). \tag{4.36}$$

Hence the Bayes estimate of $\boldsymbol{\beta}$ is the posterior mean $\boldsymbol{\beta}_n$. In particular, if $\mathbf{V}^{-1} \to \mathbf{0}$, then the posterior mean $\boldsymbol{\beta}_n$ becomes the OLS estimator. For the case where $\mathbf{z} = \mathbf{0}$ and $\mathbf{V}^{-1} = \lambda \mathbf{I}_p$, the Bayes estimator becomes $(\lambda \mathbf{I}_p + \mathbf{X}^T \mathbf{X})^{-1} \mathbf{X}^T \mathbf{Y}$, which is the *ridge regression* estimator proposed by Hoerl and Kennard (1970) and shrinks the OLS estimate toward $\mathbf{0}$.

Without assuming that σ^2 is known, a convenient prior distribution for σ^2 in Bayesian analysis is the inverted chi-square distribution, as shown in the preceding section, which also gives an inverted chi-square posterior distribution of σ^2. Since $\chi^2_m/(2\lambda)$ has the same distribution as gamma$(m/2, \lambda)$ (see Section 2.3.4), and since an inverted chi-square distribution with m degrees of freedom is equivalent to $(2\sigma^2)^{-1} \sim \chi^2_m/(2\lambda)$ for some scale parameter λ, it is more convenient to work with the gamma(g, λ) prior distribution for $(2\sigma^2)^{-1}$ instead of the inverted chi-square prior distribution for σ^2. This also removes the integer constraint on $m = 2g$. Specifically, letting $\tau = (2\sigma^2)^{-1}$, we assume that the prior distribution of $(\boldsymbol{\beta}, \tau)$ is given by

$$\tau \sim gamma(g, \lambda), \qquad \boldsymbol{\beta} | \tau \sim N\big(\mathbf{z}, \mathbf{V}/(2\tau)\big).$$

Then the posterior distribution of $(\boldsymbol{\beta}, \tau)$ given $(\mathbf{X}, \mathbf{Y})$ is also of the same form:

$$\tau \sim gamma\Big(g + \frac{n}{2}, \frac{1}{a_n}\Big), \qquad \boldsymbol{\beta} | \tau \sim N\big(\boldsymbol{\beta}_n, \mathbf{V}_n/(2\tau)\big),$$
$$a_n = \lambda^{-1} + \mathbf{z}^T \mathbf{V}^{-1} \mathbf{z} + \mathbf{Y}^T \mathbf{Y} - \boldsymbol{\beta}_n^T \mathbf{V}_n^{-1} \boldsymbol{\beta}_n. \tag{4.37}$$

Hence the Bayes estimate of $\boldsymbol{\beta}$ is still $\boldsymbol{\beta}_n (= E(\boldsymbol{\beta}|\mathbf{X}, \mathbf{Y}))$, and the Bayes estimate of σ^2 is

$$E\Big(\frac{1}{2\tau}|\mathbf{X}, \mathbf{Y}\Big) = \frac{a_n}{2(g + n/2 - 1)}. \tag{4.38}$$

4.3.5 Empirical Bayes and shrinkage estimators

In Example 4.1 and (4.29), which consider Bayesian estimation of the mean μ of a normal distribution based on $X_1, \dots X_n$ sampled from the distribution, the Bayes estimator "shrinks" the sample mean $\bar{X}_n$ toward the prior mean μ_0, and, for large n (relative to σ^2/v_0), the shrinkage has negligible effect. The following variant of Example 4.1 amplifies the shrinkage effect in the problem of Bayesian estimation of n normal means $\mu_1, \dots, \mu_n$ in the case $X_i \sim N(\mu_i, \sigma^2)$. In this setting, it is even possible to estimate the prior mean μ_0 and prior variance v_0 from $X_1, \dots, X_n$ by the method of moments, thereby yielding *empirical Bayes* estimates of μ_i.

Example 4.3. Let (μ_i, X_i), $i = 1, \dots, n$, be independent random vectors such that $X_i|\mu_i \sim N(\mu_i, \sigma^2)$ and $\mu_i \sim N(\mu_0, v_0)$. Again assuming that σ^2 is known, the Bayes estimate $E(\mu_i|X_i)$ of μ_i is given by (4.29) with $n\bar{X}$ replaced by X_i and n/σ^2 replaced by $1/\sigma^2$. Note that (4.29) requires a "subjective" choice of the "hyperparameters" μ_0 and v_0. In the present setting, we can in fact estimate them from the data since

$$E(X_i) = E\{E(X_i|\mu_i)\} = E(\mu_i) = \mu_0, \tag{4.39}$$

$$\mathrm{Var}(X_i) = E\big(\mathrm{Var}(X_i|\mu_i)\big) + \mathrm{Var}\big(E(X_i|\mu_i)\big) = \sigma^2 + v_0. \tag{4.40}$$

Since (X_i, μ_i) are i.i.d., we can use (4.39), (4.40) and the method of moments to estimate μ_0 and v_0 by

$$\widehat{\mu}_0 = \bar{X}\left(= n^{-1}\sum_{i=1}^{n} X_i\right), \qquad \widehat{v}_0 = \left(n^{-1}\sum_{i=1}^{n}(X_i - \bar{X})^2 - \sigma^2\right)_+, \tag{4.41}$$

where $y_+ = \max(0, y)$. The empirical Bayes approach replaces the hyperparameters μ_0 and v_0 in the Bayes estimate $E(\mu_i|X_i)$, yielding

$$\widehat{E}(\mu_i|X_i) = \bar{X} + \left(1 - \frac{\sigma^2}{n^{-1}\sum_{i=1}^{n}(X_i - \bar{X})^2}\right)_+ (X_i - \bar{X}). \tag{4.42}$$

The estimate (4.42) is closely related to the minimum-variance linear predictor $E(\mu|X) = E\mu + \frac{\mathrm{Cov}(\mu, X)}{\mathrm{Var}(X)}(X - EX)$ in Section 1.5; it simply replaces $E\mu$, $\mathrm{Var}(X)$, and $\mathrm{Cov}(\mu, X)$ by the method-of-moments estimates, noting that

$$E(\mu X) = E\big(\mu E(X|\mu)\big) = E(\mu^2)$$

and therefore $\mathrm{Cov}(\mu, X) = E(\mu^2) - E(\mu)E(X) = v_0$. A variant of (4.42) has been shown by James and Stein (1961) to have smaller total mean squared error $\sum_{i=1}^n E_{\mu_i}\left[(\widehat{\mu}_i - \mu_i)^2\right]$ than the MLE $\widehat{\mu}_i = X_i$ when $n \geq 3$. The James-Stein estimator has the form $\widehat{\mu}_i = \left\{1 - \sigma^2(n-2)/\sum_{i=1}^n X_i^2\right\}X_i$, whose mean squared error is analytically more tractable than (4.42). The James-Stein estimator and (4.42) are also called *shrinkage estimators*, as they "shrink" the MLE X_i towards 0 (for James-Stein) or toward $\bar{X}$ (for (4.42)). They show the advantages of using Bayesian-type shrinkage in reducing the total mean squared error when one encounters several (three or more in the case of James-Stein) similar estimation problems.

Instead of prespecifying a prior distribution for Bayesian modeling, the empirical Bayes approach basically specifies a family of prior distributions with unspecified *hyperparameters*, which are estimated from the data. This is particularly useful for problems of the type in Example 4.1 or of the change-point type that will be considered in Chapter 9 (Section 9.5.2), where parameters may undergo occasional changes (jumps) at unknown times and with unknown magnitudes. A prior distribution for the unobserved time series of parameter values, with the jump probability and jump size as hyperparameters to be estimated from the observations, is assumed in the Bayesian approach to estimating these piecewise constant parameters in Section 9.5.2.

An important property of well-chosen shrinkage estimators is their adaptability to the amount of information that the sample contains about the unknown parameters. In "data-rich" situations, the shrinkage estimator roughly behaves like the MLE because of the small posterior weight attached to the shrinkage target. On the other hand, in "data-poor" situations, the MLE is unreliable and the shrinkage estimator places much more weight on the shrinkage target, which the data can estimate considerably better.

4.4 Investment applications of shrinkage estimators and Bayesian methods

In practice, the parameters $\boldsymbol{\mu}$ and $\boldsymbol{\Sigma}$ in Markowitz's portfolio theory are unknown and have to be estimated from historical data, as noted in Section 3.5.1. Although it may seem that this should not cause difficulties because one can use a large amount of historical data and the sample mean vector and covariance matrix are consistent estimates of $\boldsymbol{\mu}$ and $\boldsymbol{\Sigma}$, in practice one can only use a relatively small sample size n compared with the database of past returns because of possible structural changes over time. In fact, some companies may not have existed or survived throughout a long window of past history. Moreover, the number p of assets under consideration may not be small compared with n, making it harder for the plug-in frontier to be near Markowitz's efficient frontier, as pointed out in Section 3.5.1, which also describes Michaud's resampled frontier that uses the average of bootstrap weights (3.38) instead of the plug-in weights that substitute the unknown $\boldsymbol{\mu}$ and $\boldsymbol{\Sigma}$ in the optimal weights by their sample counterparts. In this section, we summarize two recent approaches to portfolio theory in which $\boldsymbol{\mu}$ and $\boldsymbol{\Sigma}$ are unknown. The first approach is to replace the unknown $\boldsymbol{\mu}$ and $\boldsymbol{\Sigma}$ in Markowitz's efficient frontier by shrinkage or empirical Bayes estimates, which have been shown to be more reliable than the sample means and covariances when $p(p+1)/2$ is not small compared with n. Instead of plugging the estimated parameters into Markowitz's efficient frontier, the second approach tackles the more fundamental problem of portfolio optimization when $\boldsymbol{\mu}$ and $\boldsymbol{\Sigma}$ are random variables with specified prior distributions so that the case of known $\boldsymbol{\mu}$ and $\boldsymbol{\Sigma}$ (as in Markowitz's theory) becomes a special case (with a degenerate prior distribution). This Bayesian approach incorporates the uncertainties in $\boldsymbol{\mu}$ and $\boldsymbol{\Sigma}$ in the optimization problem via their posterior distributions. Moreover, the assumption of a specified prior distribution in this approach can also be removed by using bootstrap resampling, instead of the Bayes risk, to study performance.

4.4.1 Shrinkage estimators of $\boldsymbol{\mu}$ and $\boldsymbol{\Sigma}$ for the plug-in efficient frontier

Assuming $\mathbf{r}_t$ to be independent $N(\boldsymbol{\mu}, \boldsymbol{\Sigma})$ for $1 \le t \le n$ and $(\boldsymbol{\mu}, \boldsymbol{\Sigma})$ to have the prior distribution

$$\boldsymbol{\mu}|\boldsymbol{\Sigma} \sim N(\boldsymbol{\nu}, \boldsymbol{\Sigma}/\kappa), \qquad \boldsymbol{\Sigma} \sim IW_m(\boldsymbol{\Psi}, n_0), \tag{4.43}$$

where m is the number of assets and IW denotes the inverted Wishart distribution, the posterior distribution of $(\boldsymbol{\mu}, \boldsymbol{\Sigma})$ given $(\mathbf{r}_1, \dots, \mathbf{r}_n)$ is also of the same form and given in (4.34), yielding explicit formulas for the Bayes estimators of $(\boldsymbol{\mu}, \boldsymbol{\Sigma})$. The Bayes estimator of $\boldsymbol{\mu}$ shrinks the sample mean $\bar{\mathbf{r}}$ toward the prior mean $\boldsymbol{\nu}$. The Bayes estimator of $\boldsymbol{\Sigma}$ is a convex combination of the prior mean $\boldsymbol{\Psi}/(n_0 - m - 1)$ and an adjusted sample covariance matrix:

$$\begin{aligned}\widehat{\boldsymbol{\Sigma}} = & \frac{n_0 - m - 1}{n + n_0 - m - 1} \frac{\boldsymbol{\Psi}}{n_0 - m - 1} + \frac{n}{n + n_0 - m - 1} \left\{ \frac{1}{n} \sum_{i=1}^{n} (\mathbf{r}_t - \bar{\mathbf{r}})(\mathbf{r}_t - \bar{\mathbf{r}})^T \right. \\ & \left. + \frac{\kappa}{n+\kappa} (\bar{\mathbf{r}} - \boldsymbol{\nu})(\bar{\mathbf{r}} - \boldsymbol{\nu})^T \right\};\end{aligned} \tag{4.44}$$

see (4.34). If $\boldsymbol{\mu}$ were known, then the sample covariance estimate would be $\sum_{t=1}^{n} (\mathbf{r}_t - \boldsymbol{\mu})(\mathbf{r}_t - \boldsymbol{\mu})^T/n$. The adjusted sample covariance matrix inside the curly brackets in (4.44) not only replaces $\boldsymbol{\mu}$ by $\bar{\mathbf{r}}$ (as in the MLE) but also adds to the MLE of $\boldsymbol{\Sigma}$ the covariance matrix $\kappa(\bar{\mathbf{r}}-\boldsymbol{\nu})(\bar{\mathbf{r}}-\boldsymbol{\nu})^T/(n+\kappa)$, which accounts for the uncertainties due to replacing $\boldsymbol{\mu}$ by $\bar{\mathbf{r}}$.

Instead of using the preceding Bayesian estimator, which requires specification of the hyperparameters $\boldsymbol{\mu}$, κ, n_0, and $\boldsymbol{\Psi}$, Ledoit and Wolf (2003, 2004) propose to estimate $\boldsymbol{\mu}$ by $\bar{\mathbf{X}}$ but to shrink the MLE of $\boldsymbol{\Sigma}$ toward structured covariance matrices that can have relatively small "estimation error" in comparison with the MLE of $\boldsymbol{\Sigma}$. Let $\mathbf{S} = \sum_{t=1}^{n}(\mathbf{r}_t - \bar{\mathbf{r}})(\mathbf{r}_t - \bar{\mathbf{r}})^T/n$. Ledoit and Wolf's rationale is that $\mathbf{S}$ has a large estimation error (or, more precisely, variances of the matrix entries) when $p(p+1)/2$ is comparable with n, whereas a structured covariance matrix $\mathbf{F}$ has much fewer parameters and can therefore be estimated with much smaller variances. In particular, they consider $\mathbf{F}$ that corresponds to the single-factor model in CAPM (see Sections 3.3 and 3.4) and point out that its disadvantage is that $\boldsymbol{\Sigma}$ may not equal $\mathbf{F}$, resulting in a bias of $\widehat{\mathbf{F}}$ when the assumed structure (e.g., CAPM) does not hold. They therefore propose to estimate $\boldsymbol{\Sigma}$ by a convex combination of $\widehat{\mathbf{F}}$ and $\mathbf{S}$:

$$\widehat{\boldsymbol{\Sigma}} = \widehat{\delta}\widehat{\mathbf{F}} + (1-\widehat{\delta})\mathbf{S}, \tag{4.45}$$

where $\widehat{\delta}$ is an estimator of the *optimal shrinkage constant* δ used to shrink the MLE toward the estimated structured covariance matrix $\widehat{\mathbf{F}}$. Defining the optimal shrinkage constant as the minimizer of

$$E\left\{\sum_{1\le i,j\le p}\left[\delta f_{ij} + (1-\delta)s_{ij} - \sigma_{ij}\right]\right\}^2$$

(see Exercise 3.3(b)), they show that with p fixed as $n \to \infty$, the minimizer δ can be expressed as $\kappa/n + O(1/n^2)$ and that κ can be consistently estimated by a method-of-moments estimator $\widehat{\kappa}$; see Exercise 4.7(b) for details. Therefore they propose to use $\widehat{\delta} = \min\{1, (\widehat{\kappa}/n)_+\}$ in (4.45).

Besides $\mathbf{F}$ associated with CAPM, Ledoit and Wolf (2004) suggest using a constant correlation model for $\mathbf{F}$ in which all pairwise correlations are identical. It is easier to implement than the single-factor model, and they have found that it gives comparable performance in simulation and empirical studies. They have provided for its implementation a computer code in `MATLAB` that can be downloaded from `http://www.ledoit.net`. They advocate using this shrinkage estimate of $\boldsymbol{\Sigma}$ in lieu of $\mathbf{S}$ in implementing Markowitz's efficient frontier or Michaud's resampled efficient frontier; see Sections 3.2.4 and 3.5.1.

4.4.2 An alternative Bayesian approach

The preceding Bayes and shrinkage approaches focus primarily on Bayes estimates of $\boldsymbol{\mu}$ and $\boldsymbol{\Sigma}$ (with normal and inverted Wishart priors) and shrinkage estimators of $\boldsymbol{\Sigma}$. However, the construction of efficient portfolios when $\boldsymbol{\mu}$ and

$\boldsymbol{\Sigma}$ are unknown is more complicated than estimating them as well as possible and plugging the estimates into (3.11). Note in this connection that (3.11) involves $\boldsymbol{\Sigma}^{-1}$ instead of $\boldsymbol{\Sigma}$ and that estimating $\boldsymbol{\Sigma}$ as well as possible need not imply that $\boldsymbol{\Sigma}^{-1}$ is reliably estimated. Moreover, a major difficulty with the "plug-in" efficient frontier and its "resampled" version is that Markowitz's idea of the variance of $\mathbf{w}^T\mathbf{r}$ as a measure of the portfolio's risk cannot be captured simply by the plug-in estimate $\mathbf{w}^T\widehat{\boldsymbol{\Sigma}}\mathbf{w}$ of $\mathrm{Var}(\mathbf{w}^T\mathbf{r})$. Whereas the problem of minimizing $\mathrm{Var}(\mathbf{w}^T\mathbf{r})$ subject to a given level μ_* of the mean return $E(\mathbf{w}^T\mathbf{r})$ is meaningful in Markowitz's framework, in which both $E(\mathbf{r})$ and $\mathrm{Cov}(\mathbf{r})$ are known, the surrogate problem of constraining $\mathbf{w}^T\bar{\mathbf{r}} = \mu_*$ and minimizing $\mathbf{w}^T\widehat{\boldsymbol{\Sigma}}\mathbf{w}$ under such constraint becomes much less convincing since both $\bar{\mathbf{r}}$ and $\widehat{\boldsymbol{\Sigma}}$ have inherent errors (risks) themselves, as pointed out by Lai, Xing, and Chen (2007). Using ideas from adaptive stochastic control and optimization, they consider the more fundamental problem of minimizing $\mathrm{Var}(\mathbf{w}^T\mathbf{r}) - \lambda E(\mathbf{w}^T\mathbf{r})$ when $\boldsymbol{\mu}$ and $\boldsymbol{\Sigma}$ are unknown and treated as state variables whose uncertainties are specified by their posterior distributions given the observations $\mathbf{r}_1, \ldots, \mathbf{r}_n$ in a Bayesian framework. Note that if the prior distribution puts all its mass at $(\boldsymbol{\mu}_0, \boldsymbol{\Sigma}_0)$, then the minimization problem reduces to Markowitz's portfolio optimization problem that assumes $\boldsymbol{\mu}_0$ and $\boldsymbol{\Sigma}_0$ are given. The choice of the prior distribution is not limited to the conjugate family (4.33). In particular, we can consider covariance matrices of the form $\boldsymbol{\Sigma} = \mathbf{B}\boldsymbol{\Omega}\mathbf{B}^T + \mathbf{V}$, in which $\boldsymbol{\Omega}$ is the covariance matrix of factors given by "domain knowledge" related to the stocks in the portfolio (as in the Fama-French three-factor model of Section 3.4.4).

Lai, Xing, and Chen (2007) provide an efficient algorithm to compute the solution of the Bayesian portfolio optimization problem above. By using bootstrap resampling, they also modify the algorithm to handle the non-Bayesian problem in which the Bayesian objective function is replaced by the bootstrap estimate of the corresponding performance criterion. Letting the Lagrange multiplier λ in the objective function vary over a grid of values gives the efficient frontier when $\boldsymbol{\mu}$ and $\boldsymbol{\Sigma}$ are unknown, providing a natural extension of Markowitz's theory.

A point raised by Ledoit and Wolf (2003, 2004) is that the covariance matrix $\boldsymbol{\Sigma}$ is poorly estimated by its sample counterpart and can be greatly improved by their shrinkage estimators when p is not small in comparison with n, which is often the case in portfolio management. This is a "data-poor" situation for which prior information (subject-matter knowledge) should definitely help. Note that the Bayesian approach of Lai, Xing, and Chen (2007) does not simply plug the Bayes estimator of $(\boldsymbol{\mu}, \boldsymbol{\Sigma})$ into Markowitz's efficient frontier but rather reformulates Markowitz's portfolio optimization problem as a Bayesian statistical decision problem. The commonly used plug-in approach, however, can still work well if the estimates that are plugged into the efficient frontier are close to the actual values of $(\boldsymbol{\mu}, \boldsymbol{\Sigma})$ with respect to the metric induced by the objective function. Even when p is comparable to n, this can still happen if $\boldsymbol{\Sigma}$ satisfies certain sparsity constraints so that it can be estimated consistently by an estimator that first identifies and then exploits these constraints. Section 9.3 will consider such sparsity constraints on $\boldsymbol{\Sigma}$ and the associated estimators.

Exercises

4.1. Consider the linear regression model

$$y_i = \alpha + \beta x_i + \epsilon_i, \qquad i = 1, \dots, n,$$

in which $\epsilon_1, \dots, \epsilon_n$ are random variables with zero mean and independent of $(x_1, \cdots, x_n)$.

(a) Suppose the ϵ_i are independent normal and have common variance σ^2. Find the maximum likelihood estimates of α, β, and σ^2.

(b) Suppose $(\epsilon_1, \cdots, \epsilon_n)^T$ has a multivariate normal distribution with covariance matrix $\sigma^2\mathbf{V}$, where $\mathbf{V}$ is a known correlation matrix and σ is an unknown parameter. Find the maximum likelihood estimates of α, β, and σ^2.

4.2. Prove (4.3).

4.3. (a) Show that the linear regression model (1.3) under assumptions (A) and (C*) in Section 1.1.4 is a generalized linear model, and relate the F-tests for ANOVA in Section 1.2.2 to the GLR tests in analysis of deviance.

(b) Using analysis of deviance, carry out an analog of ANOVA and stepwise variable selection (see Section 1.3.2) for logistic regression.

4.4. Show that if $\mathbf{W} \sim W_m(\boldsymbol{\Sigma}, n)$ and $\boldsymbol{\Sigma}$ has the prior distribution $IW_m(\boldsymbol{\Psi}, n_0)$, the posterior distribution of $\boldsymbol{\Sigma}$ given $\mathbf{W}$ is $IW_m(\mathbf{W} + \boldsymbol{\Psi}, n + n_0)$.

4.5. Prove (4.37).

4.6. The file `d_tcm_var.txt` contains the variances v_τ of the yields of U.S. 1-year, 2-year, 3-year, 5-year, 7-year, 10-year, and 20-year zero-coupon bonds, calculated from the weekly Treasury constant maturity rates from August 11, 1995 to September 28, 2007. Assume the Hull-White model for the yield of a τ-year zero-coupon bond. Estimate the parameters κ and σ of the Hull-White model using nonlinear least squares, and give a 95% confidence interval for each parameter.

4.7. The file `m_ret_10stocks.txt` contains the monthly returns of ten stocks from January 1994 to December 2006. The ten stocks include Apple Computer, Adobe Systems, Automatic Data Processing, Advanced Micro Devices, Dell, Gateway, Hewlett-Packard Company, International Business Machines Corp., Microsoft Corp., and Oracle Corp. The file `m_sp500ret_3mtcm.txt` contains three columns. The second column gives the monthly returns of the S&P 500 index from January 1994 to December 2006. The third column gives the monthly rates of the 3-month Treasury bill in the secondary market, which are obtained from the Federal Reserve Bank of St. Louis and used as the risk-free rate here. Consider portfolios that consist of the ten stocks and allow short selling.

(a) Using a single-index model (see Exercise 3.3) for the structured covariance matrix $\mathbf{F}$, calculate the estimate $\widehat{\mathbf{F}}$ of $\mathbf{F}$ in (4.45).

(b) The $\widehat{\delta}$ in (4.45) suggested by Ledoit and Wolf (2003, 2004) is of the following form. Let $\widehat{f_{ij}}$ and $\widehat{\sigma}_{ij}$ denote the (i,j)th entry of $\widehat{\mathbf{F}}$ and $\mathbf{S}$, respectively, and define

$$\widehat{\gamma} = \sum_{i=1}^{p}\sum_{i=1}^{p}(\widehat{f}_{ij} - \widehat{\sigma}_{ij})^2, \qquad \bar{\sigma} = \frac{2}{p(p-1)}\sum_{i=1}^{p-1}\sum_{j=i+1}^{p}\frac{\widehat{\sigma}_{ij}}{\sqrt{\widehat{\sigma}_{ii}\widehat{\sigma}_{jj}}},$$

$$\widehat{\pi}_{ij} = n^{-1}\sum_{t=1}^{n}\left\{(r_{it} - \bar{r}_i)(r_{jt} - \bar{r}_j) - \widehat{\sigma}_{ij}\right\}^2, \qquad \widehat{\pi} = \sum_{i=1}^{p}\sum_{i=1}^{p}\widehat{\pi}_{ij},$$

$$\widehat{\theta}_{k,ij} = n^{-1}\sum_{t=1}^{n}\left\{(r_{kt} - \bar{r}_k)^2 - \widehat{\sigma}_{kk}\right\}\left\{(r_{it} - \bar{r}_i)(r_{jt} - \bar{r}_j) - \widehat{\sigma}_{ij}\right\},$$

$$\widehat{\rho} = \sum_{i=1}^{p}\widehat{\pi}_{ii} + \sum_{i=1}^{p}\sum_{j\neq i, j=1}^{p}\frac{\bar{\sigma}}{2}\left\{\sqrt{\frac{\widehat{\sigma}_{jj}}{\widehat{\sigma}_{ii}}}\widehat{\theta}_{i,ij} + \sqrt{\frac{\widehat{\sigma}_{ii}}{\widehat{\sigma}_{jj}}}\widehat{\theta}_{j,ij}\right\}, \quad \widehat{\kappa} = \frac{\widehat{\pi} - \widehat{\rho}}{\widehat{\gamma}}.$$

Then $\widehat{\delta} = \min\{1, (\widehat{\kappa}/n)_+\}$. Compute the covariance estimate (4.45) with $\widehat{\mathbf{F}}$ in (a) and the $\widehat{\delta}$ suggested by Ledoit and Wolf, and plot the estimated efficient frontier using this covariance estimate.

(c) Perform PCA on the ten stocks. Using the first two principal components as factors in a two-factor model for $\mathbf{F}$ (see Section 3.4.3), estimate $\mathbf{F}$.

(d) Using the estimated $\widehat{\mathbf{F}}$ in (c) as the shrinkage target in (4.45), compute the new value of $\widehat{\delta}$ and the new shrinkage estimate (4.45) of $\boldsymbol{\Sigma}$. Plot the corresponding estimated efficient frontier and compare it with that in (b).

5

Time Series Modeling and Forecasting

Analysis and modeling of financial time series data and forecasting future values of market variables constitute an important empirical core of quantitative finance. This chapter introduces some basic statistical models and methods of this core. Section 5.1 begins with stationary time series models that generalize the classical i.i.d. (independent and identically distributed) observations and introduces moving average (MA) and autoregressive (AR) models and their hybrid (ARMA) models. Estimation of model order and parameters and applications to forecasting are described. Section 5.2 considers nonstationary models that can be converted to stationary ones by detrending, differencing, and transformations. In particular, differencing is particularly effective for nonstationary processes with stationary increments, which generalize random walks (i.e., sums of i.i.d. random variables) to ARIMA (or integrated ARMA) models. Section 5.3 considers linear state-space models and the Kalman filter and their applications to forecasting. Volatility modeling of financial time series will be considered in the next chapter. Part II of the book will introduce more advanced topics in time series analysis. In particular, nonparametric time series modeling is introduced in Chapter 7, and time series models of high-frequency transaction data are introduced in Chapter 11. Chapter 9 considers multivariate time series, while Chapters 8 and 10 describe and strengthen the connections between empirical time series analysis and stochastic process models in the theory of option pricing and interest rate derivatives.

5.1 Stationary time series analysis

5.1.1 Weak stationarity

A time series $\{x_1, x_2, x_3, \dots\}$ is said to be *weakly stationary* (or *covariance stationary*) if Ex_t and $\mathrm{Cov}(x_t, x_{t+k})$ do not depend on $t \geq 1$. For a weakly stationary sequence $\{x_t\}$, $\mu := Ex_t$ is called its mean and $\gamma_h := \mathrm{Cov}(x_t, x_{t+h})$,

$h \geq 0$, is called the *autocovariance function*. A sequence $\{x_t\}$ is said to be *stationary* (or *strictly stationary*) if the joint distribution of $(x_{t+h_1}, \ldots, x_{t+h_m})$ does not depend on $t \geq 1$ for every $m \geq 1$ and $0 \leq h_1 < \cdots < h_m$. If $\{x_t\}$ is stationary and $E|x_1| < \infty$, then $\lim_{n\to\infty} n^{-1} \sum_{t=1}^{n} x_t = x_\infty$ with probability 1 for some random variable x_∞ such that $Ex_\infty = \mu$. Moreover, $x_\infty = \mu$ under certain conditions; see Appendix B. Note that the correlation coefficient $\mathrm{Corr}(x_t, x_{t+h})$ at lag h also does not depend on t and is given by $\rho_h := \gamma_h/\gamma_0$, $h \geq 0$, which is called the *autocorrelation function* (ACF) of $\{x_t\}$.

Spectral distribution, autocovariance, and spectral estimates

Define $\gamma_{-h} = \gamma_h$. A sequence $\{\gamma_h, -\infty < h < \infty\}$ is nonnegative definite in the sense that $\sum_{t=-\infty}^{\infty} \sum_{s=-\infty}^{\infty} a_t a_s \gamma_{t-s} \geq 0$ for any sequence $\{a_j, -\infty < j < \infty\}$. A theorem due to Herglotz says that nonnegative definite sequences are Fourier transforms of finite measures on $[-\pi, \pi]$. Therefore, for the ACF γ_h of a weakly stationary sequence $\{x_t\}$, there exists a nondecreasing function F, called the *spectral distribution function* of $\{x_t\}$, such that

$$\gamma_h = \int_{-\pi}^{\pi} e^{ih\theta} dF(\theta), \qquad -\infty < h < \infty. \tag{5.1}$$

If F is absolutely continuous with respect to Lebesgue measure, then $f = dF/d\theta$ is called the *spectral density function* of $\{x_t\}$. In particular, if $\sum_{h=0}^{\infty} |\gamma_h| < \infty$, then f exists and is given by

$$f(\theta) = \frac{1}{2\pi} \sum_{h=-\infty}^{\infty} \gamma_h e^{-ih\theta} = \frac{1}{2\pi}\left(\gamma_0 + 2\sum_{h=1}^{\infty} \gamma_h \cos(h\theta)\right). \tag{5.2}$$

Suppose $\sum_{h=1}^{\infty} |\gamma_h| < \infty$. Based on the sample $\{x_1, \ldots, x_n\}$, a consistent estimate of μ is the sample mean $\bar{x} = n^{-1} \sum_{t=1}^{n} x_n$ by the weak law of large numbers for these weakly stationary sequences (see Appendix B). The method of moments estimates γ_h by

$$\widehat{\gamma}_h = \frac{1}{n-h} \sum_{t=1}^{n-h} (x_{t+h} - \bar{x})(x_t - \bar{x}). \tag{5.3}$$

If $\{x_t x_{t+h}, t \geq 1\}$ is weakly stationary with an absolutely summable autocovariance function, then $\widehat{\gamma}_h$ is a consistent estimate of γ_h for fixed h; see Appendix B.

Replacing γ_h by $\widehat{\gamma}_h$ for $|h| < n$ in (5.2), in which $\sum_{h=-\infty}^{\infty}$ is replaced by $\sum_{|h|<n}$, provides a sample estimate of $f(\theta)$. However, this estimate of f behaves erratically because the number of observations to estimate γ_h is $n-h$, which decreases with h and does not provide enough information to estimate

γ_h well for large h. One way to get around this difficulty is to use lag weights $1 - |h|/n$ to dampen the effect of $\widehat{\gamma}_h$ for large lags in (5.2), leading to the spectral estimate

$$\widehat{f}(\theta) = \frac{1}{2\pi} \sum_{|h|<n} \left(1 - \frac{|h|}{n}\right) \widehat{\gamma}_h e^{-ih\theta}. \tag{5.4}$$

This is tantamount to replacing the sample autocovariance function (5.3) by the *window estimate*

$$\widetilde{\gamma}_h = \left(1 - \frac{h}{n}\right) \widehat{\gamma}_h = \frac{1}{n} \sum_{t=1}^{n-h} (x_{t+h} - \bar{x})(x_t - \bar{x}). \tag{5.5}$$

The sample ACF can be computed and plotted by using the `MATLAB` function `autocorr` or the `type="correlation"` option of the `R` function `acf`.

5.1.2 Tests of independence

Suppose x_t are i.i.d. and $Ex_t^2 < \infty$. Then $\rho_j = 0$ for all $j \geq 1$. Moreover, $\widehat{\rho}_h$ is asymptotically $N(0, 1/n)$ as $n \to \infty$, for any fixed $h \geq 1$, and $\widehat{\rho}_1, \ldots, \widehat{\rho}_m$ are asymptotically independent. This property has been used to test the null hypothesis $\rho_h = 0$ with approximate significance level α. The test rejects H_0 if $|\widehat{\rho}_h| \geq z_{1-\alpha/2}/\sqrt{n}$, where z_q is the qth quantile of the standard normal distribution. In addition, to test the null hypothesis $\rho_1 = \cdots = \rho_m = 0$, a widely used test statistic is the *Box-Pierce statistic* $Q^*(m)$ or the *Ljung-Box statistic* $Q(m)$, where

$$Q^*(m) = n \sum_{h=1}^{m} \widehat{\rho}_h^2, \qquad Q(m) = n(n+2) \sum_{h=1}^{m} \widehat{\rho}_h^2/(n-h). \tag{5.6}$$

Both $Q^*(m)$ and $Q(m)$ are asymptotically χ_m^2 as $n \to \infty$ when the x_t are i.i.d. Therefore the null hypothesis is rejected if $Q(m)$ or $Q^*(m)$ exceeds the $1 - \alpha$ quantile $\chi_{m;1-\alpha}^2$ of the χ_m^2-distribution. For a moderate sample size m, $Q(m)$ is better approximated by χ_m^2 than $Q^*(m)$. The `R` function `Box.test` can be used to compute the Box-Pierce or Ljung-Box statistic for the null hypothesis $\rho_1 = \cdots = \rho_m$.

Example: Time series of monthly log returns of six stocks

The stock returns in Sections 3.2.4 and 3.3.4, where they are treated as 63 i.i.d. observations to illustrate the estimated efficient frontier and CAPM, are actually time series data of monthly log returns during the period August 2000 to October 2005. Figure 5.1 plots these data over time. Note that the "dot-com bubble burst" occurred in 2000 and that the year 2001 had its historic date September 11. In spite of these events, the monthly log returns appear to be roughly stationary. To test the null hypothesis of independence of r_t, we apply the Ljung-Box test with $m = 20$. The test rejects the null hypothesis at the 0.05 level only for Citigroup. Figure 5.2 plots the estimated autocorrelations of the monthly log returns of the six stocks. Their pointwise 95% confidence limits (see preceding paragraph) are marked by dotted lines.

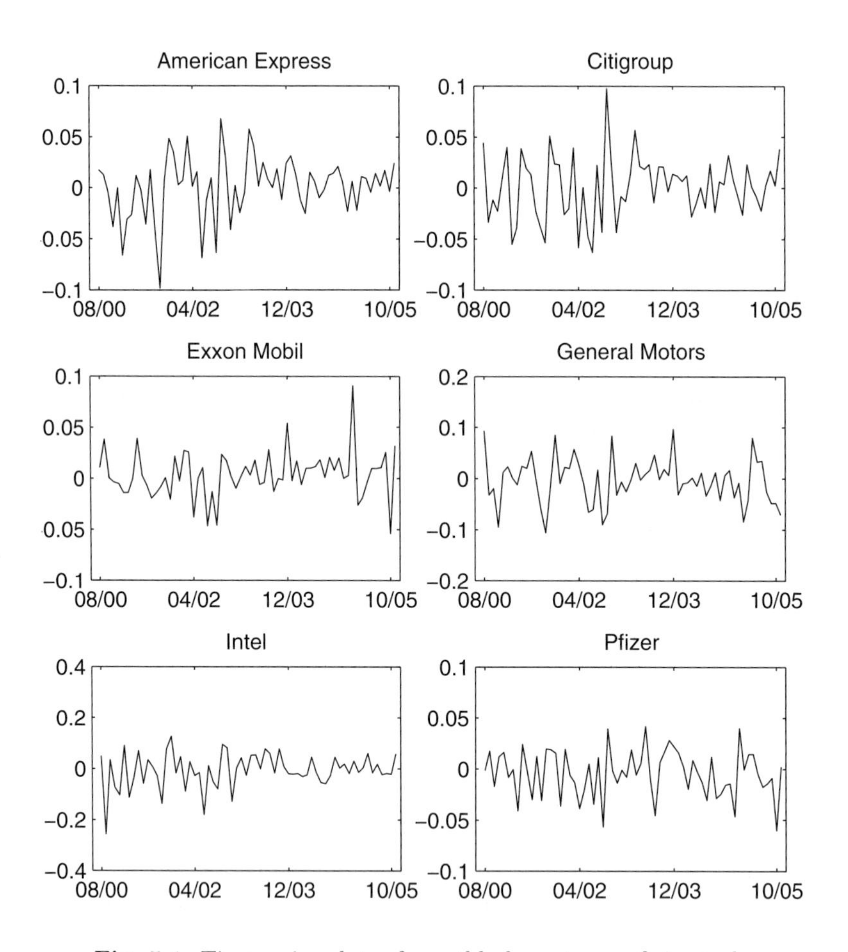

Fig. 5.1. Time series plots of monthly log returns of six stocks.

Figure 5.3 plots the time series of monthly log returns of the Dow Jones Industrial Average and the monthly rate of the 3-month U.S. Treasury bill during the sample period. These data have been used in Section 3.3.4 besides the stock returns to illustrate the statistical analysis of CAPM, which assumes

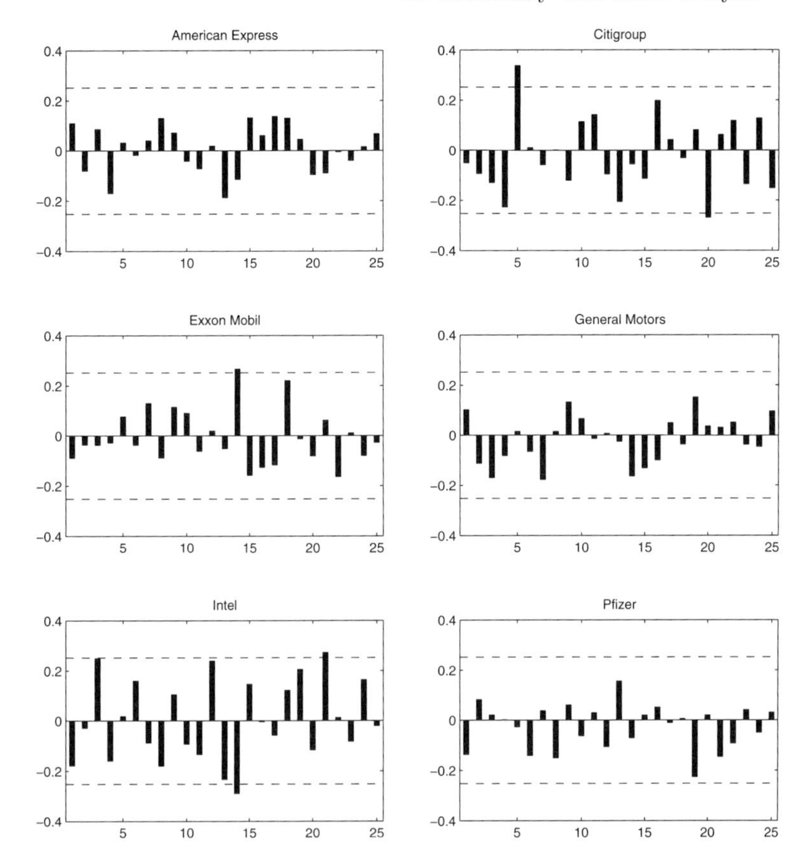

Fig. 5.2. Autocorrelations of monthly log returns. The dashed lines represent the rejection boundaries of a 5%-level test of zero autocorrelation at indicated lag.

i.i.d. excess returns. In Section 5.3.2, we apply Kalman filtering to address certain departures from the underlying assumptions of CAPM.

5.1.3 Wold decomposition and MA, AR, and ARMA models

A weakly stationary sequence $\{x_t\}$ whose spectral distribution is absolutely continuous with respect to Lebesgue measure can be expressed as

$$x_t = \mu + \sum_{j=0}^{\infty} \psi_j u_{t-j}, \tag{5.7}$$

in which $\psi_0 = 1$ and u_t are uncorrelated random variables with $Eu_t = 0$ and $\mathrm{Var}(u_t) = \sigma^2$. The representation (5.7) is called the *Wold decomposition*, and the u_t are called *innovations*. The infinite series in (5.7) is the L_2-limit of

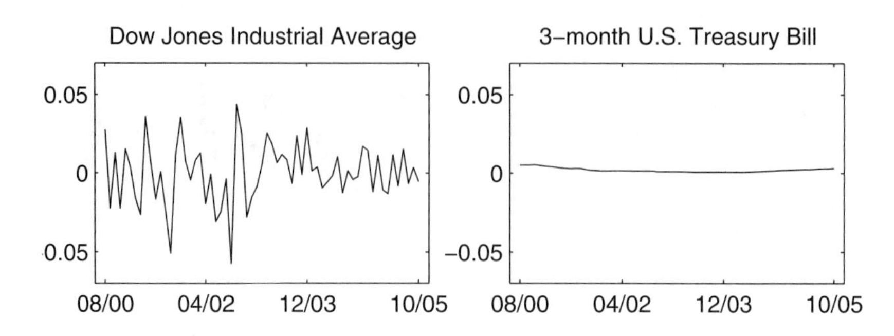

Fig. 5.3. The monthly log returns of the Dow Jones Industrial Average and the monthly rate of the 3-month U.S. Treasury bill.

$\sum_{j=0}^{q} \psi_j u_{t-j}$. In other words, defining

$$x_t = \mu + u_t + \psi_1 u_{t-1} + \cdots + \psi_q u_{t-q}, \tag{5.8}$$

the expected value of the squared difference between (5.7) and (5.8) converges to 0 as $q \to \infty$. The model (5.8) is called a *moving average* model of order q, or MA(q).

Let B denote the *backshift operator* defined by $Bx_t = x_{t-1}$. Iterating the operator yields $B^2 x_t = B(Bx_t) = x_{t-2}, \ldots, B^m x_t = x_{t-m}$. Hence we can write (5.7) as $x_t = \mu + \psi(B)u_t$, where $\psi(B) = 1 + \psi_1 B + \psi_2 B^2 + \ldots$. This operator notation has the advantage of "solving" u_t in terms of $\{x_s, s \le t\}$ as follows. Replacing x_t by $x_t - \mu$, we shall assume $\mu = 0$. We can formally rewrite $x_t = \psi(B)u_t$ as

$$u_t = \frac{1}{\psi(B)} x_t = \phi(B) x_t,$$

where $\phi(B) = 1 - \phi_1 B - \phi_2 B^2 - \ldots$ is the Taylor series of $1/\psi(B)$. Such a Taylor series expansion is valid if $\psi(z)$ satisfies the *invertibility condition* that

$$\text{the zeros of } \psi(z) \text{ are outside the unit circle;} \tag{5.9}$$

i.e., $\psi(z) \ne 0$ for complex z with $|z| \le 1$. Approximating the power series $\phi(B)$ by the polynomial $1 - \phi_1 B - \cdots - \phi_p B^p$ yields

$$x_t = \mu + \phi_1 x_{t-1} + \cdots + \phi_p x_{t-p} + u_t, \tag{5.10}$$

which is called an *autoregressive model* of order p, or AR(p). Note that whereas $Ex_t = \mu$ in the MA(q) model (5.8), $Ex_t := \widetilde{\mu}$ satisfies $(1 - \phi_1 - \cdots - \phi_p)\widetilde{\mu} = \mu$ under the *stationarity assumption* that

$$\text{the zeros of } 1 - \phi_1 z - \cdots - \phi_p z^p \text{ are outside the unit circle.} \tag{5.11}$$

We can combine the AR and MA approximations of the infinite-order moving average (5.7) into an ARMA(p, q) model of the form

$$x_t = \mu + \phi_1 x_{t-1} + \cdots + \phi_p x_{t-p} + u_t + \psi_1 u_{t-1} + \cdots + \psi_q u_{t-q}, \tag{5.12}$$

which is tantamount to using a rational (instead of polynomial) approximation $\psi(B)/\phi(B)$ to the power series $1 + \sum_{j=1}^{\infty} \psi_j B^j$. Whereas a polynomial approximation may require a polynomial of high degree, relatively small values of p and q may suffice for the rational approximation, thereby yielding a more parsimonious model.

Partial autocorrelation functions

Partial correlation coefficients have been introduced in Section 1.3.2. For a weakly stationary time series $\{x_t\}$, the *partial autocorrelation coefficient function* (PACF) at lag h is the partial correlation coefficient between x_t and x_{t-h} adjusted for $x_{t-1}, \ldots, x_{t-h+1}$. It does not depend on t and is denoted by $\rho_{h \cdot 1, \ldots, h-1}$. For an AR$(p)$ model, the PACF cuts off at lag p (i.e., it is zero for lags larger than p). For an MA(q) model, the ACF cuts off at lag q. The sample PACF can be computed and plotted by using the `MATLAB` function `parcorr` or the `type="partial"` option of the `R` function `acf`.

5.1.4 Forecasting in ARMA models

In the AR(p) model (5.10), the conditional expectation $E(x_{t+1}|x_t, x_{t-1}, \ldots)$ is $\phi_1 x_t + \cdots + \phi_p x_{t-p+1}$. It is the minimum-variance one-step-ahead forecast $\widehat{x}_{t+1|t}$ of x_{t+1} based on the current and past observations $x_t, x_{t-1}, \ldots$. For an AR(∞) model,

$$\widehat{x}_{t+1|t} = \phi_1 x_t + \phi_2 x_{t-1} + \cdots. \tag{5.13}$$

Since MA(q) and more general ARMA(p, q) models are actually AR(∞) models under the invertibility condition (5.9), we can express their one-step-ahead forecasts by (5.13). However, it is more convenient to assume the initialization

$$x_0 = \cdots = x_{-p+1} = 0 = u_0 = \cdots = u_{-q+1} \tag{5.14}$$

for an ARMA(p, q) to retrieve $u_1, \ldots, u_t$ from (5.12) so that

$$\widehat{x}_{t+1|t} = \mu + \sum_{i=1}^{p} \phi_i x_{t-i+1} + \sum_{j=1}^{q} \psi_j u_{t-j+1}. \tag{5.15}$$

In particular, for the MA(1) model $x_t = \mu + u_t + \psi_1 u_{t-1}$, setting $u_0 = 0$, we have $u_1 = x_1 - \mu$, $u_2 = x_2 - \mu - \psi_1 u_1$, ..., $u_t = x_t - \mu - \psi_1 u_{t-1}$, and $\widehat{x}_{t+1|t} = \mu + \psi_1 u_t$.

To obtain h-steps-ahead forecasts, we use the tower property of conditional expectations:

$$E(x_{t+j}|x_t, x_{t-1}, \dots) = E\Big\{E(x_{t+j}|x_{t+j-1}, x_{t+j-2}, \dots)\Big|x_t, x_{t-1}, \dots\Big\},$$

and proceed inductively on j until we reach $j = h$, starting with $j = 1$. In particular, for the AR(p) model with $p \geq 2$,

$$\begin{aligned}\widehat{x}_{t+2|t} &= \phi_1 E(x_{t+1}|x_t, x_{t-1}, \dots) + \phi_2 x_t + \dots + \phi_p x_{t-p+2} \\ &= \phi_1 \widehat{x}_{t+1|t} + \phi_2 x_t + \dots + \phi_p x_{t-p+2}.\end{aligned}$$

For the MA(1) model, since $E(u_{t+j}|x_t, x_{t-1}, \dots) = 0$ for $j \geq 1$, we have $\widehat{x}_{t+h|t} = \mu$ for $h \geq 2$.

5.1.5 Parameter estimation and order determination

Consider the ARMA(p, q) model (5.12) with $u_t \sim N(0, \sigma^2)$. Let $\boldsymbol{\theta} = (\mu, \phi_1, \dots, \phi_p, \psi_1, \dots, \psi_q)^T$. The log-likelihood function is given by

$$\begin{aligned}l(\boldsymbol{\theta}, \sigma) = -\frac{n}{2}\log(2\pi\sigma^2) - \frac{1}{2\sigma^2}\sum_{t=1}^{n}\big(x_t - \mu - \phi_1 x_{t-1} - \dots - \phi_p x_{t-p} \\ -\psi_1 u_{t-1} - \dots - \psi_q u_{t-q}\big)^2,\end{aligned} \tag{5.16}$$

in which $u_1, \dots, u_{n-1}$ can be retrieved recursively from (5.12) for a given $\boldsymbol{\theta}$ under the initial condition (5.14).

An important practical issue is the choice of the order (p, q) of an ARMA model. The information criteria AIC and BIC in (1.32) and (1.34) were originally developed to address this issue. Discarding the scaling factor $1/n$ in (1.32) and (1.34), these information criteria are of the form

$$\text{AIC}(d) = -2l(\widehat{\boldsymbol{\theta}}, \widehat{\sigma}) + 2d, \tag{5.17}$$

$$\text{BIC}(d) = -2l(\widehat{\boldsymbol{\theta}}, \widehat{\sigma}) + d\log n, \tag{5.18}$$

where n is the sample size, $d = p+q+1$, $l(\boldsymbol{\theta}, \sigma)$ is the log-likelihood function of the ARMA(p, q) model, and $(\widehat{\boldsymbol{\theta}}, \widehat{\sigma})$ is the MLE. The model selection procedure is to choose the model with the smallest AIC or BIC. Note that we can replace $d = p + q + 1$ by $d = p + q$ in (5.17) or (5.18), as is done by some software packages, since we are basically comparing the information criteria of candidate ARMA(p, q) models. Moreover, packages such as `MATLAB` use the penalized log-likelihood $2l(\widehat{\boldsymbol{\theta}}, \widehat{\sigma}) - 2d$ instead of (5.17) to define the AIC so that the selection procedure is to choose the model with the largest penalized log-likelihood. Whereas BIC is concerned with choosing the correct p and q when the data are indeed generated by a finite-order ARMA model, the AIC is designed to select the model that predicts best when the underlying model has infinite order.

5.2 Analysis of nonstationary time series

A first step in empirical analysis of time series y_t is plotting the data against time to see if the plot appears stationary, as illustrated in Figure 5.1. If the plot appears nonstationary, the next step is to explore if the data can be detrended, differenced, or transformed into stationary models.

5.2.1 Detrending

The underlying idea behind detrending is based on regression models of the form

$$y_t = f(t) + w_t, \qquad 1 \le t \le n, \tag{5.19}$$

where w_t is stationary (or weakly stationary) with $E(w_t) = 0$ and $f(t)$ is a nonrandom function of t. The *mean function* $f(t)$ is either assumed to have a parametric form that involves an unknown parameter vector $\boldsymbol{\theta}$ or to be nonparametric. The nonparametric approach will be treated in Chapter 7 as a special case of *nonparametric regression*. A widely used method in the time series literature is to estimate $f(t)$ by moving averages of the form

$$\widehat{f}_t = \left(\sum_{j=t-q}^{t+q} y_i \right) \Big/ (2q+1), \qquad q+1 \le t \le n-q. \tag{5.20}$$

For the parametric approach, we can write $f(t)$ as $f(t; \boldsymbol{\theta})$ and estimate $\boldsymbol{\theta}$ by the method of least squares, minimizing $\sum_{t=1}^{n} \{y_t - f(t; \boldsymbol{\theta})\}^2$. The consistency and asymptotic normality properties of the least squares estimator still hold when the i.i.d. assumption on w_t is generalized to weakly stationary w_t satisfying certain assumptions; see Appendix B. After performing the regression, it is important to plot the residuals $y_t - f(t; \widehat{\boldsymbol{\theta}})$ against t to see if the plot appears stationary. The stationarity assumption on w_t in (5.19) is important for consistent estimation of $f(t; \boldsymbol{\theta})$, and this issue will be explored further in Chapter 9 (Section 9.4.3) in connection with "spurious regression."

In econometric time series, the mean function $f(t)$ often includes a seasonal component that reflects certain cyclical behavior of economic activity. Thus f is decomposed as $f(t) = m(t) + s(t)$, where s is a periodic function with given period d. For identifiability, the periodic component is assumed to be centered, i.e., $\sum_{t=1}^{d} s(t) = 0$. If d is odd, let $q = (d-1)/2$ and define $\widehat{y}_t$ by the moving average in (5.19), noting that the window size in this case is $2q+1 = d$. If d is even, let $q = d/2$ and define

$$\widehat{y}_t = \frac{1}{2q}\left\{\frac{y_{t-q}}{2} + y_{t-q+1} + \cdots + y_{t+q-1} + \frac{y_{t+q}}{2}\right\}, \quad q+1 \le t \le n-q. \quad (5.21)$$

For $k = 1, \ldots, d$, let $\bar{\Delta}_k$ be the mean of the sample $\{y_{k+jd} - \widehat{y}_{k+jd} : q+1-k \le jd \le n-q-k\}$. The *method-of-moments estimate*, which replaces population moments by their sample counterparts, of the periodic function $s(t)$ is

$$\widehat{s}(k) = \bar{\Delta}_k - d^{-1}\sum_{i=1}^{d} \bar{\Delta}_i \text{ for } 1 \le k \le d, \quad \widehat{s}(t) = \widehat{s}(k) \text{ for } t = k + jd, \quad (5.22)$$

in which j is some integer. From the *deseasonalized* series $y_t - \widehat{s}(t)$, $m(t)$ can be estimated by parametric regression or moving average methods. More refined nonparametric regression techniques than simple averaging, as in (5.20) and (5.22), will be described in Chapter 7. The function `stl` in `R` (or `S`) uses these more refined techniques to estimate the decomposition $y_t = m(t) + s(t) + w_t$ of a time series into a trend $m(t)$, a seasonal component $s(t)$ and a stationary disturbance w_t; see Venables and Ripley (2002, pp. 403–404) for details and illustrations.

5.2.2 An empirical example

Figure 5.4 plots the monthly unemployment rates in Dallas County, Arizona, from January 1980 to June 2005. The data are obtained from the Website `www.Economagic.com`. The ACF and PACF of the unemployment rates are plotted in Figure 5.5. Note that all rates vary from 4% to 12% except those in the first five months, which are above 15%. We split the time series into a training sample of historical data from January 1980 to December 2004 and a second sample of "test data" from January to June 2005. The second sample is used to measure the performance of the out-of-sample forecasts developed from the training sample. Since there are obvious seasonal effects on unemployment, we use the `R` or `S` function `stl` to decompose the training sample into a trend, a seasonal component, and residual; Figure 5.6 plots the trend and seasonal components and the residuals of the decomposition, with a period of 12 months for the seasonal component. We fit an ARMA model to the deseasonalized series x_t, which consists of the trend and the residuals. Note that the trend in Figure 5.6 does not show an obvious linear or polynomial pattern and consists of fluctuations that resemble those of an ARMA model. The ACF and PACF of the deseasonalized series are plotted in Figure 5.7,

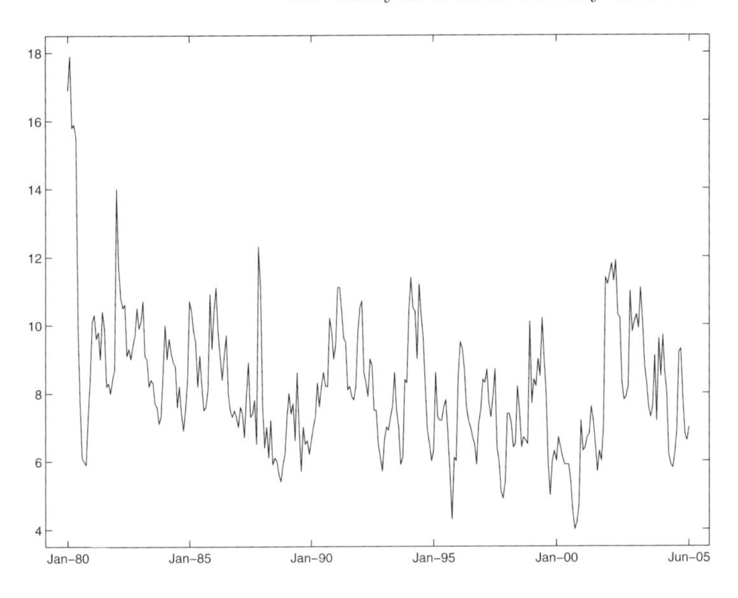

Fig. 5.4. Monthly unemployment rates (in %) in Dallas County, Arizona, from January 1980 to June 2005.

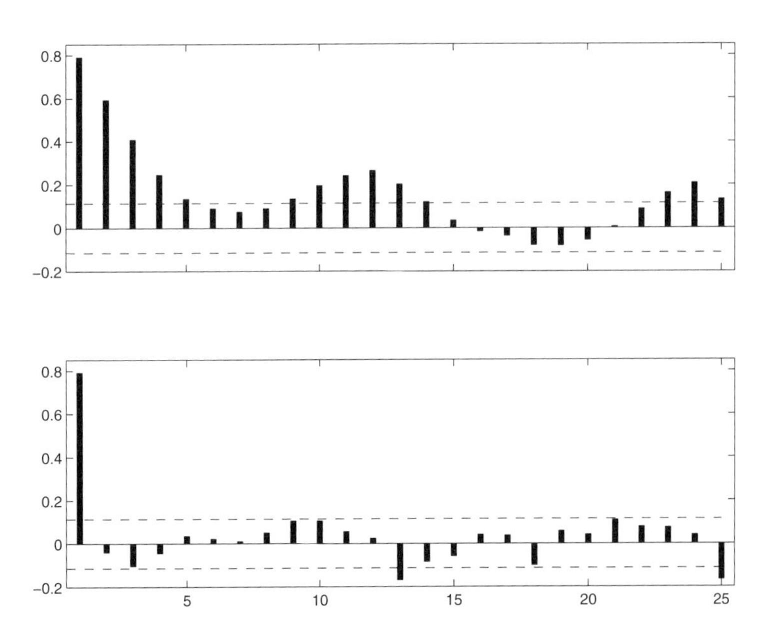

Fig. 5.5. ACF (top panel) and PACF (bottom panel) of the unemployment rate. The dashed lines represent rejection boundaries of 5%-level tests of zero ACF and PACF at indicated lag.

which shows significant autocorrelations for lags up to 12. We fit an ARMA model to the deseasonalized series by using the AIC to determine the order of the ARMA model; see Table 5.1. Based on the AIC, we fit the ARMA(1,1) model

$$\begin{aligned} x_t = \underset{(0.4417)}{8.3421} + \underset{(0.0550)}{0.8828}x_{t-1} + \epsilon_t - \underset{(0.067)}{0.099}\epsilon_{t-1} \end{aligned}$$

with the standard errors of the parameter estimates given in parentheses. The results are obtained by using the R function arima, which also gives the estimate 1.025 of the error variance $\mathrm{Var}(\epsilon_t)$ and diagnostic plots for the fitted model. Most of the standardized residuals $\epsilon_t/\sqrt{\mathrm{Var}(\epsilon_t)}$ are small, and so are their autocorrelations shown in the top panel of Figure 5.8. The bottom panel of Figure 5.8 shows the p-values of the Ljung-Box statistics $Q(m)$ in (5.6) for different values of the lag m. These p-values are well above the level 0.05 (shown by the broken line), below which the Ljung-Box test rejects the null hypothesis of zero autocorrelations up to lag m.

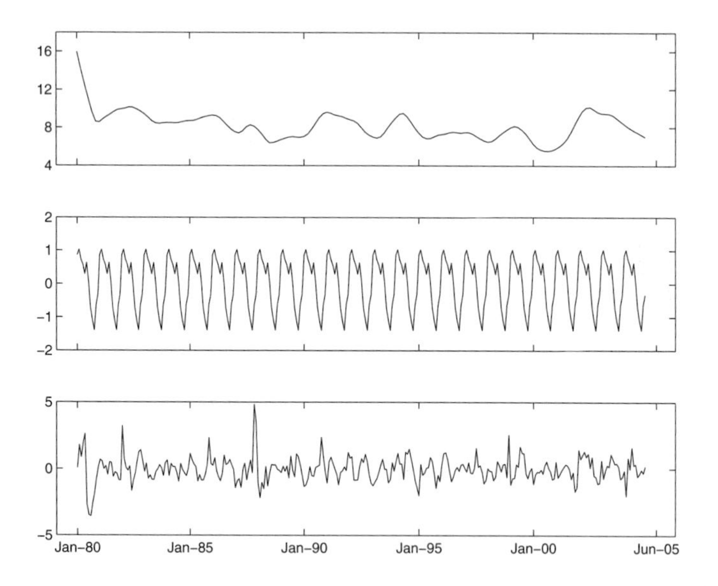

Fig. 5.6. Decomposition of the time series of unemployment rates into the trend (top panel), the seasonal component (middle panel), and residuals (bottom panel).

We can use the fitted ARMA(1, 1) model to obtain k-months-ahead forecasts as in Section 5.1.4. The R function arima can be used to calculate these forecasts and their standard errors (s.e.). Table 5.2 gives the forecast values of

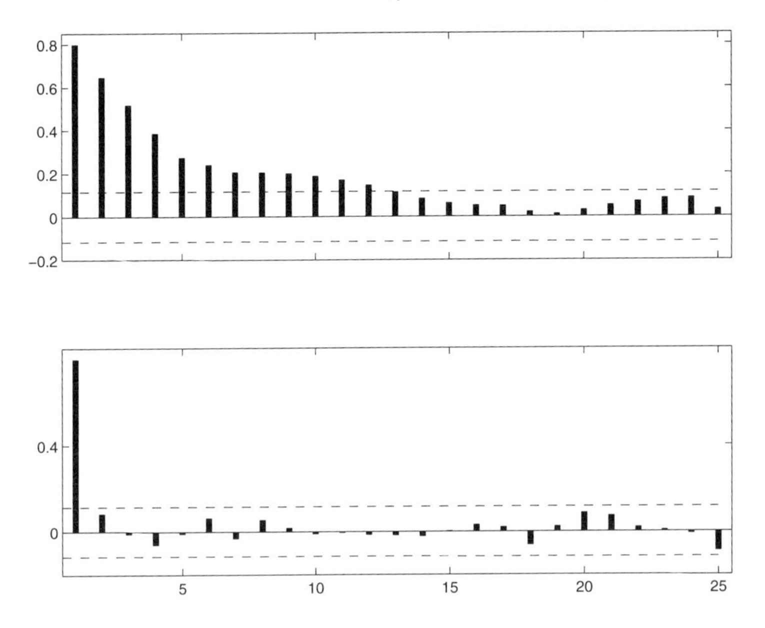

Fig. 5.7. ACF (top panel) and PACF (bottom panel) of the deseasonalized time series from the training sample. The dashed lines represent rejection boundaries of 5%-level tests of zero ACF and PACF at indicated lag.

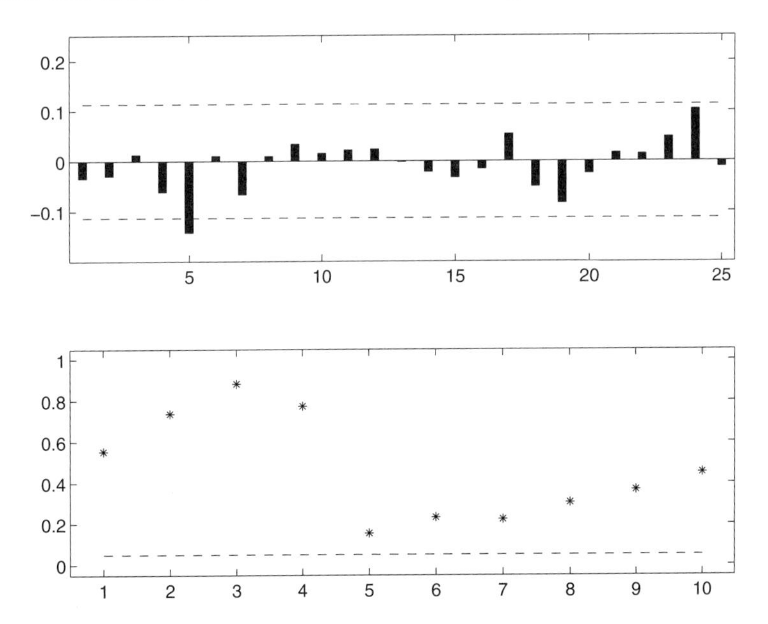

Fig. 5.8. Diagnostic plots for the fitted ARMA(1, 1) model. Top panel: ACF of residuals; bottom panel: p-values of Ljung-Box statistics $Q(m)$.

Table 5.1. AICs given by ARMA(p, q) $(1 <= p,\ q <= 3)$.

	$q = 1$	$q = 2$	$q = 3$
$p = 1$	868.090	869.766	870.569
$p = 2$	869.988	871.942	872.594
$p = 3$	871.959	871.490	871.938

Table 5.2. Forecasts of unemployment rates in 2005.

	Jan.	Feb.	Mar.	Apr.	May	June
Actual rate	9.2	9.3	7.9	6.8	6.6	7.0
Deseasonalized rate	7.21	7.48	7.65	7.79	7.90	7.98
(s.e.)	(1.14)	(1.49)	(1.75)	(1.89)	(1.98)	(2.02)
Seasonal rate	0.86	1.03	0.73	0.57	0.29	0.63
Predicted rate	8.07	8.51	8.38	8.36	8.19	8.61

the deseasonalized series from January to June 2005 based on the ARMA(1,1) model fitted to the training sample from January 1980 to December 2004. The estimated seasonal components of the unemployment rates are also given in Table 5.2. Adding them to the deseasonalized forecasts yields the predicted unemployment rates in the last row of Table 5.2. The difference between the predicted and the actual rates is within the standard error shown in parentheses.

5.2.3 Transformation and differencing

Some nonstationary time series can be converted to stationary ones by differencing and using appropriate transformations. For example, the top panel of Figure 5.9 plots the daily closing prices P_t of Intel from March 26, 1990 to December 29, 2006; the data are obtained from Yahoo! Inc. The plot shows steady growth in prices until a sharp drop in April 2000 followed by another increase before flattening out after September 2003. This suggests using the transformation $p_t = \log P_t$. The middle panel plots p_t, which has an upward trend and is still nonstationary. The bottom panel shows the differenced series $r_t = p_t - p_{t-1}$, and the plot looks stationary.

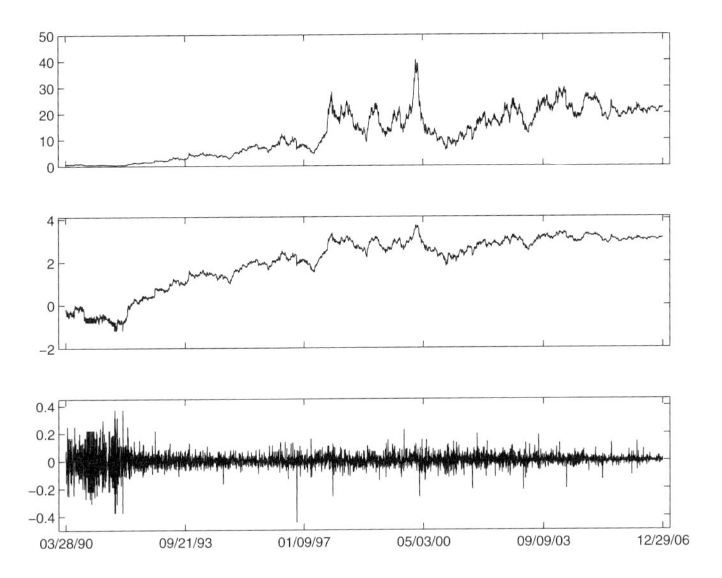

Fig. 5.9. Daily prices (top panel), log prices (middle panel), and log returns (bottom panel) of Intel.

U.S. Treasury-LIBOR swap rates

For the example in Section 2.2.3 on U.S. Treasury-LIBOR swap rates, the top panel of Figure 2.1 shows the time series of swap rates for different maturities during a 5-year period to be nonstationary. However, the differenced series for 1-year and 30-year swap rates in the middle and bottom panels of Figure 2.1 appear to be stationary. The input data for principal component analysis (PCA) should be stationary, so PCA is performed on the differenced series in Section 2.2.3 and not on the original swap rates.

5.2.4 Unit-root nonstationarity and ARIMA models

A time series x_t is said to be *unit-root nonstationary* if the differenced series $(1 - B)x_t = x_t - x_{t-1}$ is weakly stationary. More generally, it is said to be *integrated* of order d if $(1 - B)^d x_t$ is weakly stationary. It is called an *ARIMA* (autoregressive integrated moving average model) of order (p, d, q) if $(1-B)^d x_t$ is a stationary ARMA(p, q) model.

Polynomials of degree k in t are integrated of order k since

$$(1-B)^k(a_0 + a_1 t + \cdots + a_k t^k) = k! a_k. \tag{5.23}$$

Moreover, if x_t is covariance stationary (or stationary), then so is $(1-B)x_t$; see Exercise 5.9. Hence, if the trend $f(t)$ in (5.19) is a polynomial of degree k, then (5.19) is an integrated time series of order k. Therefore, for polynomial trends, an alternative to first detrending and then fitting an ARMA(p, q) model to the residuals is to fit an ARIMA(p, d, q) model directly to the time series. Whereas the detrending approach is more efficient under the assumption of a deterministic trend in (5.19), ARIMA modeling does not require such an assumption and uses differencing to achieve stationarity.

Seasonal differencing and SARIMA models

Whereas differencing (i.e., using the operator $1 - B$) can remove a linear trend, *seasonal differencing* can remove a seasonal component f_t with period s: $(1 - B^s)f_t = f_t - f_{t-s} = 0$. A *seasonal ARIMA* $(p, d, q) \times (P, D, Q)_s$ model (or SARIMA model) is of the form

$$\phi(B)\Phi(B^s)(1-B)^d(1-B^s)^D x_t = \psi(B)\Psi(B^s)u_t. \tag{5.24}$$

In (5.24), ϕ and Φ are polynomials with respective degrees p and P whose zeros lie outside the unit circle as in (5.11), and ψ and Ψ are polynomials with respective degrees q and Q. In particular, the SARIMA $(P, D, Q)_s$ model

$$\Phi(B^s)(1-B^s)^D x_t = \Psi(B^s)u_t \tag{5.25}$$

is simply a seasonal version of an ARIMA process.

The basic idea behind ARIMA (and SARIMA) modeling is to difference the series (using seasonal differencing to handle the periodic component) minimally to achieve stationarity. Graphical plots of the differenced series and its sample ACF can be used to check if it is stationary. The `R` function `arima` can still be used to fit models of this form by specifying the argument `seasonal` with the period; see Venables and Ripley (2002, pp. 405–406).

5.3 Linear state-space models and Kalman filtering

A *linear state-space model* is of the form

$$\mathbf{x}_{t+1} = \mathbf{F}_t \mathbf{x}_t + \mathbf{w}_{t+1}, \tag{5.26}$$

$$\mathbf{y}_t = \mathbf{G}_t \mathbf{x}_t + \mathbf{v}_t, \tag{5.27}$$

in which $\mathbf{x}_t, \mathbf{w}_t \in \mathbb{R}^{p\times 1}$, $\mathbf{y}_t, \mathbf{v}_t \in \mathbb{R}^{q\times 1}$, $\mathbf{w}_t$ and $\mathbf{v}_t$ are independent random vectors with $E(\mathbf{w}_t) = \mathbf{0}$, $\mathrm{Cov}(\mathbf{w}_t) = \boldsymbol{\Sigma}_t$, $E(\mathbf{v}_t) = \mathbf{0}$, and $\mathrm{Cov}(\mathbf{v}_t) = \mathbf{V}_t > \mathbf{0}$.

The $p \times p$ matrix $\mathbf{F}_t$ describes the linear dynamics in the *state equation* (5.26). The states $\mathbf{x}_t$, however, are not observable. The observations are $\mathbf{y}_t$, which are linear transformations of the states (via the $q \times p$ matrices $\mathbf{G}_t$) plus unobservable disturbance $\mathbf{v}_t$, as described in the *observation equation* (5.27). The *Kalman filter* $\widehat{\mathbf{x}}_{t|t}$ is the *minimum-variance linear estimate* of $\mathbf{x}_t$ based on the observations $\mathbf{y}_1, \dots, \mathbf{y}_t$ up to stage t. Let $\widehat{\mathbf{x}}_{t|t-1}$ denote the minimum-variance linear predictor of $\mathbf{x}_t$ based on $\mathbf{y}_1, \dots, \mathbf{y}_{t-1}$ and let $\mathbf{P}_{t|t-1} = \mathrm{Cov}(\mathbf{x}_t - \widehat{\mathbf{x}}_{t|t-1})$. The Kalman filter is derived from recursions for $(\widehat{\mathbf{x}}_{t|t-1}, \mathbf{P}_{t|t-1})$ and involves the *Kalman gain matrix*

$$\mathbf{K}_t = \mathbf{F}_t \mathbf{P}_{t|t-1} \mathbf{G}_t^T (\mathbf{G}_t \mathbf{P}_{t|t-1} \mathbf{G}_t^T + \mathbf{V}_t)^{-1}. \tag{5.28}$$

5.3.1 Recursive formulas for $\mathbf{P}_{t|t-1}, \widehat{\mathbf{x}}_{t|t-1}$, and $\widehat{\mathbf{x}}_{t|t}$

For $t \geq 1$,

$$\begin{aligned}
\widehat{\mathbf{x}}_{t+1|t} &= \mathbf{F}_t \widehat{\mathbf{x}}_{t|t-1} + \mathbf{K}_t (\mathbf{y}_t - \mathbf{G}_t \widehat{\mathbf{x}}_{t|t-1}), && (5.29)\\
\mathbf{P}_{t+1|t} &= (\mathbf{F}_t - \mathbf{K}_t \mathbf{G}_t) \mathbf{P}_{t|t-1} (\mathbf{F}_t - \mathbf{K}_t \mathbf{G}_t)^T + \boldsymbol{\Sigma}_{t+1} + \mathbf{K}_t \mathbf{V}_t \mathbf{K}_t^T, && (5.30)\\
\widehat{\mathbf{x}}_{t|t} &= \widehat{\mathbf{x}}_{t|t-1} + \mathbf{P}_{t|t-1} \mathbf{G}_t (\mathbf{G}_t \mathbf{P}_{t|t-1} \mathbf{G}_t^T + \mathbf{V}_t)^{-1} (\mathbf{y}_t - \mathbf{G}_t \widehat{\mathbf{x}}_{t|t-1}); && (5.31)
\end{aligned}$$

the recursions are initialized at $\widehat{\mathbf{x}}_{1|0} = E(\mathbf{x}_1)$ and $\mathbf{P}_{1|0} = \mathrm{Cov}(\mathbf{x}_1)$. An induction argument can be used to show that $\mathbf{x}_t - \widehat{\mathbf{x}}_{t|t-1}$ has mean $\mathbf{0}$. Therefore, by (5.31), $\mathbf{x}_t - \widehat{\mathbf{x}}_{t|t}$ also has mean $\mathbf{0}$; moreover,

$$\mathrm{Cov}(\mathbf{x}_t - \widehat{\mathbf{x}}_{t|t}) = \{\mathbf{I} - \mathbf{P}_{t|t-1} \mathbf{G}_t^T (\mathbf{G}_t \mathbf{P}_{t|t-1} \mathbf{G}_t^T + \mathbf{V}_t)^{-1} \mathbf{G}_t\} \mathbf{P}_{t|t-1}. \tag{5.32}$$

The recursive formulas (5.29)–(5.32) can be computed by the `MATLAB` function `kalman`.

Derivation of the Kalman recursions

We first consider the case of normal $\mathbf{w}_t$ and $\mathbf{v}_t$. Then $\{(\mathbf{x}_i^T, \mathbf{y}_i^T) : 1 \leq i \leq t\}$ has a multivariate normal distribution, and the conditional distribution of $\mathbf{x}_t$ given $\mathbf{y}_1, \dots, \mathbf{y}_{t-1}$ is normal. Part (i) of the following lemma follows from the result on conditional distributions in Section 2.3.2, and part (ii) of the lemma can be obtained by applying part (i) to $\mathbf{x}$ and $(\mathbf{y}^T, \mathbf{z}^T)^T$.

Basic lemma. Let $\mathbf{x}$, $\mathbf{y}$, and $\mathbf{z}$ be jointly normal such that $\mathbf{y}$ and $\mathbf{z}$ are independent. Let $\boldsymbol{\Sigma}_{\mathbf{xy}} = E\{(\mathbf{x} - E\mathbf{x})(\mathbf{y} - E\mathbf{y})^T\}$.

(i) The conditional distribution of $\mathbf{x}$ given $\mathbf{y}$ is normal with mean $E(\mathbf{x}|\mathbf{y})$ and covariance matrix $\boldsymbol{\Sigma}_{\mathbf{xx}} - \boldsymbol{\Sigma}_{\mathbf{xy}} \boldsymbol{\Sigma}_{\mathbf{yy}}^{-1} \boldsymbol{\Sigma}_{\mathbf{yx}}$. Moreover, $E(\mathbf{x}|\mathbf{y}) = E(\mathbf{x}) + \boldsymbol{\Sigma}_{\mathbf{xy}} \boldsymbol{\Sigma}_{\mathbf{yy}}^{-1} (\mathbf{y} - E\mathbf{y})$, and $\mathbf{x} - E(\mathbf{x}|\mathbf{y})$ is independent of $\mathbf{y}$ (and therefore also of $E(\mathbf{x}|\mathbf{y})$). Hence $E(\mathbf{x}|\mathbf{y})$ is a linear function of $\mathbf{y}$.

(ii) Let $\widehat{\mathbf{x}} = E(\mathbf{x}|\mathbf{y},\mathbf{z})$ and $\widetilde{\mathbf{x}} = \mathbf{x} - \widehat{\mathbf{x}}$. Then

$$\widehat{\mathbf{x}} = E\mathbf{x} + \boldsymbol{\Sigma}_{\mathbf{xy}}\boldsymbol{\Sigma}_{\mathbf{yy}}^{-1}(\mathbf{y} - E\mathbf{y}) + \boldsymbol{\Sigma}_{\mathbf{xz}}\boldsymbol{\Sigma}_{\mathbf{zz}}^{-1}(\mathbf{z} - E\mathbf{z}),$$
$$\mathrm{Cov}(\widetilde{\mathbf{x}}) = \boldsymbol{\Sigma}_{\mathbf{xx}} - \boldsymbol{\Sigma}_{\mathbf{xy}}\boldsymbol{\Sigma}_{\mathbf{yy}}^{-1}\boldsymbol{\Sigma}_{\mathbf{yx}} - \boldsymbol{\Sigma}_{\mathbf{xz}}\boldsymbol{\Sigma}_{\mathbf{zz}}^{-1}\boldsymbol{\Sigma}_{\mathbf{zx}}.$$

The proof of the recursions (5.29)–(5.32) under the additional assumption that the random disturbances $\mathbf{w}_{t+1}$ and $\mathbf{v}_t$ are normal makes use of the basic lemma and consists of three steps. The assumption of normal distributions is removed in the fourth step.

Step 1. Let $\mathcal{Y}_t = (\mathbf{y}_1, \ldots, \mathbf{y}_t)$, $\widetilde{\mathbf{x}}_{t+1|t} = \mathbf{x}_{t+1} - E(\mathbf{x}_{t+1}|\mathcal{Y}_t)$, $\widetilde{\mathbf{x}}_{t|t} = \mathbf{x}_t - E(\mathbf{x}_t|\mathcal{Y}_t)$. By (5.26), $E(\mathbf{x}_{t+1}|\mathcal{Y}_t) = \mathbf{F}_t E(\mathbf{x}_t|\mathcal{Y}_t)$ and therefore $\widetilde{\mathbf{x}}_{t+1|t} = \mathbf{F}_t\widetilde{\mathbf{x}}_{t|t} + \mathbf{w}_{t+1}$, which is a sum of two independent terms. Hence

$$\mathrm{Cov}(\widetilde{\mathbf{x}}_{t+1|t}) = \mathbf{F}_t\mathrm{Cov}(\widetilde{\mathbf{x}}_{t|t})\mathbf{F}_t^T + \boldsymbol{\Sigma}_{t+1}. \tag{5.33}$$

By part (i) of the lemma, $\widehat{\mathbf{x}}_{t|t} = E(\mathbf{x}_t|\mathcal{Y}_t)$, $\widehat{\mathbf{x}}_{t+1|t} = E(\mathbf{x}_{t+1}|\mathcal{Y}_t)$, and therefore $\mathrm{Cov}(\widetilde{\mathbf{x}}_{t+1|t}) = \mathbf{P}_{t+1|t}$, $\mathrm{Cov}(\widetilde{\mathbf{x}}_{t|t}) = \mathbf{P}_{t|t}$.

Step 2. Let $\widehat{\mathbf{y}}_{t|t-1} = E(\mathbf{y}_t|\mathcal{Y}_{t-1})$, $\widetilde{\mathbf{y}}_{t|t-1} = \mathbf{y}_t - \widehat{\mathbf{y}}_{t|t-1}$. By (5.27),

$$\widetilde{\mathbf{y}}_{t|t-1} = \mathbf{G}_t\widetilde{\mathbf{x}}_{t|t-1} + \mathbf{v}_t, \tag{5.34}$$

which is a sum of two independent terms, yielding

$$\mathrm{Cov}(\widetilde{\mathbf{y}}_{t|t-1}) = \mathbf{G}_t\mathbf{P}_{t|t-1}\mathbf{G}_t^T + \mathbf{V}_t. \tag{5.35}$$

From (5.34), it also follows that

$$E(\mathbf{x}_t\widetilde{\mathbf{y}}_{t|t-1}^T) = E(\mathbf{x}_t\widetilde{\mathbf{x}}_{t|t-1}^T)\mathbf{G}_t^T + E(\mathbf{x}_t\mathbf{v}_t^T) = \mathbf{P}_{t|t-1}\mathbf{G}_t^T. \tag{5.36}$$

To see the last equality in (5.36), note that $\mathbf{x}_t$ and $\mathbf{v}_t$ are independent and that $\mathbf{x}_t = \widehat{\mathbf{x}}_{t|t-1} + \widetilde{\mathbf{x}}_{t|t-1}$ is a sum of two independent terms, with $E(\widetilde{\mathbf{x}}_{t|t-1}) = \mathbf{0}$ and $\mathrm{Cov}(\widetilde{\mathbf{x}}_{t|t-1}) = \mathbf{P}_{t|t-1}$.

Step 3. Because of the one-to-one correspondence between $\mathcal{Y}_t$ and $(\mathcal{Y}_{t-1}, \widetilde{\mathbf{y}}_{t|t-1})$,

$$\widehat{\mathbf{x}}_{t|t} = E(\mathbf{x}_t|\mathcal{Y}_{t-1}, \widetilde{\mathbf{y}}_{t|t-1}) = \widehat{\mathbf{x}}_{t|t-1} + \left(E\mathbf{x}_t\widetilde{\mathbf{y}}_{t|t-1}^T\right)\left(\mathrm{Cov}(\widetilde{\mathbf{y}}_{t|t-1})\right)^{-1}\widetilde{\mathbf{y}}_{t|t-1}; \tag{5.37}$$

the last equality in (5.37) follows from part (ii) of the lemma, which also gives

$$\mathrm{Cov}(\mathbf{x}_t - \widehat{\mathbf{x}}_{t|t}) = \mathbf{P}_{t|t-1} - \left(E\left(\mathbf{x}_t\widetilde{\mathbf{y}}_{t|t-1}^T\right)\right)\left(\mathrm{Cov}(\widetilde{\mathbf{y}}_{t|t-1})\right)^{-1}\left(E\left(\mathbf{x}_t\widetilde{\mathbf{y}}_{t|t-1}^T\right)\right)^T. \tag{5.38}$$

Combining (5.38) with (5.35) and (5.36) yields (5.32). From (5.32) and (5.33), (5.30) follows. Since $\widehat{\mathbf{y}}_{t|t-1} = \mathbf{G}_t\widehat{\mathbf{x}}_{t|t-1}$ by (5.27), $\widetilde{\mathbf{y}}_{t|t-1} = \mathbf{y}_t - \mathbf{G}_t\widehat{\mathbf{x}}_{t|t-1}$.

Putting this and (5.35) and (5.36) into (5.37) yields (5.31). Finally, substituting (5.31) into $\widehat{\mathbf{x}}_{t+1|t} = \mathbf{F}_t\widehat{\mathbf{x}}_{t|t}$ yields (5.29).

Step 4. (Extension to non-Gaussian disturbances). In the Gaussian case, the Kalman predictor $E(\mathbf{x}_t|\mathcal{Y}_{t-1})$ only involves the mean vector $\boldsymbol{\mu}_t$ and covariance matrix $\mathbf{C}_t$ of $(\mathbf{x}_1^T, \mathbf{y}_1^T, \ldots, \mathbf{x}_t^T, \mathbf{y}_{t-1}^T)^T$. Without assuming normality, the same formulas apply to the minimum-variance linear predictor of $\mathbf{x}_t$ based on $\mathcal{Y}_{t-1}$, as has already been noted in Section 1.5.1, where the "normal equations" defining the coefficients of the minimum-variance linear predictor are given in terms of $\boldsymbol{\mu}_t$ and $\mathbf{C}_t$.

5.3.2 Dynamic linear models and time-varying betas in CAPM

The linear state-space model in (5.26) and (5.27) and its associated Kalman filter have provided a tractable method to deal with time-varying parameters in linear regression models. Whereas the regression model $y_t = \boldsymbol{\beta}^T\mathbf{x}_t + \epsilon_t$ in Chapter 1 assumes that the $p \times 1$ parameter vector $\boldsymbol{\beta}$ remains constant over the sampling period, a *dynamic linear model* (DLM) allows time-varying regression parameters that are treated as states. Specifically, a DLM is a linear state-space model of the form

$$y_t = \mathbf{x}_t^T\boldsymbol{\beta}_t + \epsilon_t, \qquad \boldsymbol{\beta}_t = \mathbf{F}\boldsymbol{\beta}_{t-1} + \mathbf{w}_t, \tag{5.39}$$

in which $\mathbf{w}_t$ are i.i.d. with mean $\mathbf{0}$ and covariance matrix $\boldsymbol{\Sigma}$ and are independent of the ϵ_t, which are i.i.d. with variance σ^2. The Kalman filter $\widehat{\boldsymbol{\beta}}_{t|t}$ can be used to estimate the time-varying regression parameters $\boldsymbol{\beta}_t$ from the observations $\mathbf{x}_1, \mathbf{y}_1, \ldots, \mathbf{x}_t, \mathbf{y}_t$, noting that here $\mathbf{G}_t$ in (5.27) reduces to $\mathbf{x}_t^T$. This model of time-varying parameters assumes that $\boldsymbol{\beta}_t$ changes continually, which may not be appropriate in many econometric applications. A better approach, which is more complicated and is treated in Chapter 9 (Sections 9.5.2 and 9.5.3), is to assume that $\boldsymbol{\beta}_t$ is piecewise constant, undergoing occasional changes or regime switches.

The $p \times p$ matrices $\mathbf{F}$ and $\boldsymbol{\Sigma}(= \mathrm{Cov}(\mathbf{w}_t))$ in (5.39) that describe the parameter dynamics are often assumed to have certain parametric forms involving an unknown parameter $\boldsymbol{\theta}$. Assuming normal ϵ_t and $\mathbf{w}_t$, the likelihood function $(\boldsymbol{\theta}, \sigma)$ can be obtained by making use of the independence of the *innovations* $y_t - \mathbf{x}_t^T\widehat{\boldsymbol{\beta}}_{t|t-1}$, which are normal with mean 0 and variance $\mathbf{x}_t^T\mathbf{P}_{t|t-1}\mathbf{x}_t + \sigma^2$ by (5.35). The maximum likelihood estimate of $(\boldsymbol{\theta}, \sigma)$ based on $\mathbf{x}_1, y_1, \ldots, \mathbf{x}_n, y_n$ therefore can be found by maximizing the likelihood function or, equivalently, by minimizing

$$S(\boldsymbol{\theta},\sigma) = \sum_{t=1}^{n} \{y_t - \mathbf{x}_t^T \widehat{\boldsymbol{\beta}}_{t|t-1}(\boldsymbol{\theta},\sigma)\}^2 \Big/ \{\mathbf{x}_t^T \mathbf{P}_{t|t-1}(\boldsymbol{\theta},\sigma)\mathbf{x}_t + \sigma^2\}$$
$$+ \sum_{t=1}^{n} \log\left(\mathbf{x}_t^T \mathbf{P}_{t|t-1}(\boldsymbol{\theta},\sigma)\mathbf{x}_t + \sigma^2\right).$$

CAPM with time-varying betas

Ferson (1989) and Ferson and Harvey (1991) have noted that the beta in CAPM (see Section 3.3.1) may vary over time. A number of authors have subsequently proposed ways to improve CAPM by including time variations in the betas. In particular, the dynamic linear model (5.39) has been considered because of the relative simplicity of the Kalman filter. We illustrate this application of Kalman filtering with the six stocks in Section 5.1.2. Figure 5.10 shows the sequential estimates of betas of the six stocks from March 2002 to October 2005. These sequential estimates, which also use the data from August 2000 to February 2002 to initialize, show that the betas of the six stocks change over time.

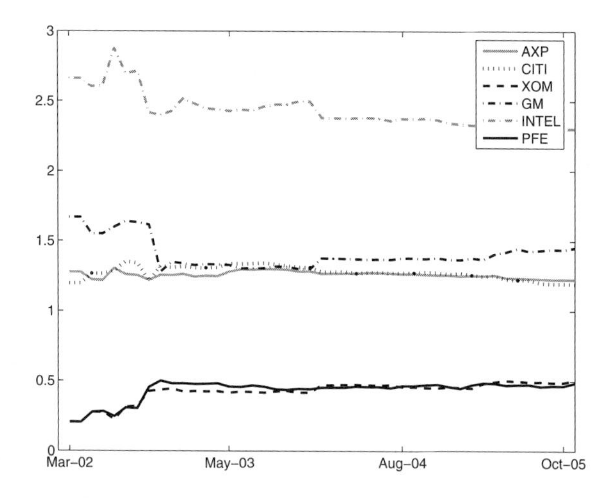

Fig. 5.10. Sequential estimates of betas of six stocks.

We next apply the DLM

$$r_t - r_{f,t} = \beta_t(r_{M,t} - r_{f,t}) + \epsilon_t, \qquad \beta_{t+1} = \beta_t + w_{t+1}, \tag{5.40}$$

with independent $\epsilon_t \sim N(0,\sigma^2)$ and $w_t \sim N(0,\sigma_w^2)$, to model the time-varying betas for each stock during the period March 2002 to October 2005. We fit CAPM to each stock for the period August 2000 to February 2002 and use the estimated beta as the initial value $\widehat{\beta}_0$ and the estimated error variance for $\widehat{\sigma}^2$ in the Kalman filter, which is used to estimate β_t sequentially from March 2002 to October 2005. Figure 5.11 shows the Kalman filter estimates (for which σ_w is chosen to be 0.1) of the time-varying betas of the six stocks during this period.

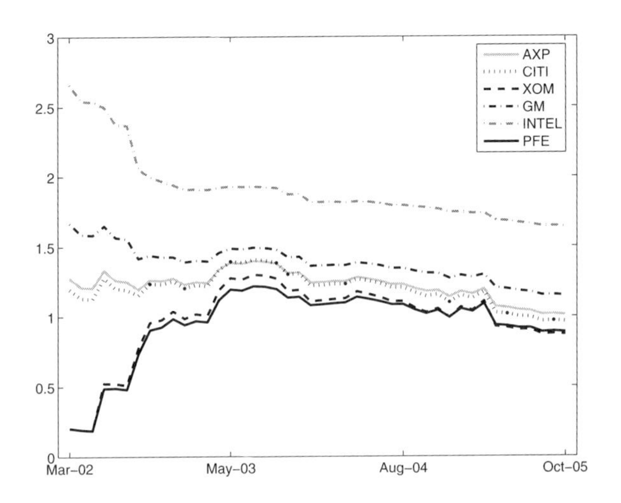

Fig. 5.11. Kalman filter estimates of betas of six stocks.

Exercises

5.1. Show that the AR(2) model $x_t = \mu + \phi_1 x_{t-1} + \phi_2 x_{t-2} + u_t$, in which u_t are i.i.d. zero-mean random variables, is stationary if $\phi_1 + \phi_2 < 1$, $\phi_1 - \phi_2 > -1$, and $\phi_2 > -1$.

5.2. Show that the AR(3) model

$$cx_t = (c-1)x_{t-1} + x_{t-3} + u_t,$$

in which u_t are i.i.d. zero-mean random variables, is nonstationary for all values of $c \neq 0$. For what of c is this autoregressive model unit-root nonstationary?

5.3. (a) Prove (5.23).
(b) Show that if x_t is covariance stationary, then so is $(1-B)x_t$.

5.4. Let $\{x_t\}$ be a weakly stationary sequence with mean 0 and autocovariance function γ_h such that $\gamma_0 > 0$. Denote the minimum-variance linear predictor of x_{n+1} based on $x_1, \ldots, x_n$ by $\widehat{x}_{n+1|n} = \phi_{n1}x_n + \cdots + \phi_{nn}x_1$,

$n \geq 1$, and denote the mean squared prediction error by $v_n = E(x_{n+1} - \widehat{x}_{n+1|n})^2$. The *Durbin-Levinson algorithm* provides a recursive method to compute v_n, ϕ_{nn}, and $\boldsymbol{\phi}_n := (\phi_{n1}, \dots, \phi_{n,n-1})^T$ in terms of $\{\gamma_h\}$:

$$\begin{aligned} \phi_{nn} &= \left(\gamma_n - \sum_{j=1}^{n-1} \phi_{n-1,j} \gamma_{n-j} \right) \Big/ v_{n-1}, \\ \boldsymbol{\phi}_n &= \boldsymbol{\phi}_{n-1} - \phi_{nn} (\phi_{n-1,n-1}, \dots, \phi_{n-1,1})^T, \\ v_n &= v_{n-1}(1 - \phi_{nn}^2). \end{aligned}$$

(a) Show that the recursive algorithm above should be initialized at $\phi_{11} = \gamma_1/\gamma_0$ and $v_0 = \gamma_0$.

(b) Show that ϕ_{nn} is the partial autocorrelation coefficient between x_{n+1} and x_1, adjusted for $x_2, \dots, x_n$.

(c) For an AR(2) model, show that $\phi_{nn} = 0$ for $n > 2$ and express ϕ_{11} and ϕ_{22} in terms of γ_0, γ_1, and γ_2.

5.5. Consider the time series of U.S. monthly unemployment rates from January 1948 to July 2007 in the file `m_us_unem.txt`. These data, which have been seasonally adjusted, are obtained from the Federal Reserve Bank of St. Louis.

(a) Plot the ACF and PACF of the rates and the differenced rates.

(b) Fit an ARMA model to the data from January 1948 to December 2006. (*Hint*: Use a model selection criterion to choose the order within a prescribed range of orders.)

(c) Use your fitted model to compute k-months-ahead forecasts ($k = 1, 2, \dots, 6$) and their standard errors, choosing December 2006 as the forecast origin. Compare your forecasts with the actual unemployment rates.

5.6. The file `q_us_gdp.txt` contains the seasonally adjusted time series of quarterly U.S. gross domestic product (GDP) from the first quarter of 1947 to the first quarter of 2007. The data are obtained from the Federal Reserve Bank of St. Louis.

(a) Plot the ACF and PACF of this series and the differenced series.

(b) Fit an ARMA model to the differenced series.

5.7. Consider the weekly log returns of Yahoo! stock from the week of April 12, 1996 to the week of June 25, 2007 in the file `w_logret_yahoo.txt`.

(a) Are there seasonal effects in the series?

(b) Are there serial correlations in the series? Use the Ljung-Box statistic $Q(10)$ to perform the test.

(c) Fit an ARMA model to the data from April 12, 1996 to April 30, 2007, and perform diagnostic checks on the fitted model.

(d) Compute k-weeks-ahead forecasts ($k = 1, 2, \dots, 8$) based on the fitted model, using April 30, 2007 as the forecast origin. Give the standard errors of your forecasts and compare them with the forecast errors, which are the differences between the predicted and actual log returns.

5.8. The file m_caus_ex.txt contains the monthly Canada/U.S. exchange rate (Canadian dollars to U.S. dollars) from January 1971 to July 2007, which are obtained from the Federal Reserve Bank of St. Louis.
   (a) Plot the time series and its ACF. Are there seasonal effects or unit-root nonstationary patterns in the series?
   (b) Using the rates from January 1971 to December 2006, build a time series model to forecast the rates in the next 6 months. You can use any of the techniques in Sections 5.1 and 5.2, and economic insights, if available, to build the forecasting model, but should explain your rationale.
   (c) Compute the k-months-ahead forecasts ($k = 1, 2, \ldots, 6$) based on your fitted model, using December 2006 as the forecast origin. Compare the forecasts with the actual exchange rates.

5.9. The first and the fifth columns of the file m_swap.txt contain the monthly swap rates (see Section 2.2.3) with 1- and 5-year maturities from July 2000 to May 2006. These data are obtained from www.Economagic.com.
   (a) Fit an ARIMA model to the 1-year swap rates using the data from July 2000 to December 2006.
   (b) For the 1-year rate, use your fitted model to compute k-months-ahead forecasts ($k = 1, \ldots, 5$) and their standard errors, choosing December 2006 as the forecast origin. Compare the forecasts with the actual swap rates, and the standard errors with the forecast errors (which are the differences between the predicted and actual rates).
   (c) Consider the spread (i.e., difference) between the 1-year and 5-year swap rates. Fit an ARIMA model to the spread using the data from July 2000 to December 2006. Compare the forecasts with the actual spreads.

5.10. The file m_sp500ret_3mtcm.txt contains three columns. The second column gives the monthly returns of the S&P 500 index from January 1994 to December 2006. The third column gives the monthly rates of the 3-month U.S. Treasury bill in the secondary market, which are obtained from the Federal Reserve Bank of St. Louis and used as the risk-free rate here. Consider the ten monthly log returns in the file m_logret_10stocks.txt.
   (a) For each stock, fit CAPM for the period from January 1994 to June 1998 and for the subsequent period from July 1998 to December 2006. Are your estimated betas significantly different for the two periods?
   (b) Consider the dynamic linear model (5.40) for CAPM with time-varying betas. Use the Kalman filter with $\sigma_w = 0.2$ to estimate β_t sequentially during the period July 1998–December 2006. The

estimated beta $\widehat{\beta}$ and error variance $\widehat{\sigma}^2$ obtained in (a) for the period from January 1994 to June 1998 can be used to initialize $\widehat{\beta}_0$ and to substitute for σ^2 in the Kalman filter.

(c) Compare and discuss your sequential estimates with the estimate of beta in (a) for the period July 1998 to December 2006.

6

Dynamic Models of Asset Returns and Their Volatilities

In Chapter 3 on single-period investment theories, historical asset returns are assumed to be i.i.d. This assumption, however, is sometimes violated, as illustrated in Section 5.1.2. On the other hand, since the historical asset returns are only used as a random sample to estimate the mean and covariance matrix of the asset returns in the single period for which investment plans are made, this simplified i.i.d. assumption is innocuous if one does not go too far back into the past, during which structural changes may have occurred.

Since subsequent chapters on finance models and applications will involve the dynamic evolution of asset returns and their volatilities, we consider here various statistical methods and models that have been developed to analyze these time series data. Although traditional time series methods such as those in Chapter 5 suffice for the dynamic levels of asset returns, they do not capture observed patterns of time-varying volatilities and volatility clustering in financial time series. Section 6.1 describes these patterns, often called *stylized facts*, of asset returns and their volatilities.

For the model of i.i.d. returns r_t in Chapter 3, volatility is defined as the standard deviation σ of r_t and can be estimated by the sample standard deviation $\widehat{\sigma}$. In the case of time-varying volatility σ_t, a simple modification of $\widehat{\sigma}$ is to use a moving average, with a sliding window or with exponentially decaying weights, instead of the usual arithmetic mean of $(r_i - \bar{r})^2$, $1 \leq i \leq n$. Section 6.2 elaborates on this idea. Section 6.3 considers conditional heteroskedastic models that are used to define σ_t^2 as the conditional variance of r_t given the past observations:

$$\sigma_t^2 = E\big\{[r_t - E(r_t|r_{t-1}, \dots, r_1)]^2 \big| r_{t-1}, \dots, r_1\big\}. \tag{6.1}$$

Section 6.4 considers joint modeling of $E(r_t|r_{t-1}, \dots, r_1)$ and σ_t^2. The conditional expectation in (6.1) can be computed from the model assumptions (which may involve unknown parameters) on the conditional distribution of

r_t given the observations up to time $t-1$. In particular, we give a detailed exposition of the conditional heteroskedastic models GARCH and EGARCH. Chapter 9 will describe other volatility models and introduce multivariate stochastic models for the conditional covariance matrix $\boldsymbol{\Sigma}_t$ of a vector of returns $\mathbf{x}_t$ on p assets at time t.

6.1 Stylized facts on time series of asset returns

Before summarizing various stylized facts from the literature on empirical analysis of asset returns and their volatilities, we begin by examining some asset returns data. Figure 6.1 plots the time series of daily log returns (adjusted for dividends) on Merck stock from January 11, 2000 to November 30, 2005. The log returns r_t appear to be stationary, fluctuating around 0, but there are several markedly large positive and negative values (especially the negative returns on September 29, 2004), while most of the other values are within ± 0.05. The top panel of Figure 6.2, which plots the weekly log returns of the NASDAQ index from the week starting on November 19, 1984 to the week starting on September 15, 2003, also shows a similar pattern that is "calm" most of the time but has big spikes at several instants, including one in the week of "Black Monday" in 1987 and the Internet bubble burst in January 2001.

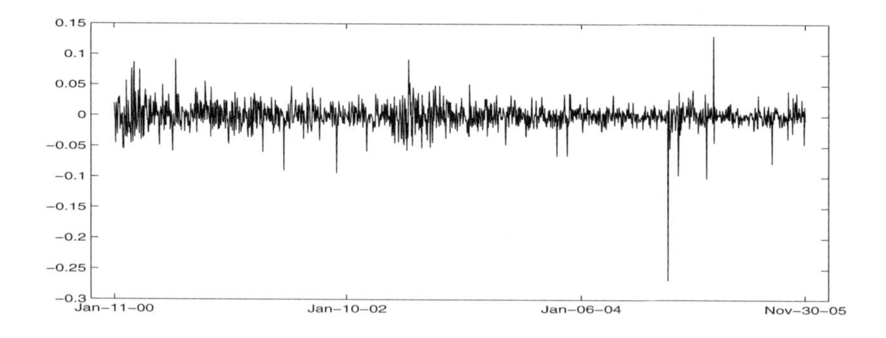

Fig. 6.1. Daily log returns on Merck stock.

The *skewness* of a random variable X with mean μ and variance σ^2 is defined as $E(X-\mu)^3/\sigma^3$, which is 0 when X is symmetric. The *kurtosis* of X is defined as $E(X-\mu)^4/\sigma^4$, which is equal to 3 when X is normal. For a sample of n independent observations from a distribution, the skewness and kurtosis of the parent distribution can be estimated by the sample skewness $\widehat{\text{sk}}$ and sample kurtosis $\widehat{\kappa}$ given by

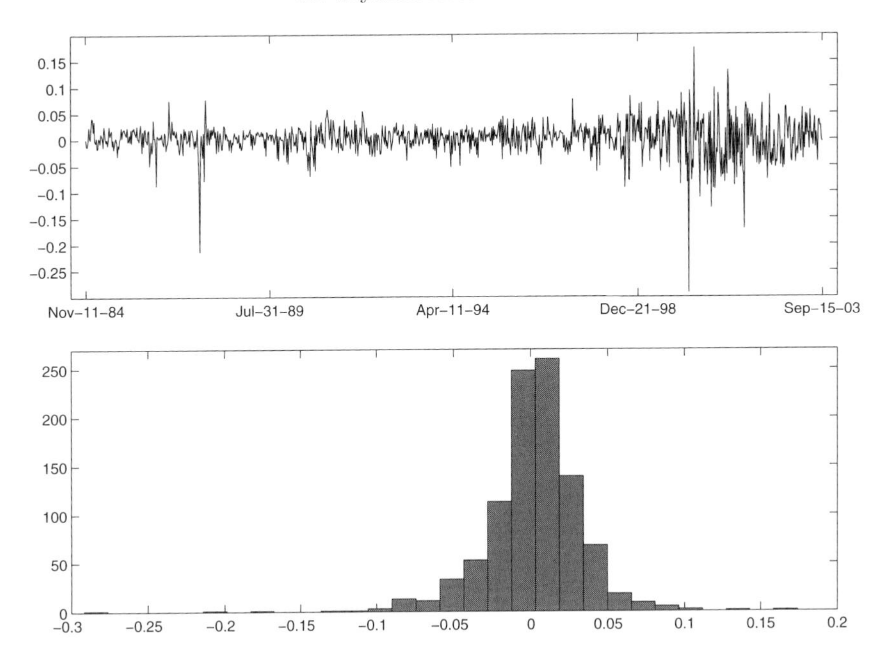

Fig. 6.2. Top: weekly log returns of NASDAQ. Bottom: histogram of weekly log returns.

$$\widehat{\text{sk}} = \frac{1}{n}\sum_{i=1}^{n}\frac{(r_i - \bar{r})^3}{\widehat{\sigma}^3}, \qquad \widehat{\kappa} = \frac{1}{n}\sum_{i=1}^{n}\frac{(r_i - \bar{r})^4}{\widehat{\sigma}^4}. \tag{6.2}$$

The bottom panel of Figure 6.2 exhibits a negatively skewed histogram of the data shown in the top panel, while Table 6.1 provides the mean $\bar{r}$, variance $\widehat{\sigma}^2$, skewness, and kurtosis of the NASDAQ returns data in Figure 6.2. A test for normality of the parent distribution based on the sample skewness $\widehat{\text{sk}}$ and sample kurtosis $\widehat{\kappa}$ is the *Jarque-Bera test* that uses the test statistic

$$\text{JB} = n\left(\frac{\widehat{\text{sk}}^2}{6} + \frac{(\widehat{\kappa} - 3)^2}{24}\right), \tag{6.3}$$

which has an approximately χ_2^2-distribution under the null hypothesis of normality. The JB statistic for the NASDAQ data in Figure 6.2 is 5839, which far exceeds 5.99, the 95th percentile of the χ_2^2-distribution. Figure 6.3 plots the autocorrelation functions for the log returns and squared log returns of the NASDAQ data. The dashed lines in both panels of Figure 6.3 represent rejection boundaries of an approximate 5%-level test of zero autocorrelation at the indicated lag; see Section 5.1.2. Note that the autocorrelations (at lags 1, 2, ...) are markedly smaller for r_t than for r_t^2. This is related to "volatility clustering," which will be explained below. The Ljung-Box statistic $Q(m)$ in Section 5.1 with $m = 25$ is 42.50 for r_t and 217.01 for r_t^2, corresponding to a p-value of 0.016 for r_t and near 0 for r_t^2 (a significant departure from the null hypothesis of independence).

Table 6.1. Summary statistics of NASDAQ weekly log returns.

Mean$\times 10^3$	Variance	Skewness	Kurtosis
2.056	0.010	−1.351	14.667

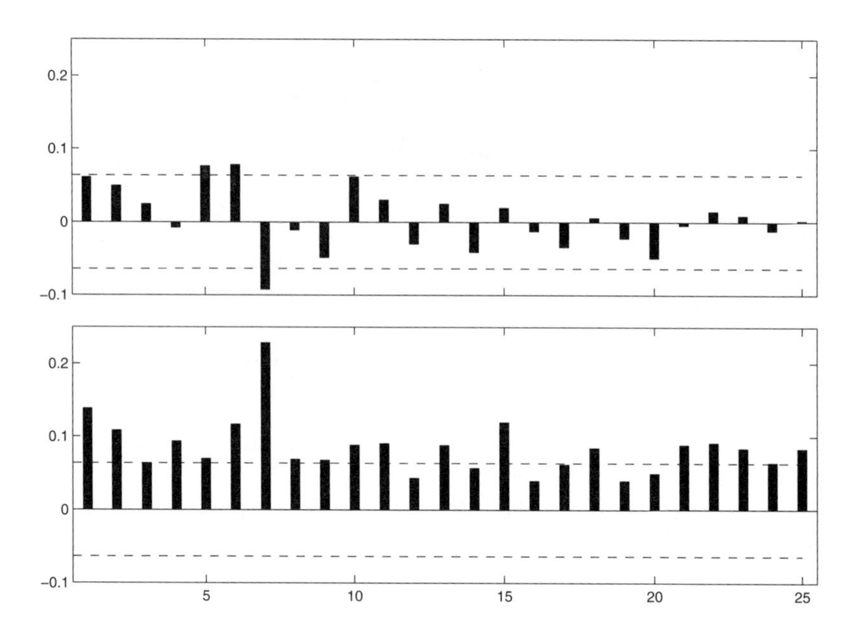

Fig. 6.3. Top: Autocorrelation function of NASDAQ log return series. Bottom: Autocorrelation function of the corresponding squared series.

Table 6.2 gives several summary statistics for the daily log returns (adjusted for dividends) of 16 U.S. stocks, including Merck, which is shown in Figure 6.1 for the period January 11, 2000 to November 30, 2005.

Volatility clustering

Many financial time series have exhibited clustering of large changes in returns, as illustrated in Figures 6.1 and 6.2. Such clustering results in much

Table 6.2. Statistics for 16 U.S. daily log returns (Jan. 11, 2000 – Nov. 30, 2005): American International Group, Inc. (AIG), American Express Company (AXP), AT&T Inc. (ATT), Boeing Aircraft Co. (BA), Walt Disney Co. (DIS), DuPont (DP), General Motors Corp. (GM), Home Depot Inc. (HD), Hewlett-Packard Co. (HP), International Business Machines Corp. (IBM), J.P. Morgan Chase (JPM), MacDonalds (MC), Merck & Co. Inc. (MRK), Microsoft Corp. (MSFT), Verizon Communications Inc. (VZ), and Wal-Mart Stores Inc. (WMT). Skew: skewness; Kurt: kurtosis; L-B$_1$: Ljung-Box statistic for r_t up to lag 10; L-B$_2$: Ljung-Box statistic for r_t^2 up to lag 10.

	Mean$\times 10^4$	SD$\times 10^2$	Skew	Kurt	L-B$_1$	L-B$_2$
AIG	1.542	1.921	0.164	6.195	12.26	334.5*
AXP	3.501	2.182	−0.055	6.062	25.18*	479.7*
ATT	−0.468	2.108	0.078	6.359	15.48	153.9*
BA	6.090	2.136	−0.415	7.679	15.16	223.6*
DIS	0.513	2.327	−0.065	9.395	12.37	59.71*
DP	−0.179	1.861	0.289	6.553	11.50	224.2*
GM	−3.617	2.313	0.199	8.550	17.65	69.63*
HD	0.858	2.507	−0.851	17.30	17.08	25.32*
HP	2.617	2.994	0.225	7.533	14.77	81.6*
IBM	0.496	2.092	0.175	9.004	19.53*	207.8*
JPM	2.141	2.416	0.329	9.250	21.43*	329.9*
MC	1.378	1.935	−0.089	7.080	6.32	87.8*
MRK	−2.536	1.985	−1.598	28.45	32.21*	2.4
MSFT	−0.921	2.360	0.105	10.53	21.50*	125.7*
VZ	−0.570	1.994	0.304	7.142	13.80	201.6*
WMT	0.092	1.959	0.267	6.140	21.27*	413.9*

*Significant at the 95% level (Ljung-Box test); $\chi^2_{10;0.95} = 18.3$.

stronger autocorrelations of the squared returns than the original returns, which are usually weakly autocorrelated. Volatility clustering is common in the intra-day, daily, and weekly returns in equity, commodity, and foreign exchange markets. Merck stock in Table 6.2 seems to be an exception to this pattern, but the nonsignificant Ljung-Box statistic L-B$_2$ for Merck's autocorrelations of r_t^2 appears to be caused by the outlier on September 29, 2004; see Figure 6.1. If we remove it from Merck's return series, then the Ljung-Box statistic L-B$_2$ increases to 30.86 and the large kurtosis of 28.45 drops to

8.01, which is much closer to the kurtosis values of most other stocks. This illustrates the well-known fact that moment statistics such as kurtosis and autocorrelations are sensitive to outliers.

Leptokurtic distributions

A distribution whose kurtosis exceeds 3 (which is the kurtosis of any normal distribution) is called *leptokurtic* and is usually associated with heavier tails than that of the normal distribution. We have pointed out in Section 3.1.2 that the classical model of asset price dynamics, namely the geometric Brownian motion model, implies normally distributed returns. One way that has been used to adjust for heavier tails in a normative market model, which assumes normality for analytic tractability in deriving pricing and risk management formulas, is to replace the normal distribution in the formula by a suitably chosen Student t-distribution, as will be illustrated in Section 6.3.

Asymmetry and leverage effect

Asymmetry of magnitudes in upward and downward movements of asset returns is of common occurrence in equity markets. In particular, the volatility response to a large positive return is considerably smaller than that to a negative return of the same magnitude. This asymmetry is sometimes referred to as a *leverage effect*. A possible cause of the asymmetry is that a drop in a stock price increases the debt-to-asset ratio of the stock, which in turn increases the volatility of returns to the equity holders. In addition, the news of increasing volatility makes the future of the stock more uncertain, and the asset price and its return therefore become more volatile.

Response to external events

As noted by Engle and Patton (2001), there are external events and exogenous variables that influence the volatility pattern of a stock return. Examples include scheduled earnings announcements of the company and macroeconomic variables.

6.2 Moving average estimators of time-varying volatilities

Historic volatility

Let σ_t be the volatility of r_t at time t. A commonly used estimate of σ_t^2 at time $t-1$ is the sample variance based on the most recent k observations:

$$\widehat{\sigma}_t^2 = \frac{1}{k-1} \sum_{i=1}^{k} (r_{t-i} - \bar{r})^2, \tag{6.4}$$

where $\bar{r} = \sum_{i=1}^{k} r_{t-i}/k$. An often-used rule of thumb to determine k is to set it equal to the number of days to which the volatility is to be applied for investment or risk management. We can convert (6.4) in the daily basis to the annual volatility by $\sqrt{A}\sigma$, where the annualizing factor A is the number of trading days, usually taken as around 252. The volatility estimate $\widehat{\sigma}_t$ given by (6.4) is called the *k-day historic volatility*. Figure 6.4 displays historic volatilities with three different window sizes k. All three estimates of σ_t capture important changes in the volatility process. In particular, they all show two big spikes, one in the week of Black Monday in 1987 and the other marking the Internet bubble burst around January 2000. Note that as the window size k increases, the time course of historic volatility becomes smoother.

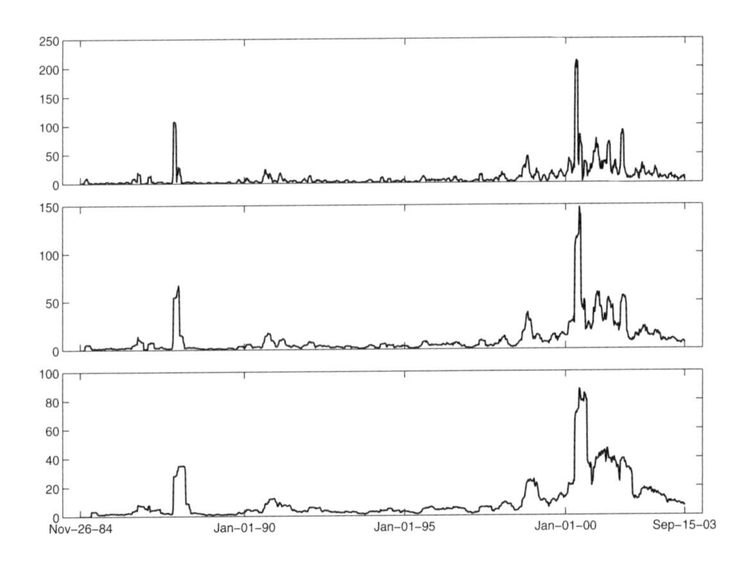

Fig. 6.4. Historic volatility estimated with different window sizes. Top: $k = 5$; middle: $k = 10$; bottom: $k = 20$.

More general weighting schemes

Historic volatility uses equal weights for the observations in the moving window of returns. Since the objective is to estimate the current level of volatility, one may want to put more weight on the most recent observations, yielding an estimate of the form

$$\widehat{\sigma}_t^2 = \sum_{i=1}^{k} \alpha_i u_{t-i}^2, \tag{6.5}$$

where $\sum_{i=1}^k \alpha_i = 1$ and u_m denotes the centered log return $r_m - \bar{r}$, or simply the log return r_m (since $\bar{r}^2$ is typically small in comparison with $k^{-1}\sum_{i=1}^k r_{m-i}^2$), or the one-period $P_m/P_{m-1} - 1$. A modification of (6.5) is to include a long-run variance rate V in the weighted sum, leading to

$$\widehat{\sigma}_t^2 = \gamma V + \sum_{i=1}^{k} \alpha_i u_{t-i}^2, \tag{6.6}$$

where $\gamma + \sum_{i=1}^k \alpha_i = 1$.

Exponentially weighted moving averages

The *exponentially weighted moving average* (EWMA) model is a particular case of (6.5) in which the weights α_i decrease exponentially fast. In this case, $\alpha_i = (1-\lambda)\lambda^{i-1}$ for some $0 < \lambda < 1$ and (6.5), with $k = \infty$ and $u_j = 0 = \widehat{\sigma}_j$ for $j \le 0$, has the recursive representation

$$\widehat{\sigma}_t^2 = \lambda \widehat{\sigma}_{t-1}^2 + (1-\lambda) u_{t-1}^2. \tag{6.7}$$

RiskMetrics, originally developed by J. P. Morgan and made publicly available in 1994, uses EWMA with $\lambda = 0.94$ for updating daily volatility estimates in its database. This value of λ was found to give forecasts of the variance rate that come closest to the equally weighted average of the u_i^2 on the subsequent 25 days. Note that (6.7) says that $\widehat{\sigma}_t^2$ is a convex combination of $\widehat{\sigma}_{t-1}^2$ and u_{t-1}^2. In analogy with (6.6), we can modify (6.7) by including also a long-run variance rate V:

$$\widehat{\sigma}_t^2 = \gamma V + \alpha u_{t-1}^2 + \beta \widehat{\sigma}_{t-1}^2 \qquad (\gamma + \alpha + \beta = 1). \tag{6.8}$$

6.3 Conditional heteroskedastic models

6.3.1 The ARCH model

Regarding the weights α_i and γ and the long-run variance rate V in the moving average scheme (6.6) as unknown parameters in a statistical model of the time-varying volatilities σ_t^2, we can estimate them by using maximum likelihood to fit the model to the observed data. One such model is Engle's (1982) *autoregressive conditional heteroskedastic* (ARCH) model, denoted by ARCH(k) and defined by

$$u_t = \sigma_t \epsilon_t, \qquad \sigma_t^2 = \omega + \sum_{j=1}^{k} \alpha_j u_{t-j}^2, \tag{6.9}$$

in which ϵ_t are i.i.d. random variables with mean 0 and variance 1 and have either the standard normal or the following standardized Student t-distribution. Let x_ν be a Student t-distribution with $\nu > 2$ degrees of freedom. Then $\epsilon_t = x_\nu / \sqrt{\nu/(\nu-2)}$ has a *standardized Student t-distribution* with variance 1 and probability density function

$$p(\epsilon) = \frac{\Gamma((\nu+1)/2)}{\Gamma(\nu/2)\sqrt{(\nu-2)\pi}} \left(1 + \frac{\epsilon^2}{\nu-2}\right)^{-(\nu+1)/2}. \tag{6.10}$$

Covariance stationary ARCH models

Since (6.9) is similar to the autoregressive model in Chapter 5, for u_t^2 to be covariance stationary, the zeros of the characteristic polynomial $1 - \alpha_1 z - \cdots - \alpha_k z^k$ are required to lie outside the unit circle. As the α_i are usually assumed to be nonnegative, this requirement is equivalent to

$$\alpha_1 + \cdots + \alpha_k < 1. \tag{6.11}$$

Under (6.11), the long-run variance V of u_t is given by

$$V = \frac{\omega}{1 - \alpha_1 - \cdots - \alpha_k}. \tag{6.12}$$

6.3.2 The GARCH model

Bollerslev (1986) introduced the *generalized ARCH* (GARCH) model, which is of the form

$$u_t = \sigma_t \epsilon_t, \quad \sigma_t^2 = \omega + \sum_{i=1}^{h} \beta_i \sigma_{t-i}^2 + \sum_{j=1}^{k} \alpha_j u_{t-j}^2, \tag{6.13}$$

where ϵ_t are i.i.d. with mean 0 and variance 1 and have either the standard normal or a standardized Student t-distribution. The GARCH(h, k) model (6.13) can be considered as an ARMA model of volatility with martingale difference innovations: Let $\beta_i = 0$ if $i > h$, $\alpha_j = 0$ if $j > k$, and $\eta_t = u_t^2 - \sigma_t^2$. Then η_t is a martingale difference sequence (see Appendix A), noting that $E(\eta_t | u_s, s \le t-1) = 0$ because $E(u_t^2 | u_s, s \le t-1) = \sigma_t^2$. Moreover,

$$u_t^2 = \omega + \sum_{j=1}^{\max(h,k)} (\alpha_j + \beta_j) u_{t-j}^2 + \eta_t - \sum_{i=1}^{h} \beta_i \eta_{t-i}. \tag{6.14}$$

Hence the same invertibility and stationarity assumptions of ARMA models apply to GARCH models. For example, to ensure that u_t is covariance stationary, it is required that all roots of $1-\sum_{j=1}^{\max(h,k)}(\alpha_j+\beta_j)z^j=0$ lie outside the unit circle; see Section 5.1.3. Since the α's and β's are usually assumed to be nonnegative, this is equivalent to

$$\sum_{j=1}^{k}\alpha_j+\sum_{i=1}^{h}\beta_i<1. \tag{6.15}$$

Under (6.15), the unconditional variance of u_t is given by

$$E(u_t^2)=\frac{\omega}{1-\sum_{i=1}^{h}\beta_i-\sum_{j=1}^{k}\alpha_j}. \tag{6.16}$$

The GARCH(1, 1) model $\sigma_t^2=\omega+\alpha u_{t-1}^2+\beta\sigma_{t-1}^2$ is often used to fit financial time series. It is closely related to the EWMA estimate (6.8), treating the α, β, and $\omega:=\gamma V$ in (6.8) as unknown parameters of the model. It implies that a large u_{t-1}^2 or σ_{t-1}^2 will lead to a large σ_t^2, which in turn will give rise to a large $u_t^2=\sigma_t^2\epsilon_t^2$. This is consistent with the volatility clustering observed in financial time series. Moreover, even when ϵ_t is standard normal, GARCH(1, 1) can still be highly leptokurtic since

$$\kappa:=\frac{E(u_t^4)}{[\mathrm{Var}(u_t)]^2}=\frac{3[1-(\alpha+\beta)^2]}{1-(\alpha+\beta)^2-2\alpha^2}>3 \tag{6.17}$$

when $(\alpha+\beta)^2+2\alpha^2<1$; see Exercise 6.1.

Forecasting future volatilities in GARCH(1, 1)

Let $\sigma^2=\omega/(1-\alpha-\beta)$. The GARCH(1, 1) model $\sigma_t^2=\omega+\alpha u_{t-1}^2+\beta\sigma_{t-1}^2$ can be written in the form

$$\sigma_t^2-\sigma^2=\alpha(u_{t-1}^2-\sigma^2)+\beta(\sigma_{t-1}^2-\sigma^2). \tag{6.18}$$

Replacing current time t by future time $t+k$ in (6.18) yields

$$\sigma_{t+k}^2-\sigma^2=\alpha(u_{t+k-1}^2-\sigma^2)+\beta(\sigma_{t+k-1}^2-\sigma^2).$$

Let $\mathcal{F}_t=\{u_s:s\le t\}$. Since ϵ_{t+k-1} is independent of $\mathcal{F}_{t+k-2}$ and has variance 1, the conditional expectation of $u_{t+k-1}^2(=\sigma_{t+k-1}^2\epsilon_{t+k-1}^2)$ given $\mathcal{F}_{t+k-2}$ is σ_{t+k-1}^2, so $E(\sigma_{t+k}^2-\sigma^2|\mathcal{F}_{t+k-2})=\lambda(\sigma_{t+k-1}^2-\sigma^2)$, where $\lambda=\alpha+\beta$. We can then take conditional expectation with respect to $\mathcal{F}_{t+k-3}$, etc., and finally arrive at

$$E(\sigma_{t+k}^2|\mathcal{F}_t)=\sigma^2+\lambda^{k-1}\{\alpha(u_t^2-\sigma^2)+\beta(\sigma_t^2-\sigma^2)\}. \tag{6.19}$$

Since $\sigma^2 = \omega/(1-\lambda)$, the k-steps-ahead forecast of σ_{t+k}^2 is given by

$$g_t(\boldsymbol{\theta}) := E(\sigma_{t+k}^2|\mathcal{F}_t) = \omega(1-\lambda^k)/(1-\lambda) + (\alpha u_t^2 + \beta\sigma_t^2)\lambda^{k-1}, \tag{6.20}$$

where $\boldsymbol{\theta} = (\omega, \alpha, \beta)^T$. Replacing the unknown $\boldsymbol{\theta}$ by the MLE $\widehat{\boldsymbol{\theta}}$ based on the observations up to time t yields the forecast $g_t(\widehat{\boldsymbol{\theta}})$.

To construct an approximately 95% confidence interval for $g_t(\widehat{\boldsymbol{\theta}})$, we can use the delta method as in Section 2.4.2 to derive for large n the standard normal approximation for

$$\left\{g_t(\widehat{\boldsymbol{\theta}}) - g_t(\boldsymbol{\theta})\right\} \Big/ \left\{\nabla^T g_t(\widehat{\boldsymbol{\theta}})(-\nabla^2 l(\widehat{\boldsymbol{\theta}}))^{-1}\nabla g_t(\widehat{\boldsymbol{\theta}})\right\}^{1/2},$$

where $l(\omega, \alpha, \beta) = -\frac{1}{2}\sum_{i=1}^n \log(2\pi\sigma_i^2) - \frac{1}{2}\sum_{i=1}^n u_i^2/\sigma_i^2$. In particular, for the special case $k = 1$ (corresponding to one-step-ahead forecasts), $g_t(\boldsymbol{\theta}) = \omega + \alpha u_t^2 + \beta\sigma_t^2$ and therefore $\nabla g_t(\boldsymbol{\theta}) = (1, u_t^2, \sigma_t^2)^T + \beta\nabla\sigma_t^2$, from which $\nabla\sigma_t^2$ can be computed recursively since $\nabla g_t(\boldsymbol{\theta}) = \nabla\sigma_{t+1}^2$.

Volatility persistence and half-life

Letting $\eta_t = u_t^2 - \sigma_t^2$ and $\lambda = \alpha + \beta$, we can express the GARCH(1, 1) model $\sigma_t^2 = \omega + \alpha u_{t-1}^2 + \beta\sigma_{t-1}^2$ in the form $(1-\lambda B)\sigma_t^2 = \omega + \alpha\eta_{t-1}$, where B is the backshift operator. Therefore, under the stationarity condition (6.15),

$$\sigma_t^2 = \omega/(1-\lambda) + \alpha[\eta_{t-1} + \lambda\eta_{t-2} + \cdots + \lambda^{j-1}\eta_{t-j} + \cdots]. \tag{6.21}$$

As λ approaches 1, the weight of the past innovation η_{t-j} in (6.21) also approaches 1, showing the persistent effect of events that occurred long ago on current volatility.

The *half-life* of current volatility is the smallest τ such that the difference between the predicted future variance $E(\sigma_{t+\tau}^2|\mathcal{F}_t)$ and the long-run variance level $\sigma^2 = \omega/(1-\lambda)$ falls below half of the difference $\sigma_{t+1}^2 - \sigma^2$. Applying the formula (6.19) for $E(\sigma_{t+\tau}^2|\mathcal{F}_t) - \sigma^2$ and setting it equal to $\frac{1}{2}(\sigma_{t+1}^2 - \sigma^2)$, we obtain that τ is the smallest positive integer such that $\lambda^{\tau-1} \le 1/2$. The half-life τ is often used as a measure of volatility persistence.

Likelihood inference and examples

Let $\boldsymbol{\theta} = (\omega, \alpha_1, \dots, \alpha_k, \beta_1, \dots, \beta_h)^T$. Assuming that the ϵ_t are i.i.d. $N(0,1)$, the log-likelihood function in the GARCH(1, 1) model (6.13) is given by

$$l(\boldsymbol{\theta}) = -\frac{1}{2}\sum_{t=1}^n \log\sigma_t^2 - \sum_{t=1}^n \frac{u_t^2}{2\sigma_t^2} - \frac{n}{2}\log(2\pi), \tag{6.22}$$

in which the σ_t^2 can be computed recursively by (6.13) when the initial values $\sigma_0^2, \ldots, \sigma_{1-h}^2$ and the value of $\boldsymbol{\theta}$ are given. Then MLE $\widehat{\boldsymbol{\theta}}$ is the solution of

$$\mathbf{0} = \nabla l(\boldsymbol{\theta}) = -\frac{1}{2}\sum_{t=1}^{n}\left[\frac{1}{\sigma_t^2} - \frac{u_t^2}{\sigma_t^4}\right]\nabla\sigma_t^2, \tag{6.23}$$

in which $\nabla\sigma_t^2$ can be evaluated recursively by

$$\nabla\sigma_t^2 = (1, u_{t-1}^2, \ldots, u_{t-k}^2, \sigma_{t-1}^2, \ldots, \sigma_{t-h}^2)^T + \sum_{i=1}^{h}\beta_i\nabla\sigma_{t-i}^2. \tag{6.24}$$

A variant of the GARCH model above assumes that the ϵ_t follow a standardized Student t-distribution with density function given by (6.10), in which $\nu \geq 2$ is treated as an unknown parameter. The underlying motivation is to allow a heavier tail in the distribution of u_t^2 and to use the data to determine how heavy the tail should be. In this case, the likelihood function is

$$\begin{aligned} l(\boldsymbol{\theta}, \nu) = n\log\left(\frac{\Gamma((\nu+1)/2)}{\Gamma(\nu/2)\sqrt{(\nu-2)\pi}}\right) - \frac{1}{2}\sum_{t=1}^{n}\log\sigma_t^2 \\ -\frac{\nu+1}{2}\sum_{t=1}^{n}\log\left(1 + \frac{u_t^2}{(\nu-2)\sigma_t^2}\right). \end{aligned}$$

We can use the function `garchfit` in the `MATLAB GARCH` toolbox to compute the components of the MLE and their standard errors in a general GARCH model with prespecified distributions of ϵ.

Example 6.1. For the daily log returns of the 16 U.S. stocks from January 2000 to November 2005 in Table 6.2, we use `garchfit` to estimate the parameters of the GARCH model

$$r_t = \mu + \sigma_t\epsilon_t, \qquad \sigma_t^2 = \omega + \beta\sigma_{t-1}^2 + \alpha(r_{t-1} - \mu)^2, \tag{6.25}$$

in which the ϵ_t are standard normal. Table 6.3 gives the estimated parameters and their standard errors. Note that the MLE of μ, which appears in both equations of (6.25), differs somewhat from the sample mean (which is a method-of-moments estimate but not the MLE of μ) given in Table 6.2. Moreover, the estimated $\lambda = \alpha + \beta$ is close to 1 in all cases except Merck, which has an outlier on September 29, 2004, as we have noted in Section 6.1. Volatility persistence is measured by the last column, HL (half-life, in days, at the end of the sample period), in Table 6.3. Some of the half-lives are over 10^3 business days (or over 4 years). Further discussion and alternatives to such "long memory" in volatility will be given in Chapter 9 (Section 9.5).

Table 6.3. Estimated GARCH parameters for 16 U.S. stocks; standard errors are given in parentheses. The symbol + denotes "larger than" the listed number.

	$10^4\mu$	$10^6\omega$	α	β	$\alpha+\beta$	HL
AIG	3.86	9.89	0.114	0.862	0.976	41.7
	(4.19)	(1.60)	(0.015)	(0.017)		
AXP	3.50	2.32	0.094	0.904	0.999	10^3+
	(5.67)	(0.564)	(0.010)	(0.010)		
ATT	2.01	0.239	0.036	0.964	0.999	10^3+
	(3.63)	(0.254)	(0.006)	(0.006)		
BA	13.8	10.4	0.101	0.878	0.979	47.6
	(4.49)	(2.60)	(0.009)	(0.013)		
DIS	5.71	4.94	0.096	0.904	0.999+	10^5+
	(4.71)	(1.15)	(0.008)	(0.007)		
DP	3.37	2.36	0.062	0.933	0.995	200
	(4.04)	(0.724)	(0.007)	(0.007)		
GM	3.39	11.0	0.064	0.918	0.982	55.6
	(5.80)	(1.00)	(0.008)	(0.007)		
HD	5.32	0.458	0.033	0.967	0.999+	10^5+
	(4.69)	(0.351)	(0.004)	(0.004)		
HP	2.62	1.78	0.013	0.984	0.997	333
	(7.78)	(0.344)	(0.002)	(0.002)		
IBM	5.57	4.17	0.125	0.874	0.999	10^3+
	(3.60)	(0.820)	(0.011)	(0.010)		
JPM	3.51	0.762	0.0633	0.937	0.999+	10^5+
	(3.75)	(0.456)	(0.007)	(0.007)		
MC	4.90	1.97	0.038	0.956	0.995	200
	(4.35)	(0.649)	(0.005)	(0.005)		
MRK	−2.50	155	0.074	0.537	0.611	2.57
	(5.29)	(45.2)	(0.024)	(0.133)		
MSFT	1.83	1.93	0.092	0.908	0.999+	10^5+
	(3.65)	(0.604)	(0.007)	(0.007)		
VZ	0.34	1.44	0.090	0.910	0.999+	10^5+
	(3.44)	(0.615)	(0.012)	(0.011)		
WM	−0.69	0.905	0.040	0.957	0.997	333
	(36.1)	(0.368)	(0.006)	(0.007)		

Table 6.4. Estimated ARCH(3) parameters for NASDAQ weekly log returns.

	$10^3\mu$	$10^4\omega$	α_1	α_2	α_3
Gaussian	4.482	1.973	0.366	0.345	0.244
	(0.622)	(0.171)	(0.045)	(0.031)	(0.047)
Student-t	4.331	2.082	0.326	0.268	0.323
	(0.644)	(0.313)	(0.070)	(0.060)	(0.078)

Example 6.2. Note that the ARCH model (6.9) is a special case of GARCH models with $h = 0$. For the NASDAQ weekly log returns in Figure 6.2, we fit the ARCH(3) model

$$r_t = \mu + \sigma_t \epsilon_t, \qquad \sigma_t^2 = \omega + \sum_{i=1}^{3} \alpha_i u_{t-i}^2,$$

by using `garchfit`, assuming the ϵ_t to be standard normal or to have a standardized Student t-distribution. Table 6.4 gives the estimated parameters and their standard errors in parentheses, with the degree of freedom of the t-distribution also treated as a parameter. The estimated degree of freedom is 6.49, and its standard error is 1.24.

6.3.3 The integrated GARCH model

Note that 12 of the 16 stocks in Table 6.3 have high volatility persistence since the $\alpha + \beta$ values in the fitted GARCH(1, 1) models are close to 1 (exceeding 0.99). If $\alpha + \beta = 1$, the GARCH(1, 1) model is of the form $\sigma_t^2 = \omega + \beta \sigma_{t-1}^2 + (1 - \beta) u_{t-1}^2$, which corresponds to the exponentially weighted moving average model (6.7) with $\lambda = \beta$ when $\omega = 0$. More generally, the *integrated GARCH* model IGARCH(h, k) is of the form (6.13) with $\sum_{j=1}^k \alpha_j + \sum_{i=1}^h \beta_i = 1$. Although IGARCH($h$, k) models have infinite unconditional variance (as can be seen by letting $\sum_{i=1}^h \beta_i + \sum_{j=1}^k \alpha_j$ approach 1 in (6.16)) and therefore cannot be covariance stationary, they are in fact strictly stationary. The k-steps-ahead forecast of σ_{t+k}^2 at time t is given by letting $\lambda \to 1$ in (6.20), yielding

$$E(\sigma_{t+k}^2 | \mathcal{F}_t) = k\omega + \beta \sigma_t^2 + (1 - \beta) u_t^2. \tag{6.26}$$

6.3.4 The exponential GARCH model

As pointed out in Section 6.1, a stylized fact of the volatility of asset returns is that the volatility response to a large positive return is considerably smaller than that of a negative return of the same magnitude. The GARCH model, which is defined by σ_t^2 and u_{t-j}^2, cannot incorporate this leverage effect. To accommodate the asymmetry, Nelson (1991) proposed the *exponential GARCH* model EGARCH(h, k) that has the form

$$u_t = \sigma_t \epsilon_t, \quad \log(\sigma_t^2) = \omega + \sum_{i=1}^{h} \beta_i \log(\sigma_{t-i}^2) + \sum_{j=1}^{k} f_j(\epsilon_{t-j}), \tag{6.27}$$

where the ϵ_t are i.i.d. with mean 0 and $f_j(\epsilon) = \alpha_j \epsilon + \gamma_j(|\epsilon| - E|\epsilon|)$. Note that the random variable $f_j(\epsilon_t)$ is the sum of two zero-mean random variables $\alpha_j \epsilon_t$ and $\gamma_j(|\epsilon_t| - E|\epsilon_t|)$. We can rewrite $f_j(\epsilon_t)$ as

$$f_j(\epsilon_t) = \begin{cases} (\alpha_j + \gamma_j)\epsilon_t - \gamma_j E|\epsilon_t|, & \text{if } \epsilon_t \geq 0, \\ (\alpha_j - \gamma_j)\epsilon_t - \gamma_j E|\epsilon_t|, & \text{if } \epsilon_t < 0, \end{cases}$$

which shows the asymmetry of the volatility response to positive and negative returns. Since (6.27) represents $\log(\sigma_t^2)$ in ARMA form with innovations $f_j(\epsilon_{t-j})$, σ_t^2 and therefore u_t also are stationary if the zeros of $1 - \beta_1 z - \cdots - \beta_h z^h$ lie outside the unit circle; see Section 5.1.3.

Forecasting future volatilities in EGARCH(1, 1)

To fix the ideas, we consider the EGARCH(1, 1) model with standard normal ϵ_t, for which $\log(\sigma_t^2) = \omega + \beta \log(\sigma_{t-1}^2) + f(\epsilon_{t-1})$, or equivalently

$$\sigma_t^2 = \left(\sigma_{t-1}\right)^{2\beta} \exp\left\{\omega + f(\epsilon_{t-1})\right\}, \tag{6.28}$$

where $f(\epsilon) = \alpha\epsilon + \gamma(|\epsilon| - \sqrt{2/\pi})$. Define $g(c) = E\big\{\exp[c(\omega + f(\epsilon))]\big\}$ for $\epsilon \sim N(0,1)$ and $c > 0$. We can evaluate $g(c)$ explicitly in terms of the standard normal distribution function Φ by

$$\begin{aligned} g(c) &= e^{c\omega} \int_{-\infty}^{\infty} e^{-\epsilon^2/2} \exp\left\{c\alpha\epsilon + c\gamma(|\epsilon| - \sqrt{2/\pi})\right\} d\epsilon / \sqrt{2\pi} \\ &= e^{c(\omega - \gamma\sqrt{2/\pi})} \Big\{ e^{(\alpha+\gamma)^2 c^2/2} \Phi\big[c(\gamma+\alpha)\big] + e^{(\alpha-\gamma)^2 c^2/2} \Phi\big[c(\gamma-\alpha)\big] \Big\}. \end{aligned} \tag{6.29}$$

Let $\mathcal{F}_t = \{\epsilon_s : s \leq t\}$. In view of (6.28), we can modify the arguments in (6.18) and (6.19) for the GARCH(1, 1) model to obtain

$$\begin{aligned} E(\sigma_{t+k}^2 | \mathcal{F}_t) &= E\Big\{ \left(\sigma_{t+k-1}^2\right)^{\beta} \big| \mathcal{F}_t \Big\} E\big\{\exp[\omega + f(\epsilon)]\big\} = \dots \\ &= g(1) g(\beta) \dots g(\beta^{k-1}) \left(\sigma_t^2\right)^{\beta^k}, \end{aligned} \tag{6.30}$$

in which $g(\cdot)$ is given by (6.29).

Likelihood inference and implementation

Let $\boldsymbol{\theta} = (\omega, \alpha_1, \dots, \alpha_k, \gamma_1, \dots, \gamma_k, \beta_1, \dots, \beta_h)^T$. Given initial values $\sigma_0^2, \dots, \sigma_{1-h}^2$ and assuming that ϵ_t are standard normal, the log-likelihood function $l(\boldsymbol{\theta})$ has the same form as (6.22), where σ_t^2 is defined recursively by (6.27) and $E|\epsilon_t| = \sqrt{2/\pi}$. The MLE $\widehat{\boldsymbol{\theta}}$ is the solution of

$$\mathbf{0} = \boldsymbol{\nabla} l(\boldsymbol{\theta}) = -\frac{1}{2} \sum_{t=1}^{n} \left[\frac{1}{\sigma_t^2} - \frac{u_t^2}{\sigma_t^4} \right] \boldsymbol{\nabla} \sigma_t^2, \tag{6.31}$$

where $\boldsymbol{\nabla}\sigma_t^2$ can be calculated recursively by the following analog of (6.24):

$$\begin{aligned} \boldsymbol{\nabla}\sigma_t^2 &= \boldsymbol{\nabla} \left\{ \exp\left(\omega + \sum_{j=1}^{k} \left[\alpha_j \epsilon_{t-j} + \gamma_j \left(|\epsilon_{t-j}| - \sqrt{\frac{2}{\pi}} \right) \right] + \sum_{i=1}^{h} \beta_i \log \sigma_{t-i}^2 \right) \right\} \\ &= \sigma_t^2 \left\{ \left(1, \epsilon_{t-1}, \dots, \epsilon_{t-k}, |\epsilon_{t-1}| - \sqrt{2/\pi}, \dots, |\epsilon_{t-k}| - \sqrt{2/\pi}, \right. \right. \\ &\qquad \left. \left. \log \sigma_{t-1}^2, \dots, \log \sigma_{t-h}^2 \right)^T + \sum_{i=1}^{h} \frac{\beta_i}{\sigma_{t-i}^2} \boldsymbol{\nabla} \sigma_{t-i}^2 \right\}. \end{aligned}$$

We can use the function `garchfit` in the `GARCH` toolbox of `MATLAB` to fit EGARCH models.

Example 6.3. Consider the weekly log returns of the closing prices of NASDAQ from the week starting on November 19, 1984 to the week starting on September 15, 2003 in Figure 6.2. The bottom panel of the figure shows that the histogram has a heavier left tail than right tail. This asymmetry between the left and right tails suggests fitting to these data the EGARCH(1, 1) model

$$y_t = \mu + \sigma_t \epsilon_t, \quad \log(\sigma_t^2) = \omega + \beta \log(\sigma_{t-1}^2) + \alpha \epsilon_{t-1} + \gamma(|\epsilon_{t-1}| - E|\epsilon_{t-1}|), \tag{6.32}$$

where ϵ_t are i.i.d. standard normal random variables. Table 6.5 shows the estimated parameters and their standard errors (in parentheses) computed by `garchfit`. Figure 6.5 plots the time series of the fitted $\widehat{\sigma}_t$, $\widehat{\epsilon}_t$, and $\widehat{\sigma}_t \widehat{\epsilon}_t$.

It is natural to ask whether a higher-order GARCH model would significantly improve the fit. In particular, consider the EGARCH(2, 1) model

$$\log(\sigma_t^2) = \omega + \beta_1 \log(\sigma_{t-1}^2) + \beta_2 \log(\sigma_{t-2}^2) + \alpha \epsilon_{t-1} + \gamma(|\epsilon_{t-1}| - E|\epsilon_{t-1}|), \tag{6.33}$$

whose parameter estimates (by maximum likelihood) are given in Table 6.5. We can use the function `lratiotest` in the `GARCH` toolbox to perform a GLR test for $H_0 : \beta_2 = 0$. The GLR statistic is 0.3316, with a p-value of 0.5647 (based on the χ_1^2 approximation to the GLR statistic). Hence EGARCH(2, 1) does not significantly improve the EGARCH(1, 1) fit to these data.

Table 6.5. Estimated EGARCH(1, 1) and EGARCH(2, 1) parameters.

	μ	ω	α	γ	β_1	β_2
EGARCH(1, 1)	0.341	0.100	−0.107	0.380	0.950	
	(0.065)	(0.019)	(0.020)	(0.033)	(0.010)	
EGARCH(2, 1)	0.340	0.102	−0.110	0.394	0.885	0.064
	(0.065)	(0.020)	(0.024)	(0.046)	(0.139)	(0.138)

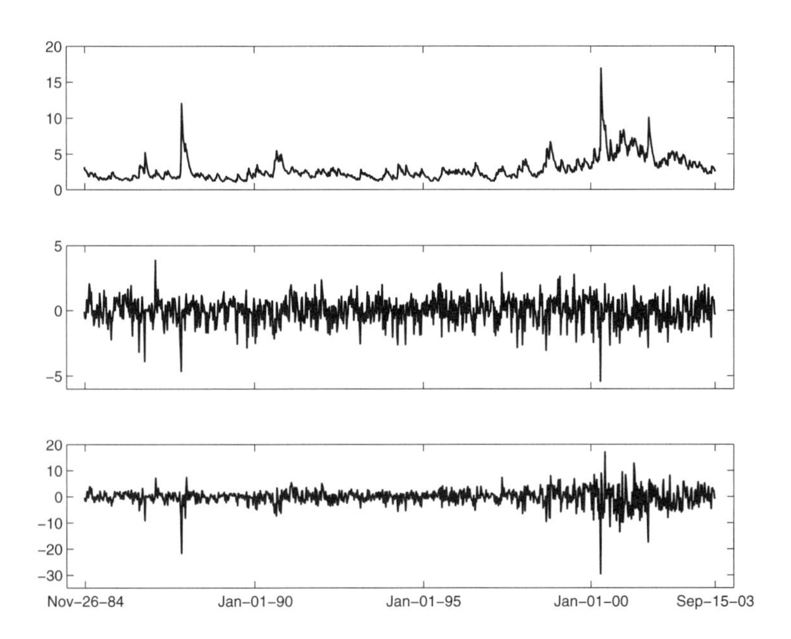

Fig. 6.5. EGARCH(1, 1) dynamics for NASDAQ weekly returns. Top panel: the fitted $\widehat{\sigma}_t$; middle panel: the fitted $\widehat{\epsilon}_t$; bottom panel: the fitted $\widehat{\sigma}_t\widehat{\epsilon}_t$.

6.4 The ARMA-GARCH and ARMA-EGARCH models

The linear time series models in Chapter 5 can be combined with GARCH or EGARCH to model the dynamics of asset returns and their volatilities. A simple special case has been considered in (6.25) that assumes a GARCH model for $r_t - \mu$, where μ is the mean of the stock returns r_t. Assuming that r_t follows an ARMA model with GARCH innovations yields the following ARMA(p,q)-GARCH(h,k) model for (r_t, σ_t):

$$r_t = \phi_0 + \sum_{i=1}^{p} \phi_i r_{t-i} + u_t + \sum_{j=1}^{q} \psi_j u_{t-j}, \qquad u_t = \sigma_t \epsilon_t,$$

$$\sigma_t^2 = \omega + \sum_{j=1}^{k} \alpha_j u_{t-j}^2 + \sum_{i=1}^{h} \beta_i \sigma_{t-i}^2. \tag{6.34}$$

The ϵ_t in (6.34) are i.i.d. standard normal or standardized Student-t random variables. The stationarity conditions for the ARMA part and the GARCH part are the same as those in Sections 5.1.3 and 6.3.2. Replacing the second equation in (6.34) by (6.27) yields the ARMA(p, q)-EGARCH(h, k) model.

6.4.1 Forecasting future returns and volatilities

Given the model parameters, the two equations in (6.34) can be used to obtain one-step-ahead forecasts of the conditional mean and conditional variance of r_t. Specifically, from Sections 5.1.3 and 6.3.2, we have the one-step-ahead forecasts

$$\widehat{r}_{t+1|t} = \phi_0 + \sum_{i=1}^{p} \phi_i r_{t+1-i} + \sum_{j=1}^{q} \psi_j u_{t+1-j}, \tag{6.35}$$

$$\widehat{\sigma}_{t+1|t}^2 = \omega + \sum_{j=1}^{k} \alpha_j u_{t+1-j}^2 + \sum_{j=1}^{h} \beta_i \sigma_{t+1-i}^2. \tag{6.36}$$

The conditional distribution of r_{t+1} given the current and past observations up to time t is $N(\widehat{r}_{t+1|t}, \widehat{\sigma}_{t+1|t}^2)$.

The k-steps-ahead forecast $\widehat{r}_{t+k|t}$ can be evaluated by using the method for the ARMA model r_t described in Section 5.1.4. To evaluate the k-steps-ahead forecast $\widehat{\sigma}_{t+k|t}^2$, we first use the MA($\infty$) representation $r_t = \mu + \sum_{i=0}^{\infty} \psi_i u_{t-i}$ with $\psi_0 = 1$ (see Section 5.1.3). Therefore the k-steps-ahead forecast of r_{t+k} can be expressed as $\widehat{r}_{t+k|t} = \mu + \sum_{i=k}^{\infty} \psi_i u_{t+k-i}$, and the corresponding forecast error is $r_{t+k} - \widehat{r}_{t+k|t} = \sum_{i=1}^{k} \psi_{k-i} u_{t+i}$. From this and the property $E(u_{t+i}^2 | \mathcal{F}_{t+i-1}) = \sigma_{t+i}^2$, which has been used to derive (6.19), it follows that

$$\widehat{\sigma}_{t+k|t}^2 = \sum_{i=1}^{k} \psi_{k-i}^2 E(\sigma_{t+i}^2 | \mathcal{F}_t). \tag{6.37}$$

We can then use the volatility forecasts in Section 6.3.2 for the GARCH model to evaluate $E(\sigma_{t+i}^2 | \mathcal{F}_t)$ in (6.37).

6.4.2 Implementation and illustration

Maximum likelihood can be used to estimate the parameters of ARMA-GARCH or ARMA-EGARCH models. The MLE can be computed by using the `garchfit` function in the `GARCH` toolbox of `MATLAB`.

Example 6.4. We fit an AR(2)-GARCH(1, 1) model to the NASDAQ weekly log returns data in Figure 6.2, assuming the ϵ_t to be standard normal or to have a standardized Student t-distribution. Table 6.6 gives the estimated parameters and their standard errors (in parentheses), with the degree of freedom of the t-distribution also treated as a parameter. The estimated degree of freedom is 6.72 and its standard error is 1.23.

Table 6.6. Estimated AR(2)-GARCH(1, 1) parameters for NASDAQ returns.

	$10^3\mu$	ϕ_1	ϕ_2	$10^5\omega$	α	β
Gaussian	3.20	0.092	0.077	2.333	0.224	0.772
	(0.623)	(0.037)	(0.036)	(0.608)	(0.019)	(0.020)
Student-t	3.32	0.096	0.070	1.466	0.133	0.856
	(0.655)	(0.033)	(0.033)	(0.615)	(0.027)	(0.028)

Exercises

6.1. Consider the GARCH(1, 1) model $u_t = \sigma_t \epsilon_t$, $\sigma_t^2 = \omega + \beta\sigma_{t-1}^2 + \alpha u_{t-1}^2$, where $\omega > 0$, $\alpha \geq 0$, and ϵ_t are i.i.d. standard normal random variables. Show that the kurtosis of u_t is given by (6.17).

6.2. Consider an AR(1)-GARCH(1, 1) model

$$r_t = \phi r_{t-1} + u_t, \quad u_t = \sigma_t \epsilon_t, \quad \sigma_t^2 = \omega + \alpha u_{t-1}^2 + \beta\sigma_{t-1}^2,$$

where ϵ_t are i.i.d. standard normal random variables.

(a) Derive the log-likelihood function of the data.

(b) Derive the skewness and kurtosis of r_t.

6.3. The file `w_logret_3stocks.txt` contains the weekly log returns of three stocks (Citigroup Inc., General Motors, and Pfizer Inc.) from the week of January 4, 1982 to the week of May 21, 2007.

(a) Compute the sample mean, variance, skewness, excess kurtosis, and Ljung-Box statistic up to lag 10 for each return series.

(b) Plot the histograms of these returns and the squared returns.

(c) For each stock, perform the Jarque-Bera test of the null hypothesis of normally distributed log returns.

(d) For each stock, plot the ACFs of the return series and the squared return series and compare them.

6.4. The file `intel_d_logret.txt` contains the daily log returns of Intel stock from July 9, 1986 to June 29, 2007.

(a) Do you find any evidence of conditional heteroskedasticity in the time series? Use the Ljung-Box statistic up to lag 10 and the ACF plots of the returns and the squared returns to draw your conclusion.

(b) Build a GARCH(1, 1) model with Gaussian innovations for the return series. Use the Ljung-Box statistic up to lag 10 to test if the estimated innovations $\widehat{\epsilon}_t$ are serially correlated. Plot the ACF of $\widehat{\epsilon}_t$ and the rejection boundaries of a 5%-level test of zero autocorrelation at each lag.

(c) Use the fitted GARCH(1, 1) model in (b) to obtain k-days-ahead volatility forecasts on June 29, 2007 as the forecast origin ($k = 1, \dots, 10$) and give 95% confidence intervals for these forecasts.

6.5. The file `ibm_w_logret.txt` contains the weekly log returns on IBM stock from the week of January 2, 1962 to the week of June 18, 2007. Build a GARCH(1, 1) model with standardized Student-t innovations with ν degrees of freedom for the time series, with ν also treated as an unknown parameter. Give standard errors of the parameter estimates.

6.6. Consider the monthly log returns on GM (General Motors) stock from January 1962 to June 2007 in the file `gm_m_logret.txt`.

(a) Fit an EGARCH(1, 1) model to the data, and provide standard errors of the parameter estimates.

(b) Compute k-months-ahead volatility forecasts ($k = 1, \dots, 10$) using June 2007 as the forecast origin, and give 95% confidence intervals for these forecasts.

6.7. The file `sp500_d_logret.txt` contains the daily log returns on the S&P 500 index from January 3 1980 to June 28, 2007.

(a) Fit an AR(1)-GARCH(1, 1) model with Gaussian innovations to the data, and give standard errors of the parameter estimates.

(b) Compute k-days-ahead forecasts ($k = 1, \dots, 5$) of the log return and its volatility, using the fitted model and June 28, 2007 as the forecast origin.

6.8. Consider the weekly log returns on the S&P 500 index from the week of January 3, 1950 to the week of June 18, 2007 in the file `sp500_w_logret.txt`.

(a) Fit the GARCH(1, 1) model

$$r_t = \mu + u_t, \quad u_t = \sigma_t \epsilon_t, \quad \sigma_t^2 = \omega + \alpha u_{t-1}^2 + \beta \sigma_{t-1}^2 \tag{6.38}$$

to the series, where ϵ_t are i.i.d. standard normal random variables. Provide the standard errors of the parameter estimates.

(b) Plot the estimated conditional volatilities $\widehat{\sigma}_t$ and innovations $\widehat{\epsilon}_t$ over time. At what times do you find atypically large values of $\widehat{\sigma}_t$? Relate these times to the historical events during the sample period.

6.9. The file `sp500_m_logret.txt` contains the monthly log returns on the S&P 500 index from the month of January 1950 to the month of June 2007.

(a) Fit the GARCH(1, 1) model (6.38), in which ϵ_t are i.i.d. standard normal, to these data. Give standard errors of the estimated parameters.

(b) Note that even though the sample period is the same as that in Exercise 6.8, a different timescale (months) is used here. Compare the results in (a) with those in Exercise 6.8(a), and discuss the implications.

6.10. The file `nasdaq_w_logret.txt` contains the weekly log returns on the NASDAQ index during the period November 11, 1984 to September 15, 2003.

(a) Fit a GARCH(1, 1) model with Gaussian innovations to the entire series.

(b) Divide the sample period into four parts: November 19, 1984 to June 16, 1987; June 9, 1987 to August 15, 1990; August 8, 1990 to March 13, 1998; and March 6, 1998 to September 15, 2003. Fit a GARCH(1, 1) model with Gaussian innovations to the NASDAQ weekly log returns during each period.

(c) Compare the results in (a) and (b), and discuss the possibility of parameter changes and their implications on volatility persistence.

Part II

Advanced Topics in Quantitative Finance

While Part I of the book can be used as a first course on statistical methods in finance, Part II is intended for a one-semester second course. As pointed out in the Preface, an important objective of this second course is to link the theory and formulas students learn from mathematical finance courses to data from financial markets. Without requiring readers to take the mathematical finance courses first, the book summarizes the key concepts and models and relates them to market data, thereby giving a focused and self-contained treatment of finance theory and statistical analysis. For readers who have not taken previous courses in quantitative finance and who would like to supplement this focused treatment with more detailed background, especially with respect to the institutional issues and the derivatives markets, Hull (2006) provides an excellent reference.

Part II begins with a substantive-empirical modeling approach to addressing the discrepancy between finance theory (subject-matter model) and empirical data by regarding the subject-matter model as one of the basis functions in a nonparametric regression model. Nonparametric regression is introduced in Chapter 7, and Chapter 9 describes advanced multivariate and time series methods in financial econometrics. The other chapters in Part II are devoted to applications to quantitative finance. Chapter 8 considers option pricing and applies the substantive-empirical modeling approach. Chapter 10 introduces the interest rate market and applies multivariate time series methods. Chapter 11 discusses statistical trading strategies and their evaluation. It also considers statistical modeling and analysis of high-frequency data in real-time trading. Chapter 12 introduces Value at Risk and other measures of market risk, describes statistical methods and models for their analysis, and considers backtesting and stress testing of internal models from which banks calculate their capital requirements under the Basel Accord.

7

Nonparametric Regression and Substantive-Empirical Modeling

The regression models in Chapters 1 and 4 are parametric, involving an unknown regression parameter $\boldsymbol{\theta} \in \mathbb{R}^p$ in $y_t = f(\boldsymbol{\theta}, \mathbf{x}_t) + \epsilon_t$. Nonparametric regression dispenses with the finite-dimensional parameter $\boldsymbol{\theta}$, replacing $f(\boldsymbol{\theta}, \mathbf{x}_t)$ by $f(\mathbf{x}_t)$, in which f itself is regarded as an unknown infinite-dimensional parameter. Since one only has a finite number of observations $(\mathbf{x}_1, y_1), \ldots, (\mathbf{x}_n, y_n)$, it is unrealistic to expect that the infinite-dimensional f can be estimated well unless f can be closely approximated by functions which involve a finite number (depending on n) of parameters that can be well estimated. Note that a similar problem also arises in Section 5.1.1 in connection with estimating the autocovariance function and the spectral density function of a weakly stationary sequence based on n successive observations from the sequence and that window estimates are used to take advantage of the summability (and hence rapid decay) of the autocorrelation coefficients.

Estimating only an approximation to f (rather than f itself) leads to bias of the estimate $\widehat{f}(\mathbf{x})$. The mean squared error $E\{\widehat{f}(\mathbf{x}) - f(\mathbf{x})\}^2$ can be decomposed as a sum of the squared bias $\left[E\big(\widehat{f}(\mathbf{x}) - f(\mathbf{x})\big)\right]^2$ and $\mathrm{Var}(\widehat{f}(\mathbf{x}))$. Section 7.3 describes the bias-variance trade-off in coming up with efficient nonparametric estimators $\widehat{f}$. An essential ingredient is the "smoothing parameter" that $\widehat{f}$ uses, but the optimal choice of the smoothing parameter depends on the unknown f. Model selection techniques that generalize those in Section 1.3 are described in Section 7.3 to address this problem.

Section 7.1 formulates the regression function $f(\mathbf{x})$ as a conditional expectation and thereby links nonparametric regression to minimum-variance prediction of y given $\mathbf{x}$, as in Section 1.5.1, which, however, is restricted only to linear predictors (i.e., linear functions of $\mathbf{x}$). Without assuming linearity or other parametric representations of the regression function f, Section 7.2 describes several methods to estimate f nonparametrically in the case of univariate x. Section 7.4 generalizes some of these methods to the case of multivariate predictors $\mathbf{x}$ and discusses the "curse of dimensionality" for nonparametric regression on multidimensional $\mathbf{x}$: Whereas in the one-dimensional case the observed values $x_1, \ldots, x_n$ can be tightly packed within an interval covering most of their range, the observed vectors $\mathbf{x}_1, \ldots, \mathbf{x}_n$ may be sparsely dispersed in multidimensional space. Choosing suitable basis functions to approximate f is an important first step in estimating f nonparametrically in the case of multivariate predictors.

Subject-matter knowledge can often help in developing a parsimonious set of basis functions to approximate f. Section 7.5 describes a combined substantive-empirical approach that uses both subject-matter knowledge and nonparametric modeling. This approach will be used in Chapter 8 for bridging the gap between option pricing theory and observed option prices.

7.1 Regression functions and minimum-variance prediction

Given a sample of n observations $(\mathbf{x}_i, y_i)$, $i = 1, \cdots, n$, in which the components of $\mathbf{x}_i = (x_{i1}, \cdots, x_{ip})^T$ are the observed values of the p-dimensional explanatory variable (or predictor) and the y_i are the observed values of the response variable, a nonparametric regression model is expressed as

$$y_i = f(\mathbf{x}_i) + \epsilon_i, \tag{7.1}$$

where the function f is assumed to be smooth but unknown and the ϵ_i are unobservable random disturbances that are i.i.d. with mean 0 and variance σ^2. In time series applications, the i.i.d. assumption is often weakened to weak stationarity; see Appendix B. The $\mathbf{x}_i$ are random vectors such that $\mathbf{x}_i$ is independent of ϵ_i but may depend on $\epsilon_1, \ldots, \epsilon_{i-1}$. Hence the regression function is the conditional expectation

$$f(\mathbf{x}) = E(y_i | \mathbf{x}_i = \mathbf{x}), \tag{7.2}$$

which is the minimum-variance predictor of y_i from $\mathbf{x}_i$.

Nonparametric regression basically consists of using a finite-dimensional approximation of the regression function f and estimating the parameters of the finite-dimensional approximation. In particular, if one uses the *linear basis approximation*

$$f(\mathbf{x}) \approx \sum_{m=1}^{M} \beta_m g_m(\mathbf{x}), \tag{7.3}$$

where $g_m : \mathbb{R}^p \longrightarrow \mathbb{R}$ is a known function, then one can perform least squares regression of y_i on $\big(g_1(\mathbf{x}_i), \ldots, g_M(\mathbf{x}_i)\big)^T$, $1 \le i \le n$, to obtain the OLS estimates $\widehat{\beta}_1, \ldots, \widehat{\beta}_M$.

Definition 7.1. Given data $(\mathbf{x}_i, y_i)$ from the regression model (7.1), a vector of estimates $\widehat{\mathbf{Y}} = \left(\widehat{f}(\mathbf{x}_1), \ldots, \widehat{f}(\mathbf{x}_n)\right)^T$ of $\left(f(\mathbf{x}_1), \ldots, f(\mathbf{x}_n)\right)^T$ is called a *linear smoother* if it can be expressed as $\widehat{\mathbf{Y}} = \mathbf{SY}$, where $\mathbf{Y} = (y_1, \ldots, y_n)^T$ and $\mathbf{S} = (S_{ij})_{1 \le i,j \le n}$ is a matrix, called the *smoother matrix*, constructed from the $\mathbf{x}_i$ such that

$$\mathbf{S1} = \mathbf{1}, \qquad \text{where } \mathbf{1} = (1, \ldots, 1)^T. \tag{7.4}$$

7.2 Univariate predictors

7.2.1 Running-mean/running-line smoothers and local polynomial regression

Definition 7.2. Let $x_1, \ldots, x_n$ denote the observed values of a univariate explanatory variable. For $x \in \mathbb{R}$, the k *nearest neighbors* of x are the k observed values x_i that are closest to x.

Consider the problem of estimating $f(x)$ in the regression model (7.1) based on $(x_i, y_i), 1 \le i \le n$. Let $N_k(x)$ denote the set of i's such that x_i, $i \in N_k(x)$, are the k nearest neighbors of x. The mean of $\{y_i : i \in N_k(x)\}$ provides an estimate $\widehat{f}(x)$ that is similar to the moving average (5.20) or (5.21), for which we have evenly spaced x_i. This mean is called a *running-mean smoother*. Alternatively we can estimate $f(x)$ by fitting the regression line $y_i = \alpha + \beta x_i + \epsilon_i$ to the data $\{(x_i, y_i) : i \in N_k(x)\}$ and using the OLS estimates $\widehat{\alpha}$ and $\widehat{\beta}$ to estimate $f(x)$ by $\widehat{\alpha} + \widehat{\beta} x$. This regression line is called a *running-line smoother*. How k should be chosen for these smoothers will be addressed in Section 7.3.

Instead of using OLS, it seems more appropriate to use GLS, weighting the observations by some function of $x_i - x$. The *locally weighted running-line smoother* (`loess` in `R`) uses GLS to fit a straight line to $\{(x_i, y_i) : i \in N_k(x)\}$ by choosing α and β to minimize

$$\sum_{i \in N_k(x)} w_i \left[y_i - (\alpha + \beta x_i)\right]^2, \tag{7.5}$$

where $w_i = K\left(|x_i - x| / \max_{j \in N_k(x)} |x_j - x|\right)$ and K is the *tri-cube function* defined by

$$K(t) = (1 - t^3)^3 \qquad \text{for } \ 0 \le t \le 1; \qquad K(t) = 0 \qquad \text{elsewhere.} \tag{7.6}$$

The tri-cube function is a popular choice for kernels; see Section 7.2.2 on the choice of kernels in kernel smoothers.

Running-line smoothers therefore estimate $f(x)$ by using locally weighted linear functions. The basic idea can be readily generalized to *local polynomials.* Local polynomial regression of any degree d involves choosing α, $\beta_1, \dots, \beta_d$ to minimize

$$\sum_{i \in N_k(x)} w_i \big[y_i - (\alpha + \beta_1 x_i + \cdots + \beta_d x_i^d)\big]^2, \qquad (7.7)$$

in which the w_i are the same as those in locally weighted running-line smoothers.

7.2.2 Kernel smoothers

A *kernel smoother* estimates $f(x)$ by the weighted average

$$\widehat{f}(x) = \sum_{i=1}^n y_i K\Big(\frac{x - x_i}{\lambda}\Big) \Big/ \sum_{i=1}^n K\Big(\frac{x - x_i}{\lambda}\Big), \qquad (7.8)$$

where $\lambda > 0$ is the *bandwidth* and K is the *kernel,* which is often chosen to be a smooth even function. Some popular choices are the tri-cube kernel (7.6), the standard normal density $K(t) = e^{-t^2/2}/\sqrt{2\pi}$, and the *Epanechnikov kernel* $K(t) = \frac{3}{4}(1 - t^2)\mathbf{1}_{\{|t| \le 1\}}$. Note that the normal density has unbounded support, whereas the Epanechnikov and tri-cube kernels have compact support (which is needed when used with the nearest-neighbor set of size k). The `R` or `Splus` function `ksmooth` computes the kernel smoother at $x \in \{x_1, \dots, x_n\}$.

Kernel density estimation

Let $x_1, \dots, x_n$ be a random sample drawn from a distribution with density function f_X. A *kernel estimate* of f_X is of the form

$$\widehat{f}_X(x) = \frac{1}{n\lambda} \sum_{i=1}^n K_\lambda\Big(\frac{x - x_i}{\lambda}\Big), \qquad (7.9)$$

where K is a probability density function (nonnegative and $\int_{-\infty}^{\infty} K(t)dt = 1$) and $\lambda > 0$ is the bandwidth.

7.2.3 Regression splines

If we partition the domain of x into subintervals and represent (7.2) by different polynomials of the same degree in different intervals, then a piecewise polynomial function can be obtained. In particular, if there are K breakpoints (called *knots*) $\eta_1 < \cdots < \eta_K$ in the domain of x, then the piecewise polynomial $f(x)$ can be written as the linear regression function

$$f(x) = \sum_{k=1}^{K+1} (\beta_{k0} + \beta_{k1}x + \cdots + \beta_{kM}x^M)\mathbf{1}_{\{\eta_{k-1} \le x < \eta_k\}}, \tag{7.10}$$

where $\eta_0 = -\infty$, $\eta_{K+1} = \infty$, and therefore the parameters of (7.10) can be estimated by the method of least squares. Figure 7.1 illustrates this by fitting piecewise linear and piecewise cubic polynomials to simulated data (shown in circles). In particular, the piecewise linear function in the left panel corresponds to using instead of x the regressor

$$\left(\mathbf{1}_{\{x<\eta_1\}},\ x\mathbf{1}_{\{x<\eta_1\}},\ \mathbf{1}_{\{\eta_1 \le x < \eta_2\}},\ x\mathbf{1}_{\{\eta_1 \le x < \eta_2\}},\ \mathbf{1}_{\{\eta_2 \le x\}},\ x\mathbf{1}_{\{\eta_2 \le x\}}\right)^T$$

in the linear regression model of Chapter 1.

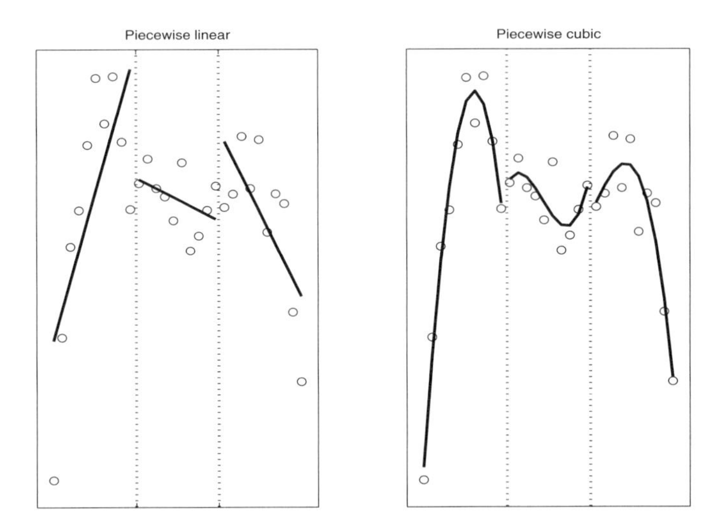

Fig. 7.1. Fitted piecewise polynomials.

In many applications to finance, the regression functions are required to be smooth, unlike those in Figure 7.1, which have discontinuities. In particular, we require $f(x)$ to have continuous derivatives up to order $M-1$; that is,

$$f^{(l)}(\eta_k+) = f^{(l)}(\eta_k-), \qquad k = 1, \cdots, K; l = 0, 1, \cdots, M-1. \tag{7.11}$$

The piecewise polynomial (7.10) that satisfies the smoothness constraint (7.11) is called a *spline* of degree M. It can be represented as a linear combination of $K + M + 1$ basis functions

$$g_j(x) = x^j,\ j = 0, \cdots, M; \quad g_{l+M}(x) = (x - \eta_l)_+^M,\ l = 1, \cdots, K. \tag{7.12}$$

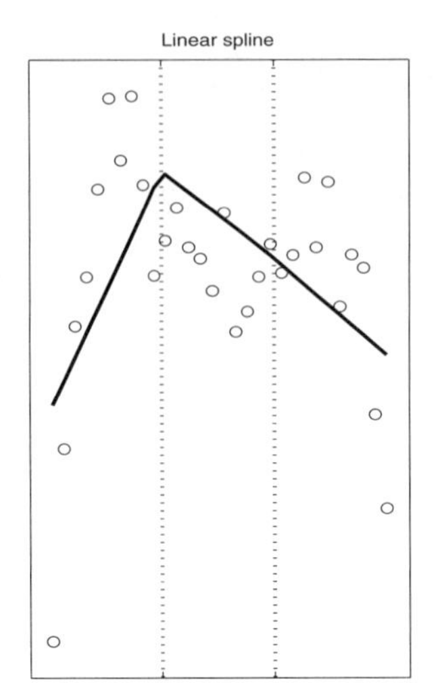

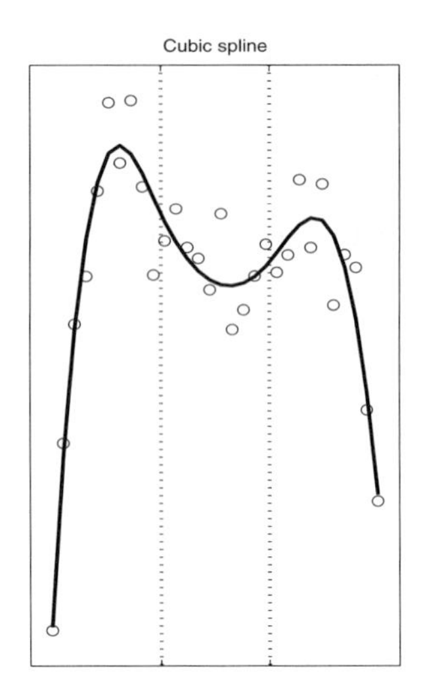

Fig. 7.2. Fitted linear and cubic splines.

In particular, for linear splines ($M = 1$) with $K = 2$ knots η_1 and η_2, the basis functions are 1, x, $(x - \eta_1)_+$, and $(x - \eta_2)_+$. For cubic splines ($M = 3$) with knots η_1 and η_2, the basis functions are 1, x, x^2, x^3, $(x - \eta_1)_+^3$, and $(x - \eta_2)_+^3$. Figure 7.2 plots the linear and cubic splines with two knots fitted to the data in Figure 7.1. Splines with specified knots are also called *regression splines*, and least squares regression can be used to estimate the coefficients associated with the basis functions. In practice, one seldom uses $M > 3$, as it is difficult to identify discontinuities in derivatives of order higher than 2.

Knot placement

A practical issue when working with regression splines is selecting the number and positions of the knots. Given the number K of knots, a simple approach to knot placement is to put two knots (called *boundary knots*) at the extremes of the observed predictor values and the remaining knots (called *interior knots*) at the $k/(K-1)$ quantiles of these data, $1 \leq k \leq K-2$. A simpler way, referred to as *cardinal splines*, is to put the $K - 2$ interior knots uniformly within the range of the observed predictor values. The choice of the number K of knots will be discussed in Section 7.3.

Natural cubic splines

A *natural cubic spline* is a cubic spline f satisfying the additional constraint $f'' = f''' = 0$ beyond the boundary knots. Therefore, instead of extrapolation to a cubic polynomial outside the range of the data, which often results in erratic behavior of the fitted spline near the extremes of the observed predictor values, a linear extrapolation is used to alleviate this problem. Whereas a cubic spline with K knots has $K+4$ basis functions (see (7.12) with $M=3$), the additional two constraints at the boundary knots for a natural cubic spline give up 4 degrees of freedom, yielding K basis functions

$$g_1(x) = 1, \quad g_2(x) = x, \quad g_k(x) = d_k(x) - d_{K-1}(x) \text{ for } 2 \le k \le K-1, \tag{7.13}$$

where $d_k(x) = \{(x-\eta_k)_+^3 - (x-\eta_K)_+^3\}/(\eta_K - \eta_k)$ and the knots are arranged in increasing order of magnitude, with η_1 and η_K being the boundary knots.

B-splines

The B-spline basis functions provide a numerically superior basis to (7.12), which is called the *truncated power basis*. Their main characteristic is that every basis function B_j is nonzero over a span of at most five distinct knots, in contrast with the basis function $(x-\eta_l)_+^M$, which is positive to the right of η_l. The `R` (or `Splus`) function `bs` generates B-spline function values that can be used for least squares estimates of the coefficients of the basis functions. The `R` (or `Splus`) function `ns` generates the B-splines that are linear beyond the boundary knots.

7.2.4 Smoothing cubic splines

Unlike the explicit regression cubic splines in Section 7.2.3, a smoothing cubic spline is defined implicitly via the optimization criterion that minimizes

$$\sum_{i=1}^{n}(y_i - f(x_i))^2 + \lambda \int_a^b [f''(u)]^2 du \tag{7.14}$$

over f, where $a = x_1 \le \cdots \le x_n = b$ and λ is called a *smoothing parameter*, which is a positive constant that penalizes the "roughness" of f. For the case $\lambda = 0$, (7.14) reduces to the residual sum of squares and f can be any function that interpolates the data. For the case $\lambda = \infty$, $f''(u)$ has to be zero everywhere so that the problem reduces to least squares linear regression. The first term in (7.14) measures the closeness of the fitted model to the data, and the second term penalizes the curvature of the function.

The criterion (7.14) is defined on an infinite-dimensional space of functions for which its second term is finite. It is remarkable that (7.14) has a unique and explicit, finite-dimensional minimizer, which is a natural cubic spline with knots $\eta_1, \ldots, \eta_K$ at the K distinct values of x_i $(1 \le i \le n)$. Hence the optimization criterion (7.14) can be expressed as

$$\mathrm{RSS}(\boldsymbol{\beta}) = (\mathbf{Y} - \mathbf{G}\boldsymbol{\beta})^T(\mathbf{Y} - \mathbf{G}\boldsymbol{\beta}) + \lambda\boldsymbol{\beta}^T\boldsymbol{\Omega}\boldsymbol{\beta}, \tag{7.15}$$

where $\mathbf{G} = (g_i(\eta_j))_{1\le i,j\le K}$, $\boldsymbol{\Omega} = (\int g_i''(u)g_j''(u)du)_{1\le i,j\le K}$, and the g_i are the natural cubic spline basis functions in (7.13). The minimizer of (7.15) is given by

$$\widehat{\boldsymbol{\beta}} = (\mathbf{G}^T\mathbf{G} + \lambda\boldsymbol{\Omega})^{-1}\mathbf{G}^T\mathbf{Y}; \tag{7.16}$$

see Exercise 7.2. The fitted smoothing spline is then $\widehat{f}(x) = \sum_{j=1}\widehat{\beta}_j g_j(x)$. The `R` (or `Splus`) function `smooth.spline` can be used to fit smoothing splines to data. It allows λ to be user-specified, with the default option of choosing λ by cross-validation, which will be considered in the next section.

7.3 Selection of smoothing parameter

The dataset $\{(\mathbf{x}_i, y_i) : 1 \le i \le n\}$ on which an estimate $\widehat{f}$ of the regression function f is based is called a *training sample.* The bias $b(\mathbf{x})$ of $\widehat{f}(\mathbf{x})$ is $E\{\widehat{f}(\mathbf{x}) - f(\mathbf{x})\}$, and the mean squared error (MSE) has the decomposition

$$\mathrm{MSE}(\mathbf{x}) = E\big[\widehat{f}(\mathbf{x}) - f(\mathbf{x})\big]^2 = b^2(\mathbf{x}) + \mathrm{Var}\big(\widehat{f}(\mathbf{x})\big). \tag{7.17}$$

When the $\mathbf{x}_i$ are sampled from a population with distribution function G, the *integrated mean squared error* (IMSE) is $\int E\big[\widehat{f}(\mathbf{x}) - f(\mathbf{x})\big]^2 dG(\mathbf{x})$. Since G is unknown, replacing it by the empirical distribution function leads to the average squared error (ASE) as a sample analog of the IMSE:

$$\mathrm{ASE} = n^{-1}\sum_{i=1}^{n}\big[\widehat{f}(\mathbf{x}_i) - f(\mathbf{x}_i)\big]^2. \tag{7.18}$$

The prediction performance of $\widehat{f}(\mathbf{x}_i)$ relates to how well it can predict a future observation $y_i^* = f(\mathbf{x}_i) + \epsilon_i^*$ from the regression model (7.1), where ϵ_i^* is independent of the training sample. The *prediction squared error* (PSE) is related to the ASE by

$$\mathrm{PSE} = n^{-1}\sum_{i=1}^{n} E\Big\{\big[y_i^* - \widehat{f}(\mathbf{x}_i)\big]^2\big|\mathbf{x}_1,\ldots,\mathbf{x}_n, y_1,\ldots,y_n\Big\} = \mathrm{ASE} + \sigma^2 \tag{7.19}$$

since ϵ_i^* is independent of $\widehat{f}(\mathbf{x}_i) - f(\mathbf{x}_i)$.

7.3.1 The bias-variance trade-off

As pointed out in Section 7.1, a nonparametric estimate of f basically consists of first approximating f by a function that involves a finite number of parameters and then estimating the parameter of the approximation. The approximation introduces bias, which decreases with increasing dimensionality of the approximation. On the other hand, the estimate based on a lower-dimensional approximation (that has fewer parameters to estimate) has smaller variance. There is therefore a trade-off between bias and variance in coming up with a suitable approximation to f. In particular, for linear smoothers in the case of univariate predictors, the squared bias increases but the variance decreases with the amount of smoothing. The smoothing parameter λ, which controls the amount of smoothing, is the bandwidth for kernel smoothers, roughness penalty for smoothing splines, or the relative neighborhood size k/n for nearest-neighbor methods. Increasing the number of knots in regression splines leads to a better approximation of f, and therefore reduces the bias, but introduces more parameters to be estimated and thereby increases the variance of $\widehat{f}(\mathbf{x}_i)$.

7.3.2 Cross-validation

The basic idea of cross-validation is to replace $(y_i^*, \widehat{f}(\mathbf{x}_i))$ by the already observed $(y_i, \widehat{f}_{(-i)}(\mathbf{x}_i))$, where $\widehat{f}_{(-i)}$ is the nonparametric regression estimate based on $\{(\mathbf{x}_j, y_j) : j \neq i\}$ (i.e., with $(\mathbf{x}_i, y_i)$ removed from the training sample). This idea is often called the *leave-one-out* or *jackknife* method. It yields the *cross-validation sum of squares*

$$\mathrm{CV} = \frac{1}{n}\sum_{i=1}^{n}\left[y_i - \widehat{f}_{(-i)}(\mathbf{x}_i)\right]^2 \tag{7.20}$$

as an estimate of the PSE. Note that CV attempts to estimate the out-of-sample error in contrast with the residual sum of squares (RSS), $\sum_{i=1}^{n}\left[y_i - \widehat{f}_i(\mathbf{x}_i)\right]^2$, which measures the in-sample error. The idea is similar to that of using jackknife (studentized) residuals in lieu of standardized residuals in Section 1.4.1. The choice of the smoothing parameter λ is often based on $\mathrm{CV}(\lambda)$, which depends on λ through $\widehat{f}_{(-i)}$, choosing the λ that gives the smallest $\mathrm{CV}(\lambda)$.

For linear smoothers, CV can be computed directly from $\widehat{f}$ without having to recompute $\widehat{f}_{(-i)}$ for every $i = 1, \ldots, n$. Let $\mathbf{S} = (S_{ij})_{1\le i,j\le n}$ be the smoother matrix. Then

$$\mathrm{CV} = \frac{1}{n}\sum_{i=1}^{n}\left(\frac{y_i - \widehat{f}(\mathbf{x}_i)}{1 - S_{ii}}\right)^2, \tag{7.21}$$

which is a consequence of the *Sherman-Morrison identity*

$$y_i - \widehat{f}_{(-i)}(\mathbf{x}_i) = \frac{y_i - \widehat{f}(\mathbf{x}_i)}{1 - S_{ii}}; \tag{7.22}$$

see Exercise 7.3.

Generalized cross-validation

For linear smoothers, the smoother matrix $\mathbf{S}$ is a projection matrix associated with linear regression. As shown in Section 1.4.1, it is of the form $\mathbf{U}(\mathbf{U}^T\mathbf{U})^{-1}\mathbf{U}^T$, whose trace is

$$\mathrm{tr}\Big(\mathbf{U}(\mathbf{U}^T\mathbf{U})^{-1}\mathbf{U}^T\Big) = \mathrm{tr}(\mathbf{I}_d) = d, \tag{7.23}$$

where $\mathbf{U}^T\mathbf{U}$ is a nonsingular $d \times d$ matrix. Therefore it is often considerably easier to compute their average, $n^{-1}\mathrm{tr}(\mathbf{S})$, than the diagonal elements S_{ii} of $\mathbf{S}$. The generalized cross-validation replaces S_{ii} in (7.21) by $\mathrm{tr}(\mathbf{S})/n$, yielding

$$\mathrm{GCV} = \frac{1}{n}\sum_{i=1}^{n}\left(\frac{y_i - \widehat{f}(\mathbf{x}_i)}{1 - \mathrm{tr}(\mathbf{S})/n}\right)^2. \tag{7.24}$$

The quantity $\mathrm{tr}(\mathbf{S})$ is the *effective number of parameters* (d in (7.23)), also called the *degrees of freedom*, of the smoother.

7.4 Multivariate predictors

7.4.1 Tensor product basis and multivariate adaptive regression splines

A simple method to extend the spline basis in Section 7.2.3 from univariate to multivariate prediction is to use tensor products of univariate basis functions. To fix the idea, consider bivariate predictors. Let $h_{1i}(x_1)$, $1 \le i \le m_1$, be the basis functions associated with the first predictor and $h_{2j}(x_2)$, $1 \le j \le m_2$, be those associated with the second predictor. Then a *tensor product basis* for the bivariate predictor $\mathbf{x}$ is defined by

$$f_{ij}(\mathbf{x}) = h_{1i}(x_1)h_{2j}(x_2), \quad 1 \le i \le m_1, 1 \le j \le m_2. \tag{7.25}$$

This idea can be readily generalized from 2 to d predictors. Note, however, that the number of basis functions grows exponentially with d, causing the "curse of dimensionality."

The *multivariate adaptive regression spline* (MARS), introduced by Friedman (1991), uses the tensor product basis formed from univariate regression splines and performs forward stepwise regression to add basis functions to the model sequentially up to a prespecified maximum number of terms. To avoid overfitting after these sequentially chosen tensor products of univariate splines are included in the model, it performs backward elimination based on a GCV criterion. It can be implemented in `R` by `mars`. Below is a summary of the procedure.

Let T_j denote the set of observed values $x_{1j}, \ldots, x_{nj}$ of the jth input variable. The forward stepwise procedure of MARS starts by including the constant function $h_0(\mathbf{x})$ in the basis set. At every stage, it adds to the current basis set $\mathcal{M}$ two basis functions of the form

$$h(\mathbf{x})(x_j - t)_+, \quad h(\mathbf{x})(t - x_j)_+ \qquad (\text{with } h \in \mathcal{M} \text{ and } t \in T_j), \tag{7.26}$$

that produce the largest decrease in the residual sum of squares when the outputs y_i are regressed on the $|\mathcal{M}|+2$ inputs, which are these basis functions evaluated at $\mathbf{x}_i$. (The notation $|\mathcal{M}|$ denotes the number of elements of $\mathcal{M}$.) An important feature of MARS therefore is that it uses a *reflected pair* of univariate linear splines, $(x_j - t)_+$ and $(t - x_j)_+$, to form the tensor product basis.

The backward elimination procedure of MARS removes terms from the model sequentially until the generalized cross-validation criterion

$$\mathrm{GCV}(\lambda) = \frac{\sum_{i=1}^n (y_i - \widehat{f}_\lambda(\mathbf{x}_i))^2}{n(1 - M(\lambda)/n)^2} \tag{7.27}$$

is minimized, where λ stands for the number of terms in the model and $M(\lambda) = r + cK$, in which r is the number of linearly independent basis functions in the model, K is the number of knots selected by the forward procedure, and c is chosen to be 3 (or 2 in certain cases) on the basis of theoretical considerations and simulation studies.

7.4.2 Additive regression models

An *additive regression model* involving p predictors has the form

$$f(\mathbf{x}) = \alpha + f_1(x_1) + \cdots + f_p(x_p), \tag{7.28}$$

where $\mathbf{x} = (x_1, \ldots, x_p)^T$ and the f_j's are unspecified smooth functions. Nonparametric estimation of the functions $f_1, \ldots, f_p$ after setting $\widehat{\alpha} = n^{-1} \sum_{i=1}^n y_i$ can be carried out by the *backfitting algorithm*, which is an iterative procedure that is initialized by setting $\widehat{f_1} = \cdots = \widehat{f_p} = 0$. At the mth iteration, for $k = 1, \ldots, p$, obtain a new estimate $\widehat{f_k}$ of f_k by applying a univariate smoother (e.g., cubic smoothing spline) to $\{(x_{ik}, y_i - \widehat{\alpha} - \sum_{j \neq k} \widehat{f_j}(x_{ij})) : 1 \le i \le n\}$. This iterative procedure is continued until the functions $\widehat{f_k}$ change by less than a prespecified threshold. The procedure is implemented by the function `gam` in `R` or `Splus`. In particular, if a cubic smoothing spline is used as the smoother, the backfitting algorithm above is a solution to the problem of minimizing

$$\mathrm{RSS}(\alpha, f_1, \ldots, f_p) := \sum_{i=1}^{n} \left\{ y_i - \alpha - \sum_{j=1}^{p} f_j(x_{ij}) \right\}^2 + \sum_{j=1}^{p} \lambda_j \int [f_j''(u)]^2 du, \tag{7.29}$$

subject to the identifiability constraints $\sum_{i=1}^{n} f_j(x_{ij}) = 0$ for $j = 1, \ldots, p$.

7.4.3 Projection pursuit regression

A *projection pursuit regression* (PPR) model assumes a regression function of the form

$$f(\mathbf{x}) = \sum_{m=1}^{M} g_m(\mathbf{w}_m^T \mathbf{x}), \tag{7.30}$$

in which g_m are unspecified smooth functions and $\mathbf{w}_m$ are unspecified unit vectors. To estimate g_m and $\mathbf{w}_m$, first consider the case $M = 1$. The function g and the direction $\mathbf{w}$ can be estimated from $(\mathbf{x}_i, y_i)$, $1 \le i \le n$, by an iterative procedure that consists of the following two steps at each iteration. For a given direction $\mathbf{w}$, apply a univariate smoother (e.g., cubic smoothing spline) to update the estimate of g_1. For a given g, we can regard $\mathbf{w}$ as the regression parameter in nonlinear least squares regression of y_i on $g(\mathbf{w}^T \mathbf{x}_i / ||\mathbf{w}||)$, ignoring the unit-length constraint on $\mathbf{w}$. This idea can be combined with stagewise regression to estimate the smooth functions and the directions in (7.30) when $M > 1$, adding a pair $(g_m, \mathbf{w}_m)$ at each stage. The $\mathbf{w}_j$'s obtained in previous stages are not adjusted, and the backfitting algorithm described in Section 7.4.2 can be used to estimate g_m and readjust the previous g_j's in the within-stage iterative step to update the function estimates. The number of terms M is usually estimated as part of the stagewise strategy. The model building stops when the next term gives little improvement to the fit. The function `ppr` in `R` or `Splus` can be used to fit the PPR model to data.

7.4.4 Neural networks

A *single-layer neural network* assumes a regression function of the form

$$f(\mathbf{x}) = \beta_0 + \sum_{m=1}^{M} \beta_m h(\alpha_m + \mathbf{w}_m^T \mathbf{x}), \tag{7.31}$$

where β_0, $\beta_1, \ldots, \beta_M$, $\alpha_1, \mathbf{w}_1, \ldots, \alpha_m, \mathbf{w}_m$ are unknown parameters and h, called the *activation function*, is usually chosen to be the sigmoid function

$$h(v) = 1/(1 + e^{-v}), \quad v \in \mathbb{R}. \tag{7.32}$$

The terms $h(\alpha_m + \mathbf{w}_m^T \mathbf{x})$ in (7.31) are called the *hidden units* of the neural network because they are not directly observable due to the unknown parameters $\alpha_m, \mathbf{w}_m$. They represent the derived features of the inputs $x_1, \ldots, x_p$, and are used to produce the output $f(\mathbf{x})$. We can think of the hidden unit as a basis function that is nonlinear in the parameters α_m and $\mathbf{w}_m$. The term "neural network" derives from the fact that it was first developed as a model of the human brain, for which each hidden unit represents a neuron. A neuron is fired when the signal passed to it exceeds a certain threshold that corresponds to choosing a step function as the activation function. Replacing the step function by a smoother threshold function for nonlinear least squares regression leads to the sigmoid function $h(\alpha v) = 1/(1 + e^{-\alpha v})$ in Figure 7.3. Note that $h(\alpha(v - v_0))$ shifts the activation threshold from 0 to v_0.

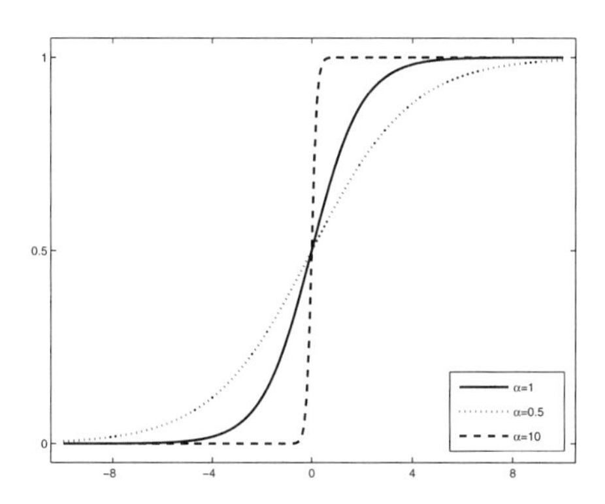

Fig. 7.3. Graph of $1/(1 + e^{-\alpha v})$.

In addition to the sigmoid function, other choices of h have been proposed. In particular, the Gaussian kernel $h(v) = e^{-v^2}$ is sometimes used. For the Gaussian kernel, a more commonly used alternative to $h(\alpha_m + \mathbf{w}_m^T \mathbf{x})$ in the neural network is $h(||\mathbf{x} - \mathbf{w}_m||/\alpha_m)$, which leads to a *radial basis network* of the form

$$f(\mathbf{x}) = \beta_0 + \sum_{m=1}^{M} \beta_m \exp\Big\{ - ||\mathbf{x} - \mathbf{w}_m||^2/\alpha_m^2 \Big\}. \tag{7.33}$$

The basis function $\exp\{ - ||\mathbf{x} - \mathbf{w}_m||^2/\alpha_m^2\}$ with nonlinear parameters α_m, $\mathbf{w}_m$ is called *radial basis function.*

Although in principle nonlinear least squares described in Section 4.2 can be used to fit neural networks or radial basis networks to data, in practice direct application of nonlinear least squares algorithms has difficulties due to nonconvergence because of the large number of parameters, resulting in overfitting. Some regularization through a penalty term, or indirectly by early stopping of a gradient descent algorithm (instead of a Gauss-Newton or Levenberg-Marquardt algorithm), is needed. The function `nnet` in `R` or `Splus` can be used to fit (7.31) to data.

7.5 A modeling approach that combines domain knowledge with nonparametric regression

The nonparametric regression methods described in the preceding sections involve different classes of basis functions of the predictors. Using basis functions such as regression splines reduces the regression problem to linear regression. Basis functions such as neural networks involve nonlinear parameters. The number of basis functions to be used can be determined by cross-validation or generalized cross-validation. Subject-matter knowledge can often help in developing a parsimonious set of basis functions and in choosing predictors.

Subject-matter knowledge has led to *substantive* models such as the Michaelis-Menten model of enzyme kinetics in Section 4.2 and the Black-Scholes model of option pricing in the next chapter. These models involve unknown parameters that can be estimated from the data by maximum likelihood or least squares and are therefore parametric models. Nonparametric regression, in contrast, does not assume a theoretical model and develops an *empirical* model from the data. The prediction performance of an empirical model depends on what predictors have been chosen to build the "black-box" model and on the amount of (stationary) data available for building the model. The prediction performance of a substantive model depends on how well the model approximates reality, which is closely related to the adequacy of the underlying theory. Unlike those in the physical sciences, models in economics are often based on assumptions that oversimplify complex market and human behavior, and therefore substantial discrepancies between theoretical models and market data are to be expected. On the other hand, these domain-knowledge models are widely adopted by economic agents and therefore can serve as a first approximation to reality, at least providing a useful set of predictor variables. A combined *substantive-empirical approach* via basis functions has been introduced by Lai and Wong (2004, 2006), in which the substantive component is associated with basis functions determined by the underlying theory and the empirical component uses flexible and computationally convenient basis functions such as regression splines. The empirical component is therefore used to correct the error in using the substantive model as a first approximation. The combined model is *semiparametric*, with a parametric substantive component and nonparametric regression for the empirical component.

Lai and Wong (2004) first came up with this modeling approach in their study of the valuation of American options, which do not have explicit formulas for the option prices, unlike their European counterparts in the Black-Scholes model; see Chapter 8. The basic idea is to apply empirical modeling to address the gap between the theoretical prices (which are computed by numerical solutions of certain optimal stopping problems) and the actual prices in the American options market. A closely related idea is to use a simple closed-form approximation to the theoretical American option price as a basis function in a combined substantive-empirical approach that uses regression splines as the other basis functions. Details will be given in Chapter 8 (see Section 8.3.5), which also describes other methods in the finance literature to address the discrepancy between actual and theoretical option prices. Lai and Wong (2006) subsequently extended this approach to time series analysis. They note that the stationary time series models in Chapter 5 and their nonlinear extensions, which use the parametric regression models of Chapter 4 or the nonparametric regression models of this chapter, are basically empirical models that relate the dynamics of a time series y_t to past observations $y_{t-1}, y_{t-2}, \ldots, y_{t-p}$ and possibly also some exogenous variables $u_{t-1}, \ldots, u_{t-p}$. In contrast, in engineering and the natural sciences, subject-matter theory suggests certain mechanistic models for the dynamics of y_t and the exogenous variables involved. To illustrate their semiparametric approach, which combines subject-matter theory with statistical modeling in time series analysis, Lai and Wong (2006) reanalyze the extensively studied time series of annual numbers of the Canadian lynx trapped for the period 1821–1934. Using a population dynamics model in the ecology literature on the Canadian lynx population to provide one of the basis functions, they apply MARS to estimate the other basis functions, which are bivariate regression splines, and demonstrate how this semiparametric approach provides improvements over previous nonlinear or nonparametric methods that have been used to analyze these data.

We next illustrate this semiparametric approach by applying it to the problem of estimating the forward rate curve of a corporate bond considered by Jarrow, Ruppert and Yu (2004).

7.5.1 Penalized spline models and estimation of forward rates

The ideas in (7.14) and (7.15) underlying smoothing cubic splines can be easily extended from the cubic spline basis to other spline bases. *Penalized spline models* impose a roughness penalty as in (7.14) or (7.15). However, unlike the linear basis representation $E(y_i|x_i) \approx \sum_{j=1}^K \beta_j g_j(x_i)$ used in smoothing cubic splines, these models allow nonlinear representation of $E(y_i|x_i)$ taking the form $\psi_i(\boldsymbol{\beta})$, where $\psi_i(\boldsymbol{\beta})$ is a nonlinear transformation (suggested by the background subject-matter theory) of $\sum_{j=1}^K \beta_j g_j(x_i)$ and $\boldsymbol{\beta} = (\beta_1, \ldots, \beta_K)^T$. Analogous to (7.15), a penalized spline model uses the penalized least squares criterion

$$\sum_{i=1}^{n} (y_i - \psi_i(\boldsymbol{\beta}))^2 + \lambda \boldsymbol{\beta}^T \boldsymbol{\Omega} \boldsymbol{\beta} \tag{7.34}$$

to estimate $\boldsymbol{\beta}$, where $\boldsymbol{\Omega} = (\int g_i''(u) g_j''(u) du)_{1 \le i,j \le K}$. The smoothing parameter λ can again be determined by cross-validation.

To see how such nonlinear transformations arise from subject-matter theory, we consider nonparametric estimation of the instantaneous forward rate curve at current time 0, based on the current prices of Treasury bonds with different maturities. Section 10.1 will introduce the fundamentals of interest rate markets, including the concept of the instantaneous forward rate $f(t, T)$ at time t for a zero-coupon bond that matures at $T(> t)$. The financial significance of $f(0, T)$ is that it is the rate one can lock in today for the future time T. As will be shown in Sections 10.1.2 and 10.1.3, the price B_i of the ith bond in the current Treasury bond market that has face value A_i and coupon payment C_{ij} at time t_{ij} ($j = 1, \dots, J_i$) and that matures at time T_i is related to the instantaneous forward rate curve $f(0, s)$ by

$$B_i = A_i \exp\left\{ - \int_0^{T_i} f(0,s)ds \right\} + \sum_{j=1}^{J_i} C_{ij} \exp\left\{ - \int_0^{t_{ij}} f(0,s)ds \right\}, \quad (7.35)$$

$i = 1, \dots, n$. With $y_i = B_i$ and $\psi_i(\boldsymbol{\beta})$ equal to the right-hand side of (7.35), the penalized spline criterion (7.34) that assumes the spline basis representation $f(0, s; \boldsymbol{\beta}) = \sum_{k=1}^{K} \beta_k g_k(s)$, with the g_k being spline basis functions, can be used to estimate $\boldsymbol{\beta}$ and thereby also the forward rate curve $f(0, s; \boldsymbol{\beta})$.

Adams and Deventer (1994) propose to estimate the forward rate curve with quadratic splines ($M = 4$ in (7.12)) and with knots at the distinct maturities and coupon-payment dates of the bonds in the sample. They call the roughness penalty constraint $\lambda \boldsymbol{\beta}^T \boldsymbol{\Omega} \boldsymbol{\beta}$ in (7.34) the "maximum smoothness constraint." Jarrow, Ruppert, and Yu (2004, pp. 60–62) provide details for fitting penalized spline models to Treasury bonds. In particular, they propose to use the spline basis (7.12) and to place interior knots at the $k/(K-1)$ quantiles of the set of coupon dates and maturities in the data, $1 \le k \le K-2$, besides the two boundary knots at the extremes of this set. They recommend choosing K sufficiently large ($K \ge 8$) to accommodate the nonlinearity in the forward rate curve. Instead of applying cross-validation (CV), they propose to modify generalized cross-validation (GCV) which applies to linear smoothers (see Section 7.3.1) by using certain linear approximations. They also consider two alternatives to CV and modified GCV for choosing λ.

7.5.2 A semiparametric penalized spline model for the forward rate curve of corporate debt

An additional source of complexity in estimating the forward rate curve of corporate bonds relative to Treasury bonds is that corporate bonds may default. The default probability and the recovery rate of a corporate bond in case of default are important factors that account for the spread between the Treasury and corporate bonds' interest rates; see Hull (2006, pp. 481–491). A major statistical difficulty is the relatively small sample size of corporate bond prices, as pointed out by Jarrow, Ruppert, and Yu (2004, p. 58): "In the estimation of the Treasury term structure, hundreds of bond prices are normally available on any given month, but for corporate term structures only a handful usually exist," as illustrated in their Tables 1 and 2 on AT&T bonds on December 31, 1995, and for the period April 1994–December 1995.

To overcome this difficulty, Jarrow, Ruppert, and Yu (2004) propose to estimate the forward rate curve $f^*(0,t)$ of a corporate bond by representing it as

$$f^*(0,t) = f(0,t;\boldsymbol{\beta}) + h(t;\boldsymbol{\alpha}), \tag{7.36}$$

where $f(0,t;\boldsymbol{\beta})$ is the forward rate curve of Treasury bonds and $h(t;\boldsymbol{\alpha})$ is a polynomial of low degree (e.g., $h(t;\boldsymbol{\alpha}) = \alpha_1 + \alpha_2 t + \alpha_3 t^2$). They propose a two-stage procedure to estimate f^*. The first stage uses Treasury bond prices to estimate $f(0,t;\boldsymbol{\beta}) = \sum_{k=1}^{K} \boldsymbol{\beta}_k g_k(t)$ in the penalized spline model described in Section 7.5.1. The second stage replaces f in (7.36) by $\widehat{f}$ estimated from the first stage and estimates $\boldsymbol{\alpha}$, and therefore also f^*, by minimizing the sum of squared differences between the actual corporate bond prices and those given by (7.36) via the nonlinear relationship (7.35) that relates the bond prices to the forward rate curve. They apply this procedure to the prices of five AT&T bonds and all U.S. Treasury STRIPS (i.e., zero-coupon bonds that are synthesized from the coupon and principal payments of Treasury bonds) available on December 31, 1995, using a quadratic spline for f and a linear function for the spread h.

Exercises

7.1. Show that the piecewise polynomial (7.10) satisfying the derivative constraints (7.11) is a linear combination of the functions $1, x, \dots, x^M, (x-\eta_1)_+^M, \dots, (x-\eta_K)_+^M$.

7.2. Show that the penalized residual sum of squares (7.15) is minimized by (7.16).

7.3. Making use of the property (7.4) of a smoother matrix, prove the Sherman-Morrison identity (7.22).

7.4. There is a connection between smoothing cubic splines and Bayesian posterior means. Consider the Bayesian regression model in which $y_i \sim N(\sum_{i=1}^{n} \beta_j g_j(x_i), \sigma^2)$ are independent and $\boldsymbol{\beta} := (\beta_1, \dots, \beta_n)^T$ has a prior $N(\mathbf{0}, \sigma^2 \lambda^{-1} \boldsymbol{\Omega}^{-1})$ distribution, where $\boldsymbol{\Omega}$ is the same as in (7.15).

   (a) Show that the posterior mean of $\boldsymbol{\beta}$ given $(x_1, y_1), \dots, (x_n, y_n)$ is the $\widehat{\boldsymbol{\beta}}$ defined in (7.16).

   (b) Show that the posterior covariance matrix of $\mathbf{G}\boldsymbol{\beta}$ is $\sigma^2 \mathbf{S}$, where $\mathbf{S}$ is the smoother matrix $\mathbf{G}(\mathbf{G}^T\mathbf{G} + \lambda\boldsymbol{\Omega})^{-1}\mathbf{G}^T$ associated with the cubic spline.

7.5. The file `ibm_intratrade_20030602.txt` contains the IBM stock transaction data on the New York Stock Exchange on June 2, 2003. The time interval between two consecutive trades is called a *duration*. Further details on durations in intraday stock trading on an exchange are given in Chapter 11 (Section 11.2.3).

(a) Plot the durations of the transactions versus the times on June 2, 2003 when they occur.

(b) Estimate the expected duration $f(t)$, as a function of time t during the trading day, by kernel smoothing. Use the Epanechnikov kernel and GCV to select the bandwidth. Plot the estimated curve $\widehat{f}(t)$.

(c) Plot the residuals $\Delta t_i - \widehat{f}(t_i)$, where $\Delta t_i = t_i - t_{i-1}$ and t_i is the time of the ith trade on June 2, 2003.

(d) Use the Epanechnikov kernel and a suitably chosen bandwidth to estimate the density function of IBM transaction durations on the trading day. Plot the histogram of the durations and the estimated density function.

7.6. The files `d_sp500f_1987.txt` and `d_sp500fopt_1987.txt` contain the daily settlement prices of S&P 500 futures and futures options (obtained from the Chicago Mercantile Exchange), respectively, for the period from January 1987 to December 1987. Each option has an expiration date T and strike price K.

(a) Use MARS to fit to the data in the first 6-month period (January–June 1987) a nonparametric regression model $c_t = Kf(S_t/K, T - t) + \epsilon_t$, where c_t is the option price and S_t the futures price on date t. Write down the estimated f and plot it. You can use `wireframe` in the `R` package `lattice` or `surf` in `MATLAB` to plot functions of two variables as surfaces, and `mars` in `R` to fit MARS. This provides an alternative nonparametric option pricing model to those used by Hutchinson, Lo and Poggio (1994); see Chapter 8 (Section 8.3.4).

(b) Use the fitted regression function in (a) to estimate the option prices during the next 6-month period. Plot the residuals.

8

Option Pricing and Market Data

Ross (1987) has noted that option pricing theory is "the most successful theory not only in finance, but in all of economics." A *call (put) option* gives the holder the right to buy (sell) the underlying asset (e.g., stock) by a certain date, known as the *expiration date* or *maturity*, at a certain price, which is called the *strike price.* "European" options can be exercised only on the expiration date, whereas "American" options can be exercised at any time up to the expiration date. The celebrated theory of Black and Scholes (1973) yields explicit formulas for the prices of European call and put options. Merton (1973) extended the Black-Scholes theory to American options. Optimal exercise of the option has been shown to occur when the asset price exceeds or falls below an exercise boundary for a call or put option, respectively. There are no closed-form solutions for the exercise boundary and American option price, but numerical methods and approximations are available, as described in Section 8.1. The Black-Scholes-Merton theory for pricing and hedging options is of fundamental importance in the development of financial derivatives and provides the foundation for financial engineering. A *derivative* is a financial instrument having a value derived from or contingent on the values of more basic underlying variables. In particular, a stock option is a derivative whose value is dependent on the price of the stock. In recent years, credit derivatives and path-dependent options have become popular, and there are emerging markets in weather, energy, and insurance derivatives; see Chapters 21–23 of Hull (2006).

The European and American call and put options considered in this chapter are often called *plain vanilla* products. They are actively traded on many exchanges throughout the world. In particular, the Chicago Board Options Exchange (CBOE) started trading options contracts in 1973. In contrast, nonstandard (such as path-dependent) options, called *exotic options*, which have been created by financial engineers to meet the needs of corporate treasurers, fund managers, and financial institutions, are traded in the *over-the-counter market*. For plain vanilla European options that are actively traded on exchanges, one can compare the actual option prices with those given explicitly by the Black-Scholes formula that involves the price of the underlying stock and the risk-free interest rate, which are directly observable, and the volatility σ of the stock's return, which has to be estimated from past data. Discrepancies between the Black-Scholes and actual prices can be used to assess the adequacy of the Black-Scholes theory and to modify it.

Instead of using historic volatility of the stock's returns to estimate σ, an approach commonly used by option traders is to solve for σ from the Black-Scholes formula, yielding the *implied volatility*. The implied volatility is used to perform delta hedging by buying or selling a suitable number (given by delta, which is defined in Section 8.1.2) of shares of the stock or to price over-the-counter (OTC) derivatives on the stock. As this approach assumes the validity of the Black-Scholes theory, market deviations from the theory have been reflected by *volatility smiles*, which are described in Section 8.2.

Section 8.3 describes several approaches in the literature to address volatility smiles and other discrepancies between the theoretical and observed option prices. In particular, it considers a combined substantive-empirical approach described in Section 7.5, whose substantive component is associated with the Black-Scholes formula and whose empirical component uses nonparametric regression to model market deviations from the Black-Scholes formula. It also discusses the applications of this approach to hedging.

As noted by Jorion (2001, pp. 12–15), derivatives markets have been growing rapidly during the past two decades and in some cases have outgrown the markets of their underlying assets. Although the prices of the derivatives are in theory determined from those of the underlying assets by the Black-Scholes formula, deviations from the theoretical assumptions in practice and separate market forces in the options market and the stock market have led to discrepancies between the theoretical and observed option prices. Whereas stock returns and their volatilities, which are the topics covered in Chapters 3 and 6, are the fundamental quantities in the stock market, the options market has option prices and implied volatilities as the basic quantities, the statistical properties of which are studied in Section 8.2.

8.1 Option prices and pricing theory

8.1.1 Options data and put–call parity

Let K denote the strike price of an option with expiration date T and let S be the current price of the underlying asset. Figures 8.1–8.3 provide examples of options data in the form of market quotes of IBM stock prices, dividends, and put and call options on IBM stock. Letting S_t denote the asset price at time t, the payoff of an option is $g(S_T)$, where

$$g(S) = \begin{cases} (K - S)_+ & \text{for put,} \\ (S - K)_+ & \text{for call,} \end{cases} \tag{8.1}$$

in which $x_+ = \max(x, 0)$. Let p_t (resp. c_t) denote the price of a European put (resp. call) on the underlying asset at time t. The *put–call parity* relates p_t and c_t to S_t and K by

$$S_t e^{-q(T-t)} + p_t - c_t = K e^{-r(T-t)}, \tag{8.2}$$

where r is the risk-free interest rate and q is the dividend rate; see Hull (2006, p. 314).

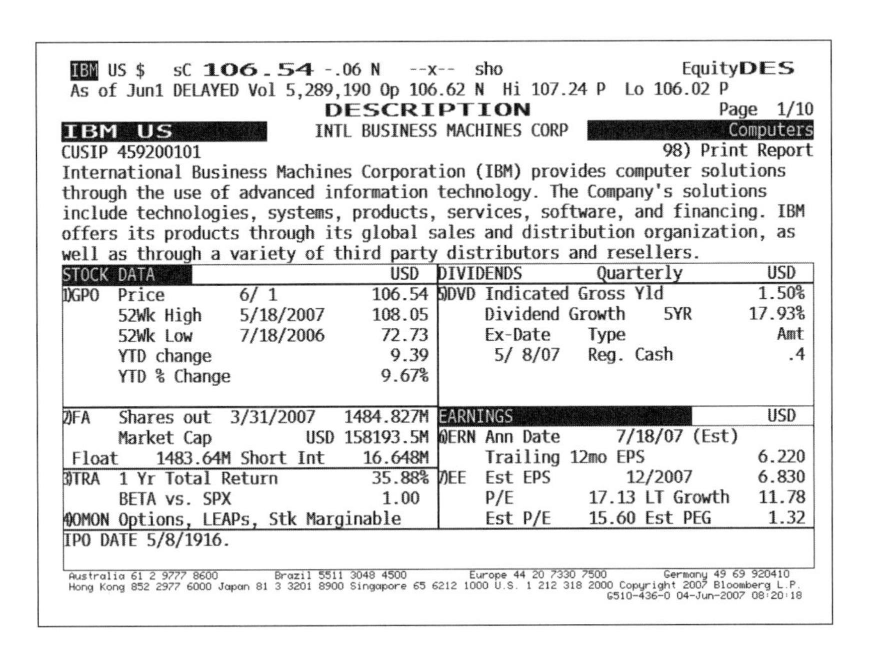
IBM US $ sC 106.54 -.06 N --x-- sho EquityDES
As of Jun1 DELAYED Vol 5,289,190 Op 106.62 N Hi 107.24 P Lo 106.02 P
DESCRIPTION Page 1/10
IBM US INTL BUSINESS MACHINES CORP Computers
CUSIP 459200101 98) Print Report
International Business Machines Corporation (IBM) provides computer solutions through the use of advanced information technology. The Company's solutions include technologies, systems, products, services, software, and financing. IBM offers its products through its global sales and distribution organization, as well as through a variety of third party distributors and resellers.

STOCK DATA			USD	DIVIDENDS		Quarterly	USD
1)GPO	Price	6/ 1	106.54	5)DVD	Indicated Gross Yld		1.50%
	52Wk High	5/18/2007	108.05		Dividend Growth	5YR	17.93%
	52Wk Low	7/18/2006	72.73		Ex-Date	Type	Amt
	YTD change		9.39		5/ 8/07	Reg. Cash	.4
	YTD % Change		9.67%				
2)FA	Shares out	3/31/2007	1484.827M	EARNINGS			USD
	Market Cap	USD	158193.5M	6)ERN	Ann Date	7/18/07 (Est)	
Float	1483.64M	Short Int	16.648M		Trailing 12mo EPS		6.220
3)TRA	1 Yr Total Return		35.88%	7)EE	Est EPS	12/2007	6.830
	BETA vs. SPX		1.00		P/E	17.13 LT Growth	11.78
4)OMON	Options, LEAPs, Stk Marginable				Est P/E	15.60 Est PEG	1.32

IPO DATE 5/8/1916.

Australia 61 2 9777 8600 Brazil 5511 3048 4500 Europe 44 20 7330 7500 Germany 49 69 920410
Hong Kong 852 2977 6000 Japan 81 3 3201 8900 Singapore 65 6212 1000 U.S. 1 212 318 2000 Copyright 2007 Bloomberg L.P.
G510-436-0 04-Jun-2007 08:20:18

Fig. 8.1. Description of IBM stock. Used with permission of Bloomberg LP.

8.1.2 The Black-Scholes formulas for European options

Assumptions in the Black-Scholes theory

(A1) S_t is a geometric Brownian motion (GBM) with drift θ and volatility σ; see (3.5).

(A2) The market has a risk-free asset with constant interest rate r.

(A3) Continuous hedging can occur and there are no transaction costs.

(A4) Short selling (see Section 3.2.1) is allowed, and the asset is perfectly divisible.

<HELP> for explanation, <MENU> for similar functions. P216 EquityOMST
Enter #<Equity><GO> for selection. Note: All values in USD

Pages: MOST ACTIVE OPTIONS ON Page 1 of 3
1 - Most Active INTL BUSINESS MACHINES CORP
2 - Most Up T T-today or P-previous
3 - Most Down V Vvolume O-open inter.

	Total	Call	Put
Volume	8,618	3,280	5,338
Open Int.	604,323	313,587	290,736

Underlying:	Last	Volume	1-Day Chg	Open	High	Low	Yest.	2-Day Ch
IBM US	106.54	5289190	-.06	106.62	107.24	106.02	106.60	-.39

Option	Symbol	Last	Chng	Vol	Option	Symbol	Last	Chng	Vol
1)Jul07 100 Puts	ST	.55	-.05	1043	17)Jan09 75 Puts	MO	1.10	-.15	150
2)Oct07 100 Puts	VT	1.88	+.08	870	18)Jun07 100 Calls	FT	6.50	-.40	72
3)Jul07 105 Puts	SA	1.69	+.04	858	19)Jun07 100 Puts	RT	.05	-.05	71
4)Oct07 105 Puts	VA	3.41	+.21	829	20)Oct07 115 Calls	JC	1.95	-.15	63
5)Jun07 105 Puts	RA	.60	-.05	580	21)Jan08 90 Puts	MR	1.15	+.07	55
6)Jun07 105 Calls	FA	2.37	-.18	568	22)Jan08 110 Calls	AB	5.60	-.40	41
7)Jul07 100 Calls	GT	7.50	-.40	526	23)Jul07 105 Calls	GA	3.84	-.26	40
8)Jun07 110 Calls	FB	.25	-.10	441	24)Jan08 75 Puts	MO	.30	+.05	40
9)Jun07 110 Puts	RB	3.50	-.10	358	25)Jan10 110 Calls	AB	15.70	+.20	35
10)Oct07 100 Calls	JT	9.87	+.27	349	26)Jan08 115 Calls	AC	3.80	+.30	32
11)Oct07 110 Calls	JB	3.60	-.20	327	27)Jun07 115 Calls	FC	.05	-.02	26
12)Jul07 110 Calls	GB	1.48	-.07	278	28)Oct07 90 Puts	VR	.55	+.04	26
13)Jan08 105 Puts	MA	4.60	+.40	201	29)Jan09 130 Calls	AF	3.80	-.20	20
14)Jan08 115 Puts	MC	10.00	-.40	200	30)Jan08 105 Calls	AA	8.10	-.44	19
15)Jan08 120 Calls	AD	2.10	-.05	179	31)Jul07 95 Puts	SS	.20	unch	14
16)Jul07 115 Calls	GC	.47	-.03	175	32)Oct07 105 Calls	JA	6.40	-.20	13

Australia 61 2 9777 8600 Brazil 5511 3048 4500 Europe 44 20 7330 7500 Germany 49 69 920410
Hong Kong 852 2977 6000 Japan 81 3 3201 8900 Singapore 65 6212 1000 U.S. 1 212 318 2000 Copyright 2007 Bloomberg L.P.
G510-436-0 04-Jun-2007 08:22:28

Fig. 8.2. Put and call options on IBM stock. Used with permission of Bloomberg LP.

IBM+JT US 10 C $ C 9.87 +.27 B 9.80/10.00 334x117 EquityDES
As of Jun1 DELAYED OpInt 5738 Vol 349 Op 10.40 P Hi 10.40 P Lo 9.60 I

EQUITY OPTION DESCRIPTION

INTL BUSINESS MACHINES CORP
IBM US 4) DES Price: 106.54

Specified Option
Bloomberg Tkr: IBM US 10 C100
9.80/ 10.00 Last: 9.87
Strke: 100 Cycle: JAN
Expiration: 10/20/07
Exercise Type: AMERICAN
Option ID: 4592009JT

Exchange Data
CBOE AMEX PHLX ISE PSE BOX
Hours, in Chicago time:
7:00 - 8:15 ; 8:30 - 15:00
Tick Size: .05 .10
Tick Val: $ 5.00 $ 10.00
Pos limit: 25000000 shares

Volatility/Risk Analysis
6) HVT Volatility 7) CALL Theo. Value
30day:18.67 Imp Vol: 19.68
60day:16.40 Delta: .7598
90day:16.33 Gamma: .0243

1) OMON Opt Cust Monitor 2) OSA Scenario Analysis 3) OPDF Default Settings

Ticker	Expiration Date	Days to Expiration	Contract Size	Multiplier
IBZ	JUN 16 2007	12	100	100.0
IBM	JUN 16 2007	12	100	100.0
IBZ	JUL 21 2007	47	100	100.0
IBM	JUL 21 2007	47	100	100.0
IBZ	OCT 20 2007	138	100	100.0
IBM	OCT 20 2007	138	100	100.0
IBZ	JAN 19 2008	229	100	100.0
IBM	JAN 19 2008	229	100	100.0
VIB	JAN 17 2009	593	100	100.0
WIB	JAN 16 2010	957	100	100.0

General Notes 8) More Details

9) CF OPTION Split History

Australia 61 2 9777 8600 Brazil 5511 3048 4500 Europe 44 20 7330 7500 Germany 49 69 920410
Hong Kong 852 2977 6000 Japan 81 3 3201 8900 Singapore 65 6212 1000 U.S. 1 212 318 2000 Copyright 2007 Bloomberg L.P.
G510-436-0 04-Jun-2007 08:23:23

Fig. 8.3. Description of an American call option on IBM stock. Used with permission of Bloomberg LP.

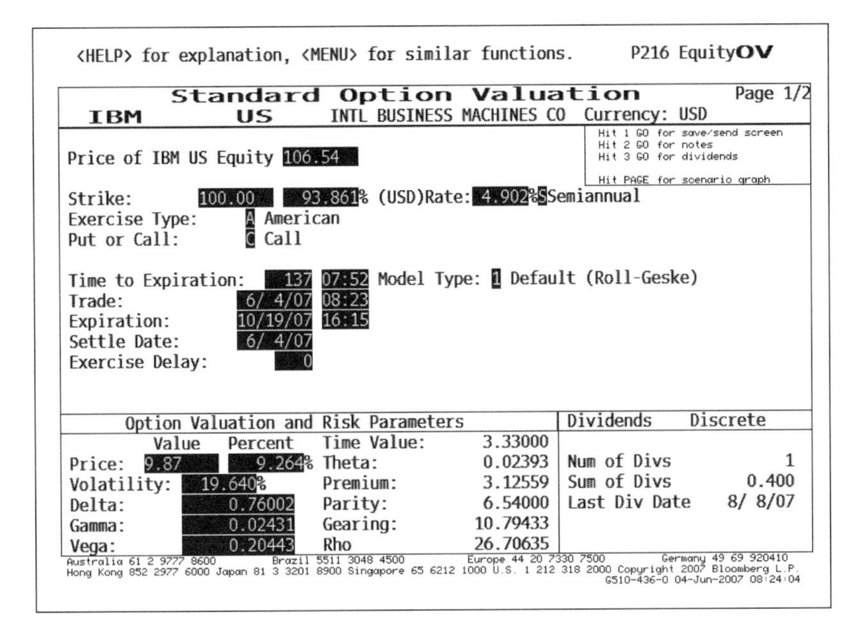

Fig. 8.4. Standard option valuation. Used with permission of Bloomberg LP.

Arbitrage-free pricing

First consider the case where the asset pays no dividends. Let $f(t, S)$ be the option price at time t when $S_t = S$. Consider a portfolio Π that at time t holds 1 unit of option and $-\Delta$ units of the asset. Then $\Pi_t = f(t, S_t) - S_t\Delta$. Since $dS_t = \theta S_t dt + \sigma S_t dw_t$, applying Ito's formula (see Appendix A) to $df(t, S_t)$ yields

$$d\Pi_t = \sigma S_t \left(\frac{\partial f}{\partial S} - \Delta \right) dw_t + \left(\theta S_t \frac{\partial f}{\partial S} + \frac{1}{2}\sigma^2 S^2 \frac{\partial^2 f}{\partial S^2} + \frac{\partial f}{\partial t} - \theta S_t \Delta \right) dt. \quad (8.3)$$

The portfolio Π_t becomes risk-free if $\Delta = \partial f / \partial S$, for which the coefficient of dw_t in (8.3) is equal to 0. In this case, Π_t should have the same return as the risk-free asset (i.e., $d\Pi_t = r\Pi_t dt$) because otherwise there are arbitrage opportunities, and (8.3) reduces to the partial differential equation (PDE)

$$\frac{\partial f}{\partial t} + rS\frac{\partial f}{\partial S} + \frac{1}{2}\sigma^2 S^2 \frac{\partial^2 f}{\partial S^2} = rf \quad \text{for} \quad 0 \leq t < T. \quad (8.4)$$

When the asset pays dividends at rate q, the preceding argument can be modified to yield the PDE

$$\frac{\partial f}{\partial t} + (r-q)S\frac{\partial f}{\partial S} + \frac{1}{2}\sigma^2 S^2 \frac{\partial^2 f}{\partial S^2} = rf \quad \text{for } 0 \le t < T. \tag{8.5}$$

The boundary condition of the PDE is $f(T, S) = g(S)$, where g is the payoff function (8.1). The PDE can be solved explicitly for f, yielding the Black-Scholes formulas for the option prices $c_t = c(t, S_t)$ and $p_t = p(t, S_t)$:

$$c(t, S) = Se^{-q(T-t)}\Phi(d_1) - Ke^{-r(T-t)}\Phi(d_2), \tag{8.6}$$

$$p(t, S) = Ke^{-r(T-t)}\Phi(-d_2) - Se^{-q(T-t)}\Phi(-d_1), \tag{8.7}$$

where $\Phi(\cdot)$ is the standard normal cumulative distribution function and

$$d_1 = \frac{\log(S/K) + (r - q + \sigma^2/2)(T-t)}{\sigma\sqrt{T-t}}, \quad d_2 = d_1 - \sigma\sqrt{T-t}.$$

Note that the preceding arguments provide not only the pricing formulas (8.6) and (8.7) but also the *delta hedging* rule in the Black-Scholes theory: A short (or long) position in an option can be made risk-free by keeping $\Delta = \partial f/\partial S$ (or $-\Delta$) units of the asset; the *short (long) position* refers to that of the option's *seller* (*buyer*). In view of (8.6), the delta of a call option is

$$\frac{\partial c}{\partial S} = e^{-q(T-t)}\Phi(d_1) \tag{8.8}$$

(see Exercise 8.1), and the put-call parity relationship (8.2) yields the delta of a put option as $\partial p/\partial S = \partial c/\partial S - 1$.

Application to currency options

For corporations that want to hedge foreign exchange exposures, foreign currency options are useful alternatives to forward contracts (see Section 1.5.2). A foreign currency is analogous to a stock that provides a dividend yield equal to the risk-free rate r_f prevailing in the foreign country, since its holder can earn interest at the rate r_f. Therefore the Black-Scholes formulas can be applied to currency options with $q = r_f$, assuming that the exchange rate process follows a GBM that has volatility σ.

Application to futures options

Options on futures contracts, or futures options, are traded on many exchanges. Consider a portfolio Π that at time t holds -1 unit of the option and $\partial f/\partial F$ units of a futures contract, as in the Black-Scholes derivation. Since it costs nothing to enter into a futures contract (see Section 1.5.2), $\Pi = -f$. Assuming $dF_t = F_t(\mu dt + \sigma dw_t)$ for the futures price and repeating the argument that led to the Black-Scholes PDE (8.5), Black (1976) obtained

$$\frac{\partial f}{\partial t} + \frac{1}{2}\frac{\partial^2 f}{\partial F^2}\sigma^2 F^2 = rf,$$

showing that a futures option has the same price as a stock option with $S_t = F_t$ and dividend rate $q = r$.

8.1.3 Optimal stopping and American options

Merton (1973) extended the Black-Scholes theory for pricing European options to American options. Optimal exercise of the option occurs when the asset price exceeds or falls below an exercise boundary $\partial\mathcal{C}$ for a call or put option, respectively. The Black-Scholes PDE still holds in the continuation region $\mathcal{C}$ of (t, S_t) before exercise, and $\partial\mathcal{C}$ is determined by the *free boundary condition* $\partial f/\partial S = 1$ (or -1) for a call (or put) option. Unlike the explicit formula (8.6) or (8.7) for European options, there is no closed-form solution of the free-boundary PDE, and numerical methods such as finite differences are needed to compute American option prices under this theory. The free-boundary PDE can also be represented probabilistically as the value function of the optimal stopping problem

$$f(t, S) = \sup_{\tau \in \mathcal{T}_{t,T}} E[e^{-r(\tau - t)} g(S_\tau) | S_t = S], \tag{8.9}$$

where $\mathcal{T}_{t,T}$ denotes the set of stopping times τ whose values are between t and T, and E is expectation with respect to the *risk-neutral measure* under which S_t is GBM with drift $r - q$ and volatility σ. Further details on risk-neutral measures and probabilistic representations of solutions of PDEs are given in Sections 10.4 and 10.5.2. Cox, Ross, and Rubinstein (1979) proposed to approximate GBM by a binomial tree with root node S_0 at time 0, so that (8.9) can be approximated by a discrete-time and discrete-state optimal stopping problem that can be computed by backward induction; see Hull (2006, pp. 391–394) for an introduction to the binomial tree method.

Denote $f(t, S)$ by $C(t, S)$ for an American call option, and by $P(t, S)$ for an American put option. Jacka (1991) and Carr, Jarrow, and Myneni (1992) derived the decomposition formula

$$\begin{aligned} P(t, S) = p(t, S) + K\rho e^{\rho u} \int_u^0 \Bigg\{ & e^{-\rho s} \Phi\left(\frac{\bar{z}(s) - z}{\sqrt{s - u}}\right) \\ & - \mu e^{-(\mu\rho s + u/2) + z} \Phi\left(\frac{\bar{z}(s) - z}{\sqrt{s - u}} - \sqrt{s - u}\right)\Bigg\} ds \end{aligned} \tag{8.10}$$

and a similar formula relating $C(t, S)$ to $c(t, S)$, where $\bar{z}(u)$ is the early exercise boundary $\partial\mathcal{C}$ under the transformation

$$\rho = r/\sigma^2, \ \mu = q/r, \ u = \sigma^2(t-T), \ z = \log(S/K) - (\rho - \mu\rho - 1/2)u. \quad (8.11)$$

Ju (1998) found that the *early exercise premium*, $P(t,S) - p(t,S)$ in (8.10), can be computed in closed form if $\partial\mathcal{C}$ is a piecewise exponential function that corresponds to a piecewise linear $\bar{z}(u)$. By using such an assumption, Ju (1998) reported numerical studies showing that his method with three equally spaced pieces substantially improves previous approximations to option prices in both accuracy and speed. AitSahlia and Lai (2001) introduced the transformation (8.11) to reduce GBM to Brownian motion, which can be approximated by a symmetric Bernoulli random walk, and to transform the early exercise boundary $\partial\mathcal{C}$ to $\bar{z}(u)$ in the new coordinate system. They developed a corrected random walk approximation to compute by backward induction the optimal stopping boundary $\bar{z}(\cdot)$, which their numerical results show can indeed be well approximated by a piecewise linear function with a few pieces. The integral obtained by differentiating that in (8.10) with respect to S also has a closed-form expression when $\bar{z}(\cdot)$ is piecewise linear, and approximating $\bar{z}(\cdot)$ by a linear spline that uses a few unevenly spaced knots gives a fast and reasonably accurate method for computing the delta $\Delta = \partial P/\partial S$ of an American put. Similar results also hold for American call options on dividend-paying stocks; American calls on stocks that do not pay dividends are optimally exercised at maturity.

8.2 Implied volatility

The interest rate r in the Black-Scholes formula (8.6) or (8.7) for the price of a European option is usually taken to be the yield of a short-maturity Treasury bill at the time when the contract is initiated. The parameter in (8.6) or (8.7) that cannot be directly observed is σ. Equating (8.6) or (8.7) to the actual price of the option yields a nonlinear equation in σ whose solution is called the *implied volatility* of the underlying asset. Traders calculate implied volatilities from actively traded options on a stock and use them to price over-the-counter options on the same stock and to calculate the option's delta for hedging applications. The implied volatilities computed from call and put options with the same strike price K and time to maturity $T-t$ should be equal because the put–call parity relationship (8.2) holds for both the Black-Scholes price pair $(p_t^{\text{BS}}, c_t^{\text{BS}})$ and the market price pair $(p_t^{\text{M}}, c_t^{\text{M}})$, from which it follows that $c_t^{\text{BS}} - c_t^{\text{M}} = p_t^{\text{BS}} - p_t^{\text{M}}$ and therefore the equation $c_t^{\text{BS}} = c_t^{\text{M}}$ gives the same solution for σ as $p_t^{\text{BS}} = p_t^{\text{M}}$.

Smiles and skews

A call option, whose payoff function is $(S-K)_+$, is said to be *in the money*, *at the money*, or *out of the money* according to whether $S_t > K$, $S_t = K$,

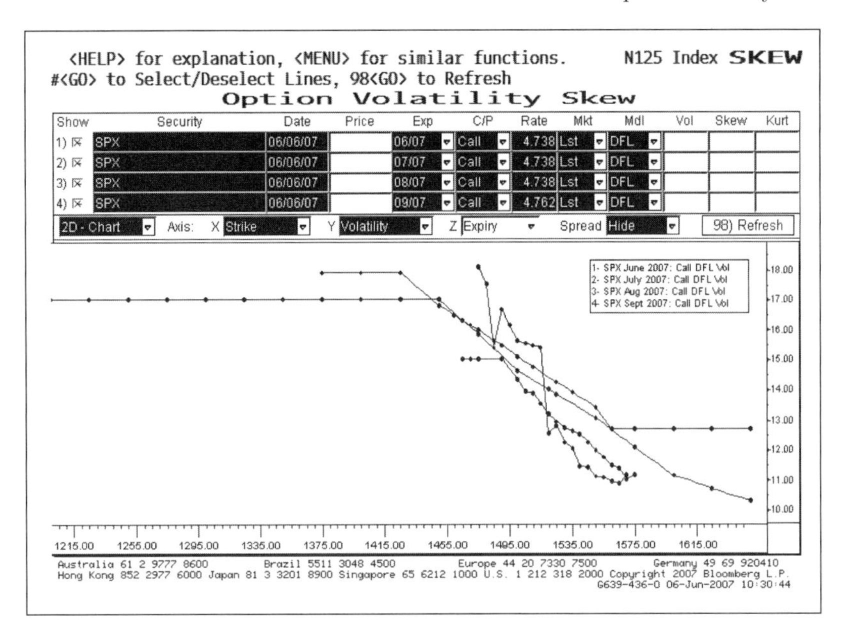

Fig. 8.5. The volatility skew of call options on the S&P 500 index. Used with permission of Bloomberg LP.

or $S_t < K$, respectively. Puts have the reverse terminology since the payoff function is $(K - S)_+$. According to the Black-Scholes theory, the σ in (8.6) and (8.7) is the volatility of the underlying asset and therefore does not vary with K and T. However, for some equity options, a *volatility skew* is observed (i.e., the implied volatility is a decreasing function of the strike price K); see Figure 8.5. The *volatility smile* is common in foreign currency options, for which the implied volatility is relatively low for at-the-money options and becomes higher as the option moves into the money or out of the money, giving the "smile" shape of the implied volatility curve as a function of K (with minimum around $K = S$). Moreover, implied volatilities also tend to vary with time to maturity.

The implied volatilities of options on an underlying asset, therefore, are often quoted as a function of K and T. *Volatility surfaces*, usually presented in the form of a table, provide the volatilities for pricing an option on the asset with any strike price and any maturity. For example, with K/S (called the *moneyness*) taking values 0.9, 0.95, 1, 1.05, and 1.1 for the columns of the table, and time to maturity listed at 1/12, 3/12, 6/12, 1, 2, and 5 years for the rows, the entries of the table give the implied volatilities of options whose market prices are available. Other volatilities outside (or left blank in) the table can be determined from these entries by linear interpolation. Figure 8.6 illustrates this by plotting the closing prices of S&P 500 futures

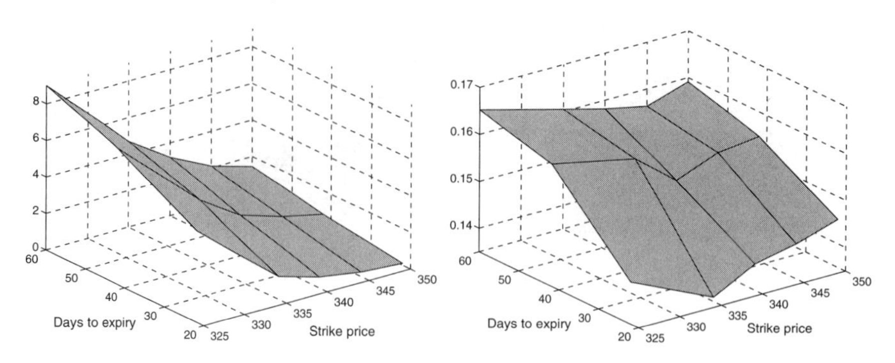

Fig. 8.6. Left panel: closing prices of call options. Right panel: implied volatility surface.

options (European) on June 20, 1989 in the left panel and the corresponding volatility surface in the right panel. The futures price on June 20, 1989 is $S = 325.85$ (dollars), and the annualized yield on the 3-month Treasury bill on June 20, 1989 is used as the risk-free interest rate $r = 0.0844$ in the Black-Scholes formula.

Time series of implied volatility surfaces

Let $m = K/S_t$ be the moneyness and $\tau = T - t$ be the time to maturity of an option whose implied volatility on date t is considered as a surface $I_t(m, \tau)$ indexed by (m, τ), thereby yielding a time series of implied volatility surfaces. Although $I_t(m, \tau)$ is available only at certain (m, τ) values and does not constitute a surface, simple kernel smoothing (see Section 7.2.2) gives a smooth surface over (m, τ). We illustrate this idea with $n = 320$ daily market quotes of (i) closing prices of the S&P 500 index and (ii) implied volatilities of European calls and puts on the S&P 500 index for the period from January 3, 2005 to April 10, 2006 that we obtained from Wharton Research Data Services. Figure 8.7 plots the mean $\bar{I}(m, \tau)$ and the standard deviation $\widehat{\sigma}(m, \tau)$ of $\{I_t(m, \tau) : 1 \le t \le n\}$. Restricting (m, τ) to a grid of evenly spaced points (m_i, τ_i) in the range of moneyness (0.5 to 1.5) and time to maturity (1 month to 2 years) of these options, we obtain a multivariate time series $\{I_t(m_i, \tau_j), 1 \le i \le 15, 1 \le j \le 20\}$. Figure 8.8 plots the time series of $I_t(m, \tau)$, $1 \le t \le n$, in the left panel for three values of (m, τ), and the transformed series

$$\Delta_t(m, \tau) = \log I_t(m, \tau) - \log I_{t-1}(m, \tau), \quad 2 \le t \le n, \tag{8.12}$$

in the right panel, showing that taking logarithms and differencing tends to result in a stationary time series.

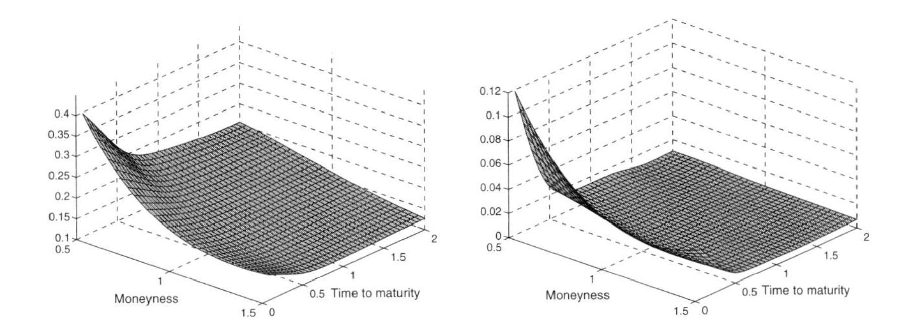

Fig. 8.7. Left panel: mean of implied volatility surfaces for S&P 500 options. Right panel: standard deviation of the implied volatilities.

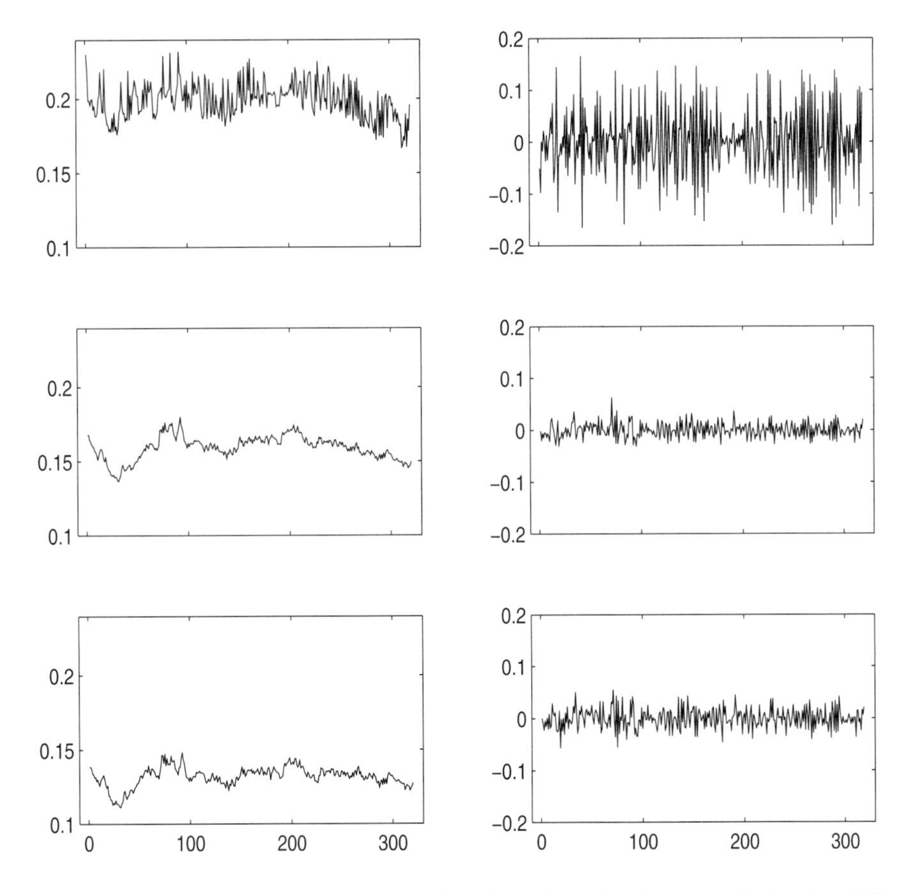

Fig. 8.8. Left panels: time series of $I_t(m, \tau)$ for $(m, \tau) = (0.714, 0.689)$, $(1, 1.496)$, $(1.286, 0.689)$. Right panels: time series $\Delta_t(m, \tau)$ for (m, τ) on the left panel.

8.3 Alternatives to and modifications of the Black-Scholes model and pricing theory

8.3.1 The implied volatility function (IVF) model

The *implied volatility* (also called the *implied tree*) model provides an exact fit to all European option prices on any given day. It assumes that the *risk-neutral process* of the asset price has the more general form

$$dS_t = (r_t - q_t)S_t dt + \sigma(t, S_t)S_t dw_t \tag{8.13}$$

rather than that described in Section 8.1.2 with $r_t \equiv r$, $q_t \equiv q$ and $\sigma(t, S_t) \equiv \sigma$. Dupire (1994) has shown that the function $\sigma(t, S)$ is given analytically by

$$\frac{\sigma^2(T,K)}{2} = \left\{ \frac{\partial c_T}{\partial T} + q_T c_T + K(r_T - q_T)\frac{\partial c_T}{\partial K} \right\} \Big/ K^2 \frac{\partial^2 c_T}{\partial K^2}, \tag{8.14}$$

where c_T is the market price of a European call option with strike price K and maturity T. To price a European option under the risk-neutral model (8.13), Andersen and Brotherton-Ratcliffe (1997) use finite difference approximations of (8.14) to recalibrate the model (8.13) daily to the market prices of standard European options. An alternative approach, proposed by Derman and Kani (1994) and Rubinstein (1994), approximates (8.13) by an *implied tree*, which is a discrete-time Markov chain approximation to (8.13) in the form of a binomial tree that is recalibrated daily to the market prices of standard options. Since this approach focuses exclusively on "in-sample" fitting, which uses the sample consisting of all European option prices on a given day, its disadvantage is that the in-sample correctness does not extend to out-of-sample forecasting of future European option prices or pricing of American options that can be exercised at any time prior to the expiration date T.

8.3.2 The constant elasticity of variance (CEV) model

Whereas the IVF model changes the constant σ in the Black-Scholes theory by a function $\sigma(t, S_t)$, the CEV model replaces σ by σS^α, imposing an additional parameter for the Black-Scholes model. Specifically, the risk-neutral process of the asset price follows the CEV model $dS_t = (r - q)S_t dt + \sigma S_t^\alpha dw_t$ introduced by Cox and Ross (1976), who showed that the formulas (8.6) and (8.7) for $c(t, S)$ and $p(t, S)$ can be modified by replacing $\Phi(d_1)$ and $\Phi(d_2)$ by the distribution functions of certain noncentral χ^2-distributions:

$$\begin{aligned} c(t,S) &= Se^{-q(T-t)}\left[1 - \chi^2(a; b+2, c)\right] - Ke^{-r(T-t)}\chi^2(c; b, a), \\ p(t,S) &= Ke^{-r(T-t)}\left[1 - \chi^2(c; b, a)\right] - Se^{-q(T-t)}\chi^2(a; b+2, c), \end{aligned} \tag{8.15}$$

in the case $\alpha > 1$ and

$$\begin{aligned} c(t,S) &= Se^{-q(T-t)}\left[1-\chi^2(c;-b,a)\right] - Ke^{-r(T-t)}\chi^2(a;2-b,c), \\ p(t,S) &= Ke^{-r(T-t)}\left[1-\chi^2(a;2-b,c)\right] - Se^{-q(T-t)}\chi^2(c;-b,c), \end{aligned} \tag{8.16}$$

in the case $0 < \alpha < 1$, where K is the strike price and T is the expiration date of the option, $\chi^2(\cdot;\nu,\lambda)$ is the distribution function of the noncentral chi-square distribution (defined below) with ν degrees of freedom and noncentrality parameter λ, and

$$v = \frac{\sigma^2}{2(r-q)(\alpha-1)}\left\{e^{2(r-q)(\alpha-1)(T-t)} - 1\right\},$$

$$a = \frac{\left[Ke^{-(r-q)(T-t)}\right]^{2(1-\alpha)}}{(1-\alpha)^2 v}, \quad b = \frac{1}{1-\alpha}, \quad c = \frac{S^{2(1-\alpha)}}{(1-\alpha)^2 v};$$

see Hull (2006, pp. 562–563). Recall that in Definition 1.2(i) on the χ^2-distribution, the Z_i are assumed to be standard normal. A generalization of this χ^2-distribution is to assume that $Z_1, \dots, Z_n$ are independent such that $Z_i \sim N(\mu_i, 1)$. Then $U = Z_1^2 + \cdots + Z_n^2$ is said to have the *noncentral chi-square distribution* with n degrees of freedom and *noncentrality parameter* $\psi^2 = \sum_{i=1}^n \mu_i^2$. The density function of U can be written as a Poisson mixture of χ^2 densities:

$$f(u) = \sum_{k=0}^{\infty} p_\psi(k) f_{n+2k}(u), \tag{8.17}$$

where $f_r(u)$ is the density function of the χ^2-distribution and $p_\psi(k) = e^{-\psi^2/2}(\psi^2/2)^k/k!$ is the Poisson density function. The parameters α and σ of the CEV model can be estimated by nonlinear least squares, minimizing over (α, σ) the sum of squared differences between the model prices and the market prices; see Exercise 8.4.

8.3.3 The stochastic volatility (SV) model

The continuous-time *stochastic volatility* (SV) model under the risk-neutral measure is $dS_t/S_t = (r-q)dt + \sigma_t dw_t$, in which $v_t := \sigma_t^2$ is modeled by

$$dv_t = \alpha(v^* - v_t)dt + \beta v_t^\xi d\widetilde{w}_t, \tag{8.18}$$

where $\widetilde{w}_t$ is Brownian motion that is independent of w_t. For this SV model, Hull and White (1987) have shown that the price of a European option is given by $\int_0^\infty b(w)g(w)dw$, where $b(w)$ is the Black-Scholes price in which σ is replaced by w, and w is the average variance rate during the life of the option, which is a random variable with density function g determined by the stochastic dynamics (8.18) for v_t. Although there is no analytic formula for

g, Hull and White (1987) have used this representation of the option price to develop closed-form approximations to the model price. The parameters α, β, v^*, and ξ in (8.18) can be estimated by minimizing the sum of squared differences between the model prices and the market prices. The SV model has been used to account for the volatility smile associated with the Black-Scholes prices; see Exercise 8.4.

8.3.4 Nonparametric methods

The methods in Sections 8.3.1–8.3.3 basically replace assumption (A1) in the Black-Scholes theory of Section 8.1.2 by other stochastic models for the asset price S_t under the risk-neutral measure to account for the observed implied volatility smile, which is incompatible with the Black-Scholes theory. Instead of specifying a particular model for S_t, Hutchinson, Lo, and Poggio (1994) proposed to use a nonparametric model that only requires S_t to have independent increments. Noting that y_t ($= c_t$ or p_t) is a function of S_t/K and $T-t$ with r and σ being constant, they assume $y_t = Kf(S_t/K, T-t)$ and approximate f by taking $\mathbf{x}_t = (S_t/K, T-t)^T$ in one of the following nonparametric regression models described in Section 7.4: (i) projection pursuit regression, (ii) neural networks, and (iii) radial basis networks. It should be noted that the transformation of S_t to S_t/K above can be motivated not only from the assumption on S_t but also from the special feature of options data. Although the strike price K could in principle be any positive number, an options exchange only sets strike prices as multiples of a fundamental unit. For example, the Chicago Board Options Exchange sets strike prices at multiples of \$5 for stock prices in the \$25 to \$200 range. Moreover, only those options with strike prices closest to the current stock price are traded and yield option prices. Since S_t is nonstationary, the observed K_t is also nonstationary. Such features create sparsity of data in the space of $(S_t, K_t, T-t)$. A nonparametric regression model of the form $f(S, K_t, T-t)$ tends only to interpolate and fails to produce good prediction because (S_t, K_t) in the future can be very different from the data used in estimating f. Choosing a regression function of the form $f(S/K, T-t)$ can make use of the fact that all observed and future S_t/K_t are close to 1, thereby circumventing the sparsity problem. Another point that Hutchinson, Lo, and Poggio (1994) highlighted is the measure of performance of the estimated pricing formula. According to their simulation study, even a linear function $f(S/K, T-t)$ can give $R^2 \approx 90\%$ (see their Table I). However, such a linear f implies a constant delta hedging scheme, which gives poor hedging results. Since the primary function of options is to hedge the risk created by changes in the price of the underlying asset, Hutchinson, Lo, and Poggio (1994) suggested using, instead of R^2, the hedging error measures $\xi = e^{-rT}E[|V(T)|]$ and $\eta = e^{-rT}[EV^2(T)]^{1/2}$, where $V(T)$ is the value of the hedged portfolio at the expiration date T. If all the Black-Scholes assumptions hold, $V(T)$ should be 0 when one uses the Black-Scholes formula. Hutchinson, Lo, and Poggio (1994) reported that the nonparametric delta gives hedging measures comparable to those of Black-Scholes in their simulation study, which assumes the Black-Scholes assumptions with the exception that time is discrete (for daily closing prices) so that daily rebalancing with the Black-Scholes delta still gives $\xi > 0$ and $\eta > 0$. They have also carried out an empirical study of out-of-sample hedging performance for S&P 500 futures options from January 1987 to December 1991 and report that the nonparametric delta performs better than the Black-Scholes delta.

For American vanilla options, Broadie et al. (2000) used kernel smoothers to estimate the option pricing formula of an American option. Using a training sample of daily closing prices of American calls on the S&P 100 index from January 3, 1984 to March 30, 1990, they compared the nonparametric estimates of American call option prices at a set of $(S_t/K, T-t)$ values with those obtained by approximations to (8.9) due to Broadie and Detemple (1996) and reported significant differences between the parametric and nonparametric estimates.

8.3.5 A combined substantive-empirical approach

For European call options, instead of using nonparametric modeling of $f(S/K, T-t)$ as in Hutchinson, Lo, and Poggio (1994), an alternative approach is to express the option price as $c + Ke^{-r(T-t)}f(S/K, T-t)$, where c is the Black-Scholes price (8.6), because the Black-Scholes formula has been widely used by option traders. This is tantamount to including $c(t, S)$ as one of the basis functions (with prescribed weight 1) to come up with a more parsimonious approximation to the actual option price. The usefulness of this idea is even more apparent in the case of American options, which do not have explicit pricing and delta-hedging formulas even under the Black-Scholes assumptions.

Consider the decomposition (8.10), which expresses an American put option price as the sum of a European put price p and the early exercise premium, which is typically small relative to p. This suggests that p should be included as one of the basis functions (with prescribed weight 1). Lai and Wong (2004) propose to use additive regression splines after the change of variables $u = -\sigma^2(T-t)$ and $z = \log(S/K)$. Specifically, for small $T-t$ (say within 5 trading days prior to expiration; i.e., $T - t \le 5/253$ under the assumption of 253 trading days per year), they approximate P by p. For $T - t > 5/253$ (or equivalently $u < -5\sigma^2/253$), they approximate P by

$$\begin{aligned} P = p + Ke^{\rho u}\Bigg\{ & \alpha + \alpha_1 u + \sum_{j=1}^{J_u} \alpha_{1+j}(u - u^{(j)})_+ + \beta_1 z + \beta_2 z^2 \\ & + \sum_{j=1}^{J_z} \beta_{2+j}(z - z^{(j)})_+^2 + \gamma_1 w + \gamma_2 w^2 + \sum_{j=1}^{J_w} \gamma_{2+j}(w - w^{(j)})_+^2 \Bigg\}, \end{aligned} \tag{8.19}$$

where $\rho = r/\sigma^2$ as in (8.11), α, α_j, β_j, and γ_j are regression parameters to be estimated by least squares from the training sample, and

$$w = |u|^{-1/2}\{z - (\rho - \theta\rho - 1/2)u\} \qquad (\theta = q/r) \tag{8.20}$$

is an "interaction" variable derived from z and u. The motivation behind the centering term $(\rho - \theta\rho - 1/2)u$ in (8.20) comes from (8.11), which transforms GBM into Brownian motion, whereas that behind the normalization $|u|^{-1/2}$ comes from the formulas for d_1 and d_2 in (8.7). The knots $u^{(j)}$ (respectively $z^{(j)}$ or $w^{(j)}$) of the linear (respectively quadratic) spline in (8.19) are the $\frac{j}{J_u}$-th (respectively $\frac{j}{J_z}$-th and $\frac{j}{J_w}$-th) quantiles of $\{u_1, \ldots, u_n\}$ (respectively $\{z_1, \ldots, z_n\}$ or $\{w_1, \ldots, w_n\}$). The choice of J_u, J_z and J_w is over all possible integers between 1 and 10 to minimize the generalized cross-validation (GCV, see Section 7.3.2) criterion, which can be expressed as

$$\mathrm{GCV}(J_u, J_z, J_w) = \sum_{i=1}^{n}(P_i - \hat{P}_i)^2 \Bigg/ \left\{ n\left(1 - \frac{J_u + J_z + J_w + 6}{n}\right)^2 \right\},$$

where the P_i are the observed American option prices in the past n periods and the $\widehat{P}_i$ are the corresponding fitted values given by (8.19), in which the regression coefficients are estimated by least squares.

In the preceding, we have assumed prescribed constants r and σ as in the Black-Scholes model; these parameters appear in (8.19) via the change of variables (8.11). In practice, σ is unknown and may also vary with time. Lai and Wong (2004) replace it in (8.19) by the standard deviation $\widehat{\sigma}_t$ of the most recent asset prices, say, during the past 60 trading days prior to t as in Hutchinson, Lo, and Poggio (1994, p. 881). This is tantamount to incorporating the asset prices $S_{t-1}, \ldots, S_{t-60}$ in the formula $P(t, S; S_{t-1}, \ldots, S_{t-60})$ with $S_t = S$. Moreover, the risk-free rate r may also change with time and can be replaced by the yield $\widehat{r}_t$ of a short-maturity Treasury bill on the close of the month before t. The same remark also applies to the dividend rate. Lai and Wong (2004) report a simulation study of the performance, measured by $E|\widehat{P} - P|$ and $E\{e^{-r\tau}|V(\tau)|\}$ (which corresponds to changing T to the early exercise time τ in the measure proposed by Hutchinson, Lo, and Poggio (1994) for European options), of their pricing formula $\widehat{P}$ and the Black-Scholes-Merton formula (8.10) when the portfolio is rebalanced daily. The simulation study shows that whereas (8.10) can perform quite poorly when the Black-Scholes assumptions are violated, $\widehat{P}$ does not encounter such difficulties.

Assumption (A3) is clearly violated in practice, and transaction costs make it prohibitively expensive to rebalance the portfolio continuously as in Black and Scholes (1973) or even daily as in Hutchinson, Lo, and Poggio (1994). When the pricing formula can be "learned" from the market, how should the

Black-Scholes delta hedging strategy be modified in the presence of transaction costs? Lai and Lim (2007) and Lai, Lim, and Chen (2007) have recently studied this problem by using (a) the connections between singular stochastic control and optimal stopping and (b) ideas from stochastic adaptive control. Here "singular" means that buying or selling shares of the underlying stock (regarded as a control action) only occurs occasionally, and "adaptive" means that the parameters of the stochastic system are unknown and have to be estimated from past information that consists not only of historical option and stock prices but also the past performance of the hedging strategy. Lai, Lim, and Chen (2007) have also pointed out certain difficulties and shortcomings with the implied volatility approach in hedging applications and have proposed to use the combined substantive-empirical approach to circumvent these difficulties.

Exercises

8.1. Prove (8.8).

8.2. The price of a European call option with strike price K and maturity T is $e^{-rT}E\{(S_T - K)_+\}$, where E is expectation under the risk-neutral measure. This representation is applicable not only to the case where $\{S_t, 0 \le t \le T\}$ is geometric Brownian motion but also to more general stock price processes, and the call option price can be expressed more generally as

$$c = e^{-rT}\int_K^\infty (x-K)_+ g(x)dx, \tag{8.21}$$

where g is the density function of S_T under the risk-neutral measure.

(a) Use (8.21) to show that g can be represented by

$$g(K) = e^{rT}\frac{\partial^2 c}{\partial K^2}. \tag{8.22}$$

(b) Note that (8.22) assumes that the option price $c = c(K, S_0, T, r)$ has an explicit formula, which is usually not the case beyond the Black-Scholes model. Discuss how you would estimate the density function g from a sample $\{(c_i, S_{0,i}, T_i, K_i, r_i) : 1 \le i \le n\}$.

8.3. The file `impvol_call010305.txt` contains the strike prices, implied volatilities, and maturity dates of call options on the S&P 500 index on January 3, 2005. The closing price of the S&P 500 index was \$1.2091, and the 1-month U.S. Treasury constant maturity rate (which we take as the risk-free rate in the Black-Scholes model) was 1.99%.

(a) Recover the prices of the corresponding European call options.

(b) There are two expiration dates in these data. Plot for each expiration date the implied volatilities versus their strike prices and describe the smiles. Use kernel smoothing to estimate the smile (i.e., the implied volatility versus strike price relationship) and plot it for each expiration date.

(c) Plot the difference between the two estimated implied volatility curves (smiles) in (b). Comment on the patterns you find.

8.4. The European call option price $c(t, S)$ and European put option price $p(t, S)$ under the CEV model for the asset price S_t (with respect to the risk-neutral measure) in Section 8.3.2 are given by (8.15) in the case $\alpha > 1$ and by (8.16) in the case $0 < \alpha < 1$. In the case $\alpha = 1$, $c(t, S)$ and $p(t, S)$ are given by the Black-Scholes formulas (8.6) and (8.7). Use the data in Exercise 8.3 to estimate the parameters σ and α of the CEV model by minimizing the sum of squared differences between the model price and the actual call option price (i.e., nonlinear least squares in Section 4.2). Note that $q = 0$. You can use the `R` function `pchisq` or the `MATLAB` function `ncx2cdf` to compute the distribution function $\chi^2(\cdot; \nu, \lambda)$.

8.5. The file `impvol_strike1_tom1.txt` contains the implied volatilities of call options on the S& P500 index from January 3, 2005 to April 10, 2006 with 1 year to maturity and strike price \$1. Fit an ARIMA model to the time series of logarithms of implied volatilities.

8.6. Consider the dataset in Exercise 7.6. The file `m_tbill_3m.txt` contains the yield of the 3-month U.S. Treasury bill (obtained as a secondary market rate in the Federal Reserve Board H.15 publication) on the closing date of each month during the period.

(a) Using the standard deviation of the 60 most recent daily log returns of the S&P 500 futures prices, estimate the volatility σ on date t in the Black-Scholes model for the period from March to December 1987.

(b) Estimate the Black-Scholes price on each date t during this period by using the estimate of σ obtained in (a) and the yield of the 3-month U.S. Treasury bill on the closing date of the month prior to or at t.

(c) Plot the time series of differences between the Black-Scholes prices in (b) and the actual option prices.

(d) Compute the implied volatilities on every trading day from March to December 1987. Plot the time series of historic volatilities in (a) and of the implied volatilities. Discuss the patterns you find from your plot.

9

Advanced Multivariate and Time Series Methods in Financial Econometrics

The importance of multivariate statistical methods, for which Chapter 2 provides an introduction, has been demonstrated in Chapters 3 and 4 in connection with portfolio optimization. As noted in Section 4.4, more advanced methods for high-dimensional multivariate data are needed to handle large portfolios. This chapter introduces several advanced multivariate statistical techniques. Section 9.1 begins with *canonical correlation analysis*, which analyzes the correlation structure between two random vectors $\mathbf{x}$ and $\mathbf{y}$, not necessarily of the same dimension, through *canonical variate pairs* of the form $(\boldsymbol{\alpha}^T\mathbf{x}, \boldsymbol{\beta}^T\mathbf{y})$, where the linear combinations $\boldsymbol{\alpha}^T\mathbf{x}$ and $\boldsymbol{\beta}^T\mathbf{y}$ are chosen in a way similar to how principal components are defined. Section 9.2 generalizes regression analysis to the case where output variables are $k \times 1$ vectors $\mathbf{y}_t$. The multivariate regression model is of the form $\mathbf{y}_t = \mathbf{B}\mathbf{x}_t + \boldsymbol{\epsilon}_t$, in which the regressors $\mathbf{x}_t$ are $p \times 1$ vectors, as in Chapter 1. The $k \times p$ coefficient matrix $\mathbf{B}$ involves kp parameters, which may be too many to estimate well for the sample size n often used in empirical studies. Of particular importance is a technique, called *reduced-rank regression*, that addresses this problem by assuming that $\text{rank}(\mathbf{B}) = r \leq \min(p, k)$ and applying canonical correlation analysis of $\mathbf{x}_i$, $\mathbf{y}_i$ to find a reduced-rank regression model. Section 9.3 introduces *modified Cholesky decompositions* of covariance matrices and their applications to the analysis of high-dimensional covariance matrices associated with large portfolios.

Multivariate time series modeling is of particular interest in financial markets where one considers prices of k different assets or interest rates with k different maturities. Although extension of the univariate time series models in Chapters 5 and 6 to the k-variate setting seems to involve only straightforward substitution of random variables by random vectors and scalar parameters by parameter matrices, there are two important issues that complicate multivariate time series modeling. The first one is the "curse of dimensionality," as the number of parameters to be estimated increases at a polynomial rate with k.

Another issue is that care has to be taken in applying multivariate analysis techniques to analyze multivariate time series that are unit-root nonstationary. Even standard methods such as regression analysis can give spurious results for unit-root nonstationary time series unless they are "cointegrated" in the sense that they have some linear combination that is stationary. Cointegration tests and statistical inference for cointegrated time series are important topics in econometrics. Section 9.4 addresses these topics and issues in multivariate time series modeling, points out the connection between reduced-rank regression and cointegration, and introduces unit-root tests that are widely used in econometric time series analysis.

Another important topic in econometric time series is regime switching and the possibility of parameter jumps. When such structural changes are ignored, models fitted to the time series data often exhibit long memory. Sections 9.5 and 9.6 introduce long-memory time series models, regime-switching models, and autoregressive models with piecewise constant autoregressive and volatility parameters. Section 9.6 also considers other advanced topics in volatility modeling, including stochastic volatility and multivariate GARCH models.

A widely used estimation method in econometrics is the *generalized method of moments*. It deals with economic models in which the stochastic mechanisms generating the data are only specified by certain moment restrictions that hold at the true value of the parameter vector. Section 9.7 describes the method and gives some applications to financial time series.

9.1 Canonical correlation analysis

9.1.1 Cross-covariance and correlation matrices

Definition 9.1. (i) The *cross-covariance matrix* $\boldsymbol{\Sigma}_{\mathbf{xy}}$ of two random vectors $\mathbf{x} = (X_1, \dots, X_p)^T$ and $\mathbf{y} = (Y_1, \dots, Y_q)^T$ is a $p \times q$ matrix defined by

$$\boldsymbol{\Sigma}_{\mathbf{xy}} = \Big(\mathrm{Cov}(X_i, Y_j)\Big)_{1\le i\le p, 1\le j\le q}. \tag{9.1}$$

(ii) The *cross-correlation matrix* $\mathbf{R}_{\mathbf{xy}}$ between $\mathbf{x}$ and $\mathbf{y}$ is

$$\mathbf{R}_{\mathbf{xy}} = \Big(\mathrm{Corr}(X_i, Y_j)\Big)_{1\le i\le p, 1\le j\le q} = \boldsymbol{\Sigma}_{\mathbf{xx}}^{-1/2}\boldsymbol{\Sigma}_{\mathbf{xy}}\boldsymbol{\Sigma}_{\mathbf{yy}}^{-1/2}. \tag{9.2}$$

(iii) For an $r \times 1$ random vector $\mathbf{z}$, the *partial covariance matrix* of $\mathbf{x}$ and $\mathbf{y}$ adjusted for $\mathbf{z}$ is defined by

$$\boldsymbol{\Sigma}_{\mathbf{xy}.\mathbf{z}} = \boldsymbol{\Sigma}_{\mathbf{xy}} - \boldsymbol{\Sigma}_{\mathbf{xz}}\boldsymbol{\Sigma}_{\mathbf{zz}}^{-1}\boldsymbol{\Sigma}_{\mathbf{zy}}. \tag{9.3}$$

From (9.1), it follows that $\boldsymbol{\Sigma}_{\mathbf{xy}} = \boldsymbol{\Sigma}_{\mathbf{yx}}^T$. The second equality in (9.2) is a matrix generalization of $\mathrm{Corr}(X, Y) = \mathrm{Cov}(X, Y)/\sigma_X\sigma_Y$; see Section 2.2.2

for the definition of the square root of a covariance matrix. Let $\mathbf{A}$ and $\mathbf{B}$ be $p \times r$ and $q \times r$ nonrandom matrices. We can generalize (2.4) to

$$\text{Cov}(\mathbf{Az}, \mathbf{Bz}) = \mathbf{A}\boldsymbol{\Sigma}_{\mathbf{zz}}\mathbf{B}^T. \tag{9.4}$$

Consequently, if $\text{Cov}(\mathbf{z}) = \sigma^2\mathbf{I}$, then

$$\mathbf{Az} \text{ and } \mathbf{Bz} \text{ are uncorrelated if } \mathbf{AB}^T = \mathbf{0}. \tag{9.5}$$

9.1.2 Canonical correlations

Let $\mathbf{x}$ and $\mathbf{y}$ be $p \times 1$ and $q \times 1$ random vectors, respectively. The objective of canonical correlation analysis is to measure the strength of association between the two sets of variables $\{x_1, \ldots, x_p\}$ and $\{y_1, \ldots, y_q\}$ via the correlations between the linear combinations $\boldsymbol{\alpha}^T\mathbf{x}$ and $\boldsymbol{\beta}^T\mathbf{y}$ by first choosing $\boldsymbol{\alpha}$ and $\boldsymbol{\beta}$ that give the largest correlation, then choosing the pair of linear combinations with the largest correlation among all pairs that are uncorrelated with the initially selected pair, and so on. These linear combinations are called *canonical variates*, and their correlation coefficients are called *canonical correlations*. The basic underlying idea is to maximize the correlation between the two sets of variables by using a few pairs of canonical variables (or "canonical variate pairs").

Specifically, *the first canonical variate pair* is the pair of linear combinations $\boldsymbol{\alpha}^T\mathbf{x}$ and $\boldsymbol{\beta}^T\mathbf{y}$ that maximizes

$$\text{Corr}(\boldsymbol{\alpha}^T\mathbf{x}, \boldsymbol{\beta}^T\mathbf{y}) = \frac{\boldsymbol{\alpha}^T\boldsymbol{\Sigma}_{\mathbf{xy}}\boldsymbol{\beta}}{\sqrt{\boldsymbol{\alpha}^T\boldsymbol{\Sigma}_{\mathbf{xx}}\boldsymbol{\alpha}}\sqrt{\boldsymbol{\beta}^T\boldsymbol{\Sigma}_{\mathbf{yy}}\boldsymbol{\beta}}}, \tag{9.6}$$

where

$$\text{Cov}\begin{pmatrix}\mathbf{x}\\ \mathbf{y}\end{pmatrix} = \begin{pmatrix}\boldsymbol{\Sigma}_{\mathbf{xx}} & \boldsymbol{\Sigma}_{\mathbf{xy}}\\ \boldsymbol{\Sigma}_{\mathbf{yx}} & \boldsymbol{\Sigma}_{\mathbf{yy}}\end{pmatrix}, \quad \boldsymbol{\Sigma}_{\mathbf{xx}} = \text{Cov}(\mathbf{x}),\ \boldsymbol{\Sigma}_{\mathbf{yy}} = \text{Cov}(\mathbf{y}).$$

This is equivalent to maximizing $\boldsymbol{\alpha}^T\boldsymbol{\Sigma}_{\mathbf{xy}}\boldsymbol{\beta}$ subject to $\text{Var}(\boldsymbol{\alpha}^T\mathbf{x}) = \boldsymbol{\alpha}^T\boldsymbol{\Sigma}_{\mathbf{xx}}\boldsymbol{\alpha} = 1$ and $\text{Var}(\boldsymbol{\beta}^T\mathbf{y}) = \boldsymbol{\beta}^T\boldsymbol{\Sigma}_{\mathbf{yy}}\boldsymbol{\beta} = 1$. In general, the kth ($k \geq 2$) canonical variate pair is the pair of linear combinations $\boldsymbol{\alpha}_k^T\mathbf{x}$ and $\boldsymbol{\beta}_k^T\mathbf{y}$, having unit variances, that maximize (9.6) among $\boldsymbol{\alpha}^T\mathbf{x}$ and $\boldsymbol{\beta}^T\mathbf{y}$ having unit variances and being uncorrelated with the linear combination in the previous $k-1$ canonical variate pairs $(\boldsymbol{\alpha}_i, \boldsymbol{\beta}_i), 1 \leq i \leq k-1$. The kth canonical variate pair, therfore, maximizes $\boldsymbol{\alpha}^T\boldsymbol{\Sigma}_{\mathbf{xy}}\boldsymbol{\beta}$ subject to

$$\boldsymbol{\alpha}^T\boldsymbol{\Sigma}_{\mathbf{xx}}\boldsymbol{\alpha} = \boldsymbol{\beta}^T\boldsymbol{\Sigma}_{\mathbf{yy}}\boldsymbol{\beta} = 1 \text{ and } \boldsymbol{\alpha}_i^T\boldsymbol{\Sigma}_{\mathbf{xx}}\boldsymbol{\alpha} = \boldsymbol{\beta}_i^T\boldsymbol{\Sigma}_{\mathbf{yy}}\boldsymbol{\beta} = 0, \quad i = 1, \cdots, k-1.$$

Canonical correlation is similar in spirit to PCA in Chapter 2, except that it considers the cross-covariance matrix $\boldsymbol{\Sigma}_{\mathbf{xy}}$ while PCA considers the covariance matrix $\boldsymbol{\Sigma}_{\mathbf{xx}}$. Therefore, we have to use singular-value decompositions for $p \times q$ matrices (Venables and Ripley, 2002, p. 62) instead of those for symmetric matrices.

To determine the canonical variate pairs, we use the $p \times q$ matrix

$$\mathbf{K} = \boldsymbol{\Sigma}_{\mathbf{xx}}^{-1/2} \boldsymbol{\Sigma}_{\mathbf{xy}} \boldsymbol{\Sigma}_{\mathbf{yy}}^{-1/2}$$

and consider the nonzero eigenvalues of $\mathbf{K}\mathbf{K}^T$. Suppose that the number of nonzero eigenvalues of $\mathbf{K}\mathbf{K}^T$ is r. Let $\lambda_1 \geq \cdots \geq \lambda_r > 0$ be those eigenvalues, which can be shown to be the same as the ordered positive eigenvalues of $\mathbf{K}^{\mathbf{T}}\mathbf{K}$. Then $\mathbf{K}$ has the singular-value decomposition

$$\mathbf{K} = (\mathbf{a}_1, \cdots, \mathbf{a}_r) \cdot \text{diag}(\sqrt{\lambda_1}, \cdots, \sqrt{\lambda_r}) \cdot (\mathbf{b}_1, \cdots, \mathbf{b}_r)^T, \tag{9.7}$$

in which $\mathbf{a}_i$ $(1 \leq i \leq p)$ and $\mathbf{b}_j$ $(1 \leq j \leq q)$ are the eigenvectors of $\mathbf{K}\mathbf{K}^T$ and $\mathbf{K}^T\mathbf{K}$, respectively, that are normalized so that $||\mathbf{a}_i|| = 1 = ||\mathbf{b}_j||$, with $\mathbf{a}_i$ and $\mathbf{b}_i$ corresponding to the positive eigenvalue λ_i for $1 \leq i \leq r$. The ith canonical correlation is given by

$$\text{Corr}(\boldsymbol{\alpha}_i^T \mathbf{x}, \boldsymbol{\beta}_i^T \mathbf{y}) = \sqrt{\lambda_i}, \tag{9.8}$$

and the ith canonical variate pair is

$$\boldsymbol{\alpha}_i = \boldsymbol{\Sigma}_{\mathbf{xx}}^{-1/2} \mathbf{a}_i, \qquad \boldsymbol{\beta}_i = \boldsymbol{\Sigma}_{\mathbf{yy}}^{-1/2} \mathbf{b}_i. \tag{9.9}$$

Since the eigenvalues of $\mathbf{K}\mathbf{K}^T$ are the same as those of $\boldsymbol{\Sigma}_{\mathbf{xx}}^{-1} \boldsymbol{\Sigma}_{\mathbf{xy}} \boldsymbol{\Sigma}_{\mathbf{yy}}^{-1} \boldsymbol{\Sigma}_{\mathbf{yx}}$, the latter matrix is often used to determine the eigenvalues λ_i and the corresponding eigenvectors. The eigenvector corresponding to λ_i in this case is already $\boldsymbol{\alpha}_i$ and is $\boldsymbol{\beta}_i$ if $\boldsymbol{\Sigma}_{\mathbf{yy}}^{-1} \boldsymbol{\Sigma}_{\mathbf{yx}} \boldsymbol{\Sigma}_{\mathbf{xx}}^{-1} \boldsymbol{\Sigma}_{\mathbf{xy}}$ is used instead; see Exercise 9.1.

Canonical correlation analysis can likewise be performed on the sample cross-covariance matrix $\widehat{\boldsymbol{\Sigma}}_{\mathbf{x},\mathbf{y}}$ based on the multivariate sample

$$\begin{pmatrix} \mathbf{x}_1 \\ \mathbf{y}_1 \end{pmatrix}, \ldots, \begin{pmatrix} \mathbf{x}_n \\ \mathbf{y}_n \end{pmatrix},$$

yielding the eigenvalues $\widehat{\lambda}_1, \ldots, \widehat{\lambda}_r$ and the corresponding canonical variate pairs $(\widehat{\boldsymbol{\alpha}}_i^T, \widehat{\boldsymbol{\beta}}_i^T), 1 \leq i \leq r$.

The following functions in MATLAB, R, or Splus can be used to perform canonical correlation analysis:

```
R or Splus:   cancor,
MATLAB:       canoncorr.
```

9.2 Multivariate regression analysis

9.2.1 Least squares estimates in multivariate regression

Consider the multivariate linear regression model

$$\mathbf{y}_k = \mathbf{B}\mathbf{x}_k + \boldsymbol{\epsilon}_k, \quad k = 1, \cdots, n, \tag{9.10}$$

with response variable $\mathbf{y}_k = (y_{k1}, \cdots, y_{kq})^T \in \mathbb{R}^q$, predictor variable $\mathbf{x}_k = (x_{k1}, \cdots, x_{kp})^T \in \mathbb{R}^p$, and random error $\boldsymbol{\epsilon}_k \in \mathbb{R}^q$ such that $E(\boldsymbol{\epsilon}_k) = \mathbf{0}$ and $\mathrm{Cov}(\boldsymbol{\epsilon}_k) = \boldsymbol{\Sigma}$. Let $\mathbf{X} = (\mathbf{x}_1, \cdots, \mathbf{x}_n)^T$, $\mathbf{Y} = (\mathbf{y}_1, \cdots, \mathbf{y}_n)^T$, and $\boldsymbol{\epsilon} = (\boldsymbol{\epsilon}_1, \cdots, \boldsymbol{\epsilon}_n)^T$. Then the regression model (9.10) can be written as

$$\mathbf{Y} = \mathbf{X}\mathbf{B}^T + \boldsymbol{\epsilon}. \tag{9.11}$$

Similar to univariate linear regression, the method of least squares can be used to estimate the regression coefficient matrix $\mathbf{B}$ and the error covariance matrix $\boldsymbol{\Sigma}$. Assume that $\mathbf{X}$ is of full rank; i.e., $p = \mathrm{rank}(\mathbf{X})$, and $p < n$. Then, minimizing the sum of squares of the components of $\mathbf{Y} - \mathbf{X}\mathbf{B}^T$ over $\mathbf{B}$ yields the least squares estimate

$$\widehat{\mathbf{B}}^T = (\mathbf{X}^T\mathbf{X})^{-1}\mathbf{X}^T\mathbf{Y} = \left(\sum_{t=1}^n \mathbf{x}_t\mathbf{x}_t^T\right)^{-1} \sum_{t=1}^n \mathbf{x}_t\mathbf{y}_t^T. \tag{9.12}$$

An unbiased estimate of $\boldsymbol{\Sigma}$ when the $\mathbf{x}_t$ are nonrandom is given by

$$\widehat{\boldsymbol{\Sigma}}_{\boldsymbol{\epsilon}} = \frac{1}{n-p}\sum_{t=1}^n (\mathbf{y}_t - \widehat{\mathbf{B}}\mathbf{x}_t)(\mathbf{y}_t - \widehat{\mathbf{B}}\mathbf{x}_t)^T, \tag{9.13}$$

which is a natural generalization of (1.10) in the case $q = 1$.

Since the multivariate regression model (9.11) is an extension of the linear regression model (1.1), various results in Chapter 1 can be generalized to the multivariate case. In fact, (9.12) is equivalent to applying OLS to each component $y_{kj} = \mathbf{x}_k^T\boldsymbol{\beta}^{(j)} + \epsilon_{kj}$ of $\mathbf{y}_k$, where $\boldsymbol{\beta}^{(j)}$ is the jth column vector of $\mathbf{B}^T, 1 \le j \le q$. Moreover, the least squares estimate is the same as the maximum likelihood estimate of $\mathbf{B}$ when the $\boldsymbol{\epsilon}_k$ in (9.10) are i.i.d. $N(\mathbf{0}, \boldsymbol{\Sigma})$, for which the MLE of $\boldsymbol{\Sigma}$ is (9.13) with $1/(n-p)$ replaced by $1/n$.

9.2.2 Reduced-rank regression

The $q \times p$ coefficient matrix $\mathbf{B}$ in (9.10) has pq parameters. Reduced-rank regression assumes that $\mathrm{rank}(\mathbf{B}) = r \le \min(p, q)$ or, equivalently, that $\mathbf{B} = \mathbf{A}\mathbf{C}$, where $\mathbf{A}$ and $\mathbf{C}$ are $q \times r$ and $r \times p$ matrices, respectively, so that the number of unknown parameters becomes $(p+q)r$. It often arises in econometric modeling via "latent variables" $\mathbf{y}_k^*$ that are related to the predictors $\mathbf{x}_k$ by $\mathbf{y}_k^* = \mathbf{C}\mathbf{x}_k + \boldsymbol{\eta}_k$. These latent variables are unobserved, and the observations are $\mathbf{y}_k = \mathbf{A}\mathbf{y}_k^* + \boldsymbol{\omega}_k$. The "reduced-form" model is

$$\mathbf{y}_k = \mathbf{A}\mathbf{C}\mathbf{x}_k + (\mathbf{A}\boldsymbol{\eta}_k + \boldsymbol{\omega}_k) = \mathbf{B}\mathbf{x}_k + \boldsymbol{\epsilon}_k.$$

We next show how the least squares estimate (9.12) can be modified to have reduced rank r. Note that (9.12) is obtained by minimizing $\sum_{k=1}^n \operatorname{tr}\left[(\mathbf{y}_k - \mathbf{B}\mathbf{x}_k)(\mathbf{y}_k - \mathbf{B}\mathbf{x}_k)^T\right]$ over $\mathbf{B}$. More generally, we can consider minimizing

$$\sum_{k=1}^n \operatorname{tr}\left[\boldsymbol{\Gamma}^{1/2}(\mathbf{y}_k - \mathbf{A}\mathbf{C}\mathbf{x}_k)(\mathbf{y}_k - \mathbf{A}\mathbf{C}\mathbf{x}_k)^T \boldsymbol{\Gamma}^{1/2}\right] \tag{9.14}$$

over $q \times r$ and $r \times p$ matrices $\mathbf{A}$ and $\mathbf{C}$, where $\boldsymbol{\Gamma}$ is a given $q \times q$ positive definite matrix. The solution of the minimization problem turns out to be closely related to the formulas for canonical correlation in Section 9.1.2. Let

$$\mathbf{S}_{\mathbf{xx}} = \frac{1}{n}\sum_{k=1}^n \mathbf{x}_k\mathbf{x}_k^T, \quad \mathbf{S}_{\mathbf{yy}} = \frac{1}{n}\sum_{k=1}^n \mathbf{y}_k\mathbf{y}_k^T, \quad \mathbf{S}_{\mathbf{xy}} = \frac{1}{n}\sum_{k=1}^n \mathbf{x}_k\mathbf{y}_k^T, \quad \mathbf{S}_{\mathbf{yx}} = \mathbf{S}_{\mathbf{xy}}^T,$$

$$\mathbf{H} = \boldsymbol{\Gamma}^{1/2}\mathbf{S}_{\mathbf{yx}}\mathbf{S}_{\mathbf{xx}}^{-1}\mathbf{S}_{\mathbf{xy}}\boldsymbol{\Gamma}^{1/2}.$$

Note the resemblance between $\mathbf{H}$ and $\mathbf{K}^{\mathbf{T}}\mathbf{K}$ in Section 9.1.2 when $\boldsymbol{\Gamma} = \mathbf{S}_{\mathbf{yy}}^{-1}$. Note that the matrices $\mathbf{A}$ and $\mathbf{C}$ are not uniquely determined. In particular, if $\mathbf{A}$ and $\mathbf{C}$ minimize (9.14) and $\mathbf{P}$ is an $r \times r$ orthogonal matrix (i.e., $\mathbf{P}\mathbf{P}^T = \mathbf{I}$), then $\mathbf{A}\mathbf{P}$ and $\mathbf{P}^T\mathbf{C}$ also minimize (9.14). To make the solution $\mathbf{A}$ and $\mathbf{C}$ unique for the given $\boldsymbol{\Gamma}$, let $\mathbf{b}_1, \cdots, \mathbf{b}_r$ be the normalized eigenvectors of $\mathbf{H}$, with $\mathbf{b}_j$ corresponding to the jth-largest eigenvalue of $\mathbf{H}$. Then the $\widehat{\mathbf{A}}$ and $\widehat{\mathbf{C}}$ that minimize (9.14) are given by

$$\widehat{\mathbf{A}} = \boldsymbol{\Gamma}^{-1/2}(\mathbf{b}_1, \ldots, \mathbf{b}_r), \qquad \widehat{\mathbf{C}} = (\mathbf{b}_1, \ldots, \mathbf{b}_r)^T \boldsymbol{\Gamma}^{1/2}\mathbf{S}_{\mathbf{yx}}\mathbf{S}_{\mathbf{xx}}^{-1}, \tag{9.15}$$

and $\mathbf{B} = \mathbf{A}\mathbf{C}$ can be estimated by

$$\widehat{\mathbf{B}} = \left\{\boldsymbol{\Gamma}^{-1/2}\left(\sum_{j=1}^r \mathbf{b}_j\mathbf{b}_j^T\right)\boldsymbol{\Gamma}^{1/2}\right\}\mathbf{S}_{\mathbf{yx}}\mathbf{S}_{\mathbf{xx}}^{-1}; \tag{9.16}$$

see Reinsel and Velu (1998, pp. 28–29). Note that $\mathbf{S}_{\mathbf{yx}}\mathbf{S}_{\mathbf{xx}}^{-1}$ is the transpose of the right-hand side of (9.12). Hence the reduced-rank estimate (9.16) simply modifies the OLS estimate (9.12) by multiplying it on the left by $\boldsymbol{\Gamma}^{-1/2}\left(\sum_{j=1}^r \mathbf{b}_j\mathbf{b}_j^T\right)\boldsymbol{\Gamma}^{1/2}$.

In the regression context, a natural choice of $\boldsymbol{\Gamma}$ is $\boldsymbol{\Sigma}^{-1}$, for which the minimizer of (9.14) is the GLS estimate of the reduced-rank parameter matrix

$\mathbf{B} = \mathbf{AC}$. Although $\boldsymbol{\Sigma}$ is unknown, we can estimate it by using (9.13), which is based on the OLS estimate (9.12) without the rank constraint.

9.3 Modified Cholesky decomposition and high-dimensional covariance matrices

Definition 9.2. Let $\mathbf{V}$ be a $p \times p$ nonnegative definite matrix.

(i) The decomposition $\mathbf{V} = \mathbf{P}\mathbf{P}^T$, with $\mathbf{P}$ being a $p \times p$ lower-triangular matrix, is called the *Cholesky decomposition* of $\mathbf{V}$.

(ii) The decomposition $\mathbf{V} = \mathbf{L}\mathbf{D}\mathbf{L}^T$, with diagonal matrix $\mathbf{D} = \text{diag}(d_1, \ldots, d_p)$ and lower-triangular matrix $\mathbf{L}$ whose diagonal elements are 1, is called the *modified Cholesky decomposition* of $\mathbf{V}$.

For a positive definite matrix $\mathbf{V}$, the elements of $\mathbf{L}$ and $\mathbf{D}$ in its modified Cholesky decomposition can be computed inductively, one row at a time, beginning with $d_1 = V_{11}$:

$$L_{ij} = \left(V_{ij} - \sum_{k=1}^{j-1} L_{ik} d_k L_{jk} \right) \Big/ d_j, \qquad j = 1, \ldots, i-1; \tag{9.17}$$

$$d_i = V_{ii} - \sum_{k=1}^{i-1} d_k L_{ik}^2, \qquad i = 2, \ldots, p. \tag{9.18}$$

The existence of modified Cholesky decomposition can be established more generally for symmetric matrices. If $\mathbf{V}$ is nonnegative definite, then the diagonal elements d_i in its Cholesky decomposition are nonnegative and therefore $\mathbf{D}^{1/2} = \text{diag}(\sqrt{d_1}, \ldots, \sqrt{d_p})$ is well defined. Therefore we can write $\mathbf{V} = (\mathbf{L}\mathbf{D}^{1/2})(\mathbf{L}\mathbf{D}^{1/2})^T$, and $\mathbf{P} := \mathbf{L}\mathbf{D}^{1/2}$ is a lower-triangular matrix, yielding the Cholesky decomposition $\mathbf{V} = \mathbf{P}\mathbf{P}^T$.

If $\mathbf{L}$ is lower diagonal with unit diagonal elements, then so is $\mathbf{L}^{-1}$. Therefore the modified Cholesky decomposition of a positive definite matrix $\mathbf{V} = \mathbf{L}\mathbf{D}\mathbf{L}^T$ yields a similar decomposition for its inverse, $\mathbf{V}^{-1} = (\mathbf{L}^T)^{-1}\mathbf{D}^{-1}\mathbf{L}^{-1}$, but with its left factor being upper triangular.

Regression interpretation of modified Cholesky decomposition and an application

Let $\mathbf{V}$ be the covariance matrix of a random vector $\mathbf{y} = (y_1, \ldots, y_p)^T$ having mean $\boldsymbol{\mu}$. If $\mathbf{V}$ is positive definite, then the matrices $\mathbf{L}$ and $\mathbf{D}$ in its modified Cholesky decomposition of $\mathbf{L}\mathbf{D}\mathbf{L}^T$ have the following regression interpretation. The below-diagonal entries of $\mathbf{L}^{-1}$ are $-\beta_j^{(i)}$, where the $\beta_j^{(i)}$ are the coefficients of the minimum-variance linear predictor

$$\widehat{y}_i = \mu_i + \sum_{j=1}^{i-1} \beta_j^{(i)} (y_j - \mu_j) \tag{9.19}$$

of y_i based on $y_1, \ldots, y_{i-1}$. Moreover, the diagonal elements of $\mathbf{D}$ are $d_1 = \mathrm{Var}(y_1)$ and

$$d_i = \mathrm{Var}(y_i - \widehat{y}_i) = V_{ii} - (V_{1i}, \ldots, V_{i-1,i})^T \mathbf{V}_{i-1}^{-1} (V_{1i}, \ldots, V_{i-1,i}) \tag{9.20}$$

for $2 \le i \le p$, where $\mathbf{V}_{i-1} = (V_{kl})_{1 \le k,l \le i-1}$; see Exercise 9.2. Hence we can impose sparsity constraints on $\mathbf{V}$ via negligibility of many of the regression coefficients $\beta_j^{(i)}$; see Bickel and Lavina (2007), Huang et al. (2006) and Ing, Lai and Chen (2007).

9.4 Multivariate time series

9.4.1 Stationarity and cross-correlation

A k-dimensional time series $\mathbf{x}_t = (x_{t1}, \cdots, x_{tk})^T$ is called *weakly stationary* (or *stationary in the weak sense*) if its mean vectors and covariance matrices are time-invariant and therefore can be denoted by

$$\boldsymbol{\mu} = E(\mathbf{x}_t), \quad \boldsymbol{\Gamma}_h = E[(\mathbf{x}_t - \boldsymbol{\mu}_t)(\mathbf{x}_{t+h} - \boldsymbol{\mu}_{t+h})^T] = \big(\Gamma_h(i,j)\big)_{1 \le i,j \le k}. \tag{9.21}$$

Letting $\mathbf{D} = \mathrm{diag}\{\sqrt{\Gamma_0(1,1)}, \cdots, \sqrt{\Gamma_0(k,k)}\}$, the *cross-correlation matrix* of $\mathbf{x}_t$ is given by

$$\boldsymbol{\rho}_0 = \big(\rho_0(i,j)\big)_{1 \le i,j \le k} = \mathbf{D}^{-1} \boldsymbol{\Gamma}_0 \mathbf{D}^{-1},$$

where $\rho_0(i,j)$ is the correlation coefficient between x_{ti} and x_{tj}. Similarly, the lag-h cross-correlation matrix of $\mathbf{x}_t$ is given by

$$\boldsymbol{\rho}_h = \big(\mathrm{Corr}(x_{t,i}, x_{t+h,j})\big)_{1 \le i,j \le k} = \mathbf{D}^{-1} \boldsymbol{\Gamma}_h \mathbf{D}^{-1}.$$

Given observations $\{\mathbf{x}_1, \ldots, \mathbf{x}_n\}$ from a weakly stationary time series, we can estimate its mean $\boldsymbol{\mu}$ by $\bar{\mathbf{x}} = n^{-1} \sum_{t=1}^n \mathbf{x}_t$ and its lag-h cross-covariance matrix $\boldsymbol{\Gamma}_h$ by

$$\widehat{\boldsymbol{\Gamma}}_h = n^{-1} \sum_{t=h+1}^{n} (\mathbf{x}_t - \bar{\mathbf{x}})(\mathbf{x}_{t-h} - \bar{\mathbf{x}})^T. \tag{9.22}$$

9.4.2 Dimension reduction via PCA

Let $\mathbf{x}_t = (x_{t1}, \ldots, x_{tk})^T$, $1 \le t \le n$, be a weakly stationary multivariate time series. Let $\mathbf{X}_j = (x_{1j}, \ldots, x_{nj})^T$, $1 \le j \le k$. Principal component analysis (PCA) is introduced in Section 2.2 to represent $\mathbf{X}_j$ in terms of normalized eigenvectors of the sample covariance matrix. For the special case of i.i.d. $\mathbf{x}_t$ considered in Section 2.2, $\widehat{\boldsymbol{\Gamma}}_0$ in (9.22) is a consistent estimate of the covariance matrix $\boldsymbol{\Gamma}_0$ in (9.21). Consistency still holds when the i.i.d. assumption is replaced by weak stationarity and some additional regularity conditions (see Appendix B). The following example, which further analyzes the time series of implied volatility surfaces in Section 8.2, uses PCA and the representation (2.16) to approximate the k-dimensional time series $\mathbf{x}_t$ by a linear combination of a few time series of principal components.

Example: Implied volatility surfaces

For the multivariate time series of implied volatilities $\{I_t(m_i, \tau_j), 1 \le i \le 15, 1 \le j \le 20\}$ plotted in the left panels of Figure 8.8, we have used the transformation $\Delta_t(m, \tau) = \log I_t(m, \tau) - \log I_{t-1}(m, \tau)$ to obtain a stationary time series. Let $\boldsymbol{\Delta}_t$ denote the 300-dimensional vector with components $\Delta_t(m_i, \tau_j)$, $1 \le i \le 15, 1 \le j \le 20$. Letting $k = (m_i, \tau_j)$, denote the k'th component of the kth eigenvector by $a_k(k')$. Then (2.16) can be used to express $\Delta_t(m_i, \tau_j)$ in the form

$$\Delta_t(m_i, \tau_j) = \sum_{k=1}^{300} y_k(t) a_k(m_i, \tau_j). \tag{9.23}$$

Combining this with (8.12) gives the representation

$$I_t(m, \tau) = I_1(m, \tau) \exp\left\{ \sum_{k=1}^{300} y_k(t) a_k(m_i, \tau_j) \right\} \tag{9.24}$$

for the time series of implied volatility surfaces. Figure 9.1 plots the normalized eigenvectors (factor loadings) corresponding to the four largest eigenvalues of the sample covariance matrix as surfaces over the (m, τ) plane. Since these four eigenvalues account for over 98% of the total variance, we can approximate $\sum_{k=1}^{300}$ in the representation (9.24) of $I_t(m, \tau)$ by $\sum_{k=1}^{4}$ or even by $\sum_{k=1}^{2}$.

9.4.3 Linear regression with stochastic regressors

In Section 1.5.3, we have considered the case of stochastic regressors $\mathbf{x}_t$ in the linear regression model $y_t = \boldsymbol{\beta}^T \mathbf{x}_t + \epsilon_t$, $1 \le t \le n$, in which ϵ_t satisfies the martingale difference assumption (1.47). The theory in Section 1.5.3 extends to multivariate linear regression models (9.10), which include vector autoregressive models, described below as special cases. An important condition in this theory is (1.49); i.e., that $\left(\sum_{t=1}^{n} \mathbf{x}_t \mathbf{x}_t^T\right)/c_n$ converges in probability to some nonrandom matrix for some nonrandom constants c_n such that

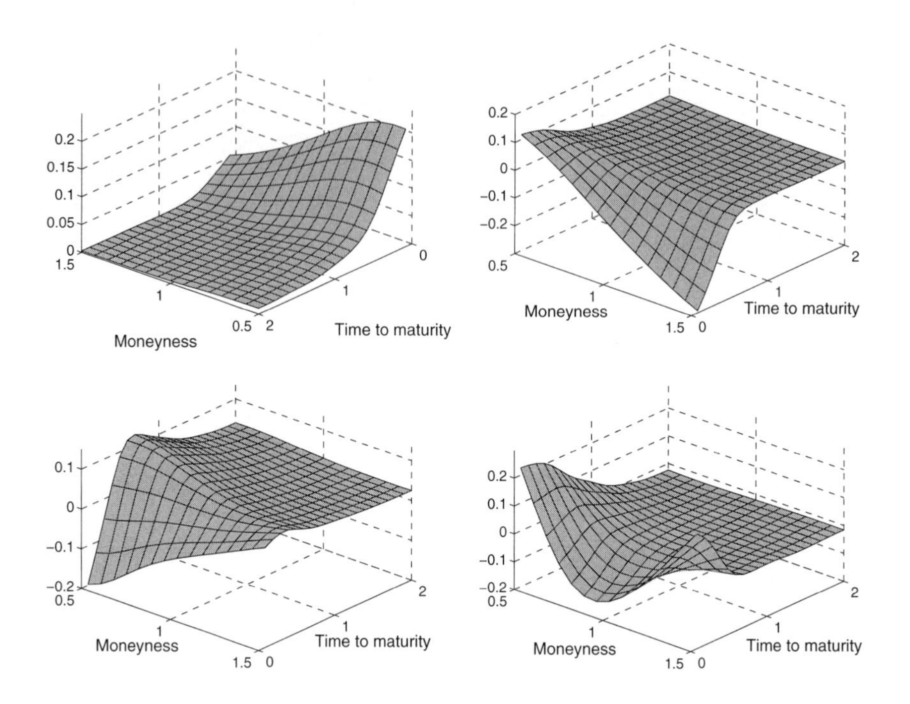

Fig. 9.1. The first four eigensurfaces of $\{\Delta_t(m, \tau)\}$, corresponding to the four largest eigenvalues $\lambda_1, \lambda_2, \lambda_3, \lambda_4$ of the sample covariance matrix. Top left: first eigensurface, with λ_1 accounting for 80.48% of the daily variance of implied volatility. Top right: second eigensurface, with λ_2 accounting for 13.14% of the daily variance of implied volatility. Bottom left: third eigensurface, with λ_3 accounting for 3.58% of the daily variance of implied volatility. Bottom right: fourth eigensurface, with λ_4 accounting for 1.25% of the daily variance of implied volatility.

$\lim_{n\to\infty} c_n = \infty$. In other words, $\sum_{t=1}^n \mathbf{x}_t\mathbf{x}_t^T$ should be asymptotically nonrandom for this theory to hold. If $\mathbf{x}_t$ is a weakly stationary time series, then one would expect this condition to be satisfied with $c_n = n$. In fact, under certain regularity conditions (see Appendix B), $n^{-1}\sum_{t=1}^n \mathbf{x}_t\mathbf{x}_t^T$ converges in probability to $E(\mathbf{x}_t\mathbf{x}_t^T)$. However, many econometric time series are unit-root nonstationary, as noted in Section 5.2.4. When the regressors $\mathbf{x}_t$ are unit-root nonstationary time series, condition (1.49) no longer holds and the least squares estimate becomes inconsistent, as was first observed by Granger and Newbold (1974).

Spurious regression

Granger and Newbold consider regressing y_t on x_t, where x_t and y_t are two independent Gaussian random walks; i.e., $x_t = x_{t-1} + u_t$, $y_t = y_{t-1} + v_t$, with i.i.d. standard normal u_t and v_t. The actual regression model therefore is $y_t = \beta x_t + \epsilon_t$ with $\beta = \text{Cov}(y_t, x_t)/\text{Var}(x_t) = 0$ and $\epsilon_t = y_t \sim N(0, t)$. Note that ϵ_t satisfies (1.47) but not (1.48). Moreover, as shown in Appendix C, $n^{-2}\sum_{t=1}^n x_t^2$ has a nondegenerate limiting distribution as $n \to \infty$; a distribution is said to be *degenerate* if it puts all its mass at a single point and to be *nondegenerate* otherwise. Therefore condition (1.49) cannot hold, as the limit of $n^{-2}\sum_{t=1}^n x_t^2$ is a random variable. Appendix C also shows that the least squares estimate $\widehat{\beta}$ has a nondegenerate limiting distribution as $n \to \infty$ in this case, and therefore $\widehat{\beta}$ is inconsistent. Granger and Newbold (1974) carried out simulation studies of $\widehat{\beta}$ in this model, reporting for example a rejection rate of 76% when using the conventional t-test (based on $\widehat{\beta}$) of the correct null hypothesis $\beta = 0$. These results led them to conclude that conventional significance tests are severely biased toward acceptance of a spurious regression relationship when the x_t and y_t are unit-root nonstationary time series.

For another illustration, consider the daily swap rates $r_{k,t}$ for maturities $k = 1, 5$, and 10 years from July 3, 2000 to July 15, 2005, discussed in Section 2.2.3. The top panel of Figure 2.1 exhibits high correlations among these rates. Regressing $r_{5,t}$ (5-year maturity) on the two other rates gives the fitted model

$$r_{5,t} = 0.305 r_{1,t} + 0.692 r_{10,t} + \epsilon_t, \tag{9.25}$$

with $R^2 = 99.9\%$, where the standard errors of the two coefficients are 0.003 and 0.002, respectively. However, the model is seriously inadequate, as shown by Figure 9.2. The sample ACF of the residuals in (9.25) decays slowly, showing the pattern of a unit-root nonstationary time series.

One way to circumvent spurious regression of y_t on x_t when both series are unit-root nonstationary is to perform regression for the differenced series. For the three interest rates considered above, let $\Delta r_{k,t} = r_{k,t} - r_{k,t-1}$. Regressing $\Delta_{5,t}$ on $\Delta_{1,t}$ and $\Delta_{10,t}$ gives the fitted model

$$\Delta r_{5,t} = 0.422 \Delta r_{1,t} + 0.804 \Delta r_{10,t} + w_t, \tag{9.26}$$

with $R^2 = 93.5\%$, where the standard errors of the two coefficients are 0.016 and 0.011, respectively. Figure 9.3 shows the residual series and its sample ACF. In contrast with the plots in Figure 9.2 for the residuals of linear regression of $r_{5,t}$ on $(r_{1,t}, r_{10,t})$, the residual series in the top panel of Figure 9.3 appears stationary and the ACF in the bottom panel resembles that of the stationary ARMA model considered in Section 5.1. An alternative approach to differencing is to include lagged values for both the response and the predictor variables in the regression model $y_t = \alpha + \beta x_t + \epsilon_t$ considered by Granger and Newbold:

$$y_t = \alpha + \phi y_{t-1} + \beta x_t + \psi x_{t-1} + \epsilon_t.$$

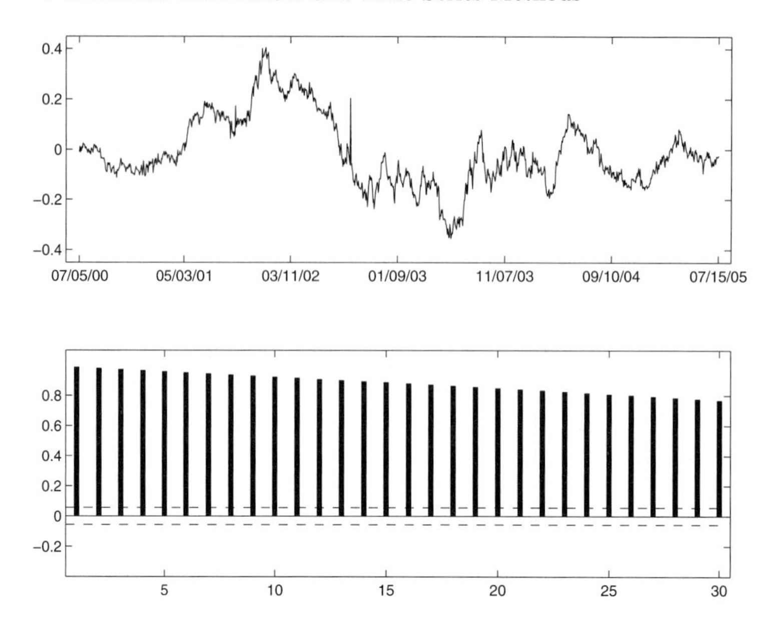

Fig. 9.2. Residual series (top) in linear regression model (9.25) and its ACF (bottom). The dashed lines in the bottom panel represent the rejection boundaries of a 5%-level test of zero autocorrelation at indicated lag.

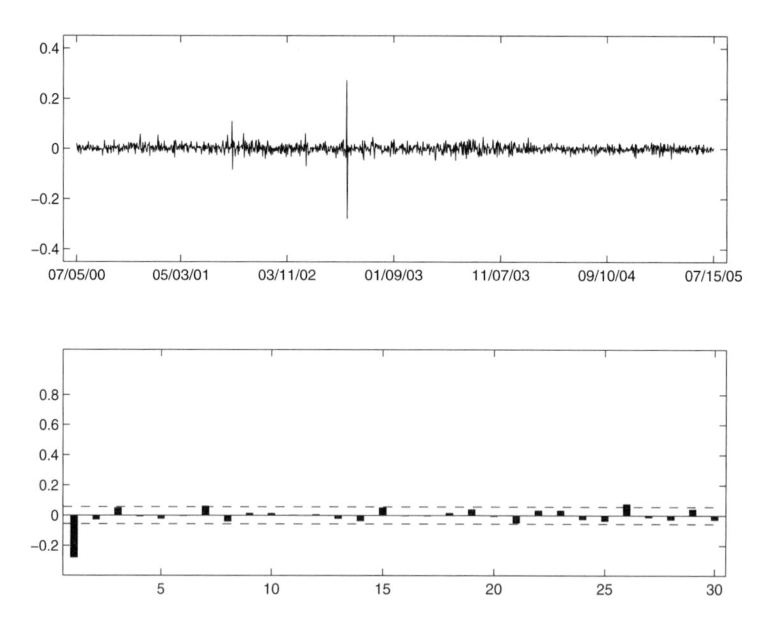

Fig. 9.3. Residual series (top) in linear regression model (9.26) and its ACF (bottom). The dashed lines in the bottom panel represent the rejection boundaries of a 5%-level test of zero autocorrelation at indicated lag.

It can be shown that the OLS estimates of the parameters α, ϕ, β, and ψ are consistent. On the other hand, without differencing or including lagged variables even though x_t and y_t are unit-root nonstationary, the regression relationship can still hold if the two series are cointegrated, which will be explained in Section 9.4.5. It is therefore important to perform diagnostic checks by examining the time series plot of the residuals and their ACF when one has time series regressors to see if the pattern is like Figure 9.2 (suggesting spurious regression) or like Figure 9.3 (suggesting the validity of the regression relationship).

Vector autoregressive (VAR) models

An important class of multivariate regression models with stochastic regressors is the autoregressive model

$$\mathbf{y}_t = \boldsymbol{\mu} + \boldsymbol{\Phi}_1 \mathbf{y}_{t-1} + \cdots + \boldsymbol{\Phi}_p \mathbf{y}_{t-p} + \boldsymbol{\epsilon}_t, \tag{9.27}$$

where $\boldsymbol{\epsilon}_t$ are i.i.d. $k \times 1$ random vectors with mean $\mathbf{0}$ and covariance matrix $\boldsymbol{\Sigma}$. Let $\boldsymbol{\Phi}(B) = \mathbf{I} - \boldsymbol{\Phi}_1 B - \cdots - \boldsymbol{\Phi}_p B^p$, where B is the backshift operator defined by $B\mathbf{y}_t = \mathbf{y}_{t-1}$. Then (9.27) can be expressed as $\boldsymbol{\Phi}(B)\mathbf{y}_t = \boldsymbol{\mu} + \boldsymbol{\epsilon}_t$. The model (9.27) is covariance stationary if all the roots of $\det\big(\boldsymbol{\Phi}(\lambda)\big) = 0$ lie outside the unit circle. Lai and Wei (1983) have shown that the least squares estimate of $(\boldsymbol{\Phi}_1, \ldots, \boldsymbol{\Phi}_p)$ is consistent when (9.27) is stationary and when there is no intercept term $\boldsymbol{\mu}$ in the case of nonstationarity.

9.4.4 Unit-root tests

As we have noted in Section 9.4.3, one way to circumvent spurious regression of y_t on x_t when they are both unit-root nonstationary is to difference both series so that the resultant series are stationary. However, one must first check that y_t and x_t are actually unit-root nonstationary (rather than stationary) because differencing of stationary x_t and y_t can lead to inconsistent OLS estimates of the regression parameters. For example, consider the AR(1) model $y_t = \mu + \phi y_{t-1} + \epsilon_t$, where $|\phi| < 1$ and ϵ_t are i.i.d. random variables with mean 0 and variance $\sigma^2 > 0$. Here $x_t = y_{t-1}$, and differencing yields the model $\Delta_t = \phi u_t + \eta_t$, where $\Delta_t = y_t - y_{t-1}$, $u_t = y_{t-1} - y_{t-2}$ and $\eta_t = \epsilon_t - \epsilon_{t-1}$. Note that η_t and u_t are correlated since $E(u_t \eta_t) = -E(\epsilon_{t-1} y_{t-1}) = -\sigma^2 \neq 0$, which causes the OLS estimate of ϕ to be inconsistent; see Section 9.7.1 for further discussion. On the other hand, if it is known that $\phi = 1$ (unit-root stationarity), then there is no need to estimate ϕ and the model can be written as $\Delta_t = \mu + \epsilon_t$, which yields the consistent OLS estimate $n^{-1} \sum_{t=1}^{n} \Delta_t$ of μ.

Augmented Dickey-Fuller test

Consider the AR(p) model $\phi(B)y_t = \mu + \epsilon_t$, where $\phi(B) = 1 - \phi_1 B - \cdots - \phi_p B^p$. Letting $\Delta y_t = y_t - y_{t-1}$, the AR($p$) model can be rewritten as

$$y_t = \mu + \beta_1 y_{t-1} - \sum_{j=1}^{p-1} \beta_{j+1} \Delta y_{t-j} + \epsilon_t, \qquad \text{where } \beta_j = \sum_{i=j}^{p} \phi_i. \tag{9.28}$$

Since $\phi(1) = 0$ if and only if $\beta_1 = 1$, a unit-root test can be formulated as testing the null hypothesis $H_0 : \beta_1 = 1$. The *augmented Dickey-Fuller* test uses the test statistic

$$\text{ADF} = (\widehat{\beta}_1 - 1)/\widehat{\text{se}}(\widehat{\beta}_1), \tag{9.29}$$

in which $\widehat{\beta}_1$ is the OLS estimate of β_1 in the regression model (9.28) and $\widehat{\text{se}}(\widehat{\beta}_1)$ is its estimated standard error given by the denominator in (1.14). The limiting distribution of $\widehat{\beta}_1 - 1$ under H_0 is nonnormal and the limiting distribution of ADF under H_0 is considerably more complicated than the t-distribution and is given in Appendix C. The R function `adf.test` (or `unitroot` in `Splus`) can be used to implement the unit-root test.

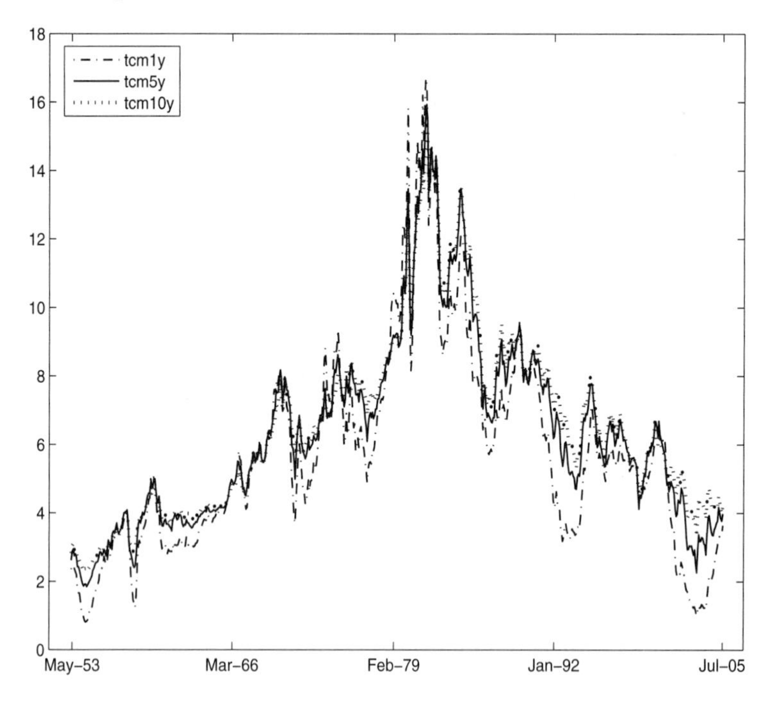

Fig. 9.4. The 1-year, 5-year, and 10-year US Treasury constant maturity (monthly average) rates from April 1953 to July 2005.

As an illustration, consider the monthly averages of the 1-year, 5-year, and 10-year U.S. Treasury constant maturity rates from April 1953 to July 2005, as shown in Figure 9.4. The series are obtained from the Federal Reserve Bank of St. Louis. We first perform the augmented Dickey-Fuller test for these three rates. The ADF statistic and the p-value for each rate are summarized in Table 9.1. The p-values are larger than 0.5, indicating that the null hypothesis cannot be rejected.

Table 9.1. Tests for unit-root nonstationarity of Treasury rates.

	1-year	5-year	10-year
ADF statistic	−2.124	−1.698	−1.492
p-value	0.526	0.706	0.794

Phillips-Perron test

The ϵ_t in (9.28) are assumed to be i.i.d., and the estimate of $\mathrm{se}(\widehat{\beta}_1)$ in (9.29) is based on this assumption. Phillips and Perron (1988) have relaxed this assumption by allowing ϵ_t to have a wide range of serial correlation patterns. They have introduced a modified test statistic

$$\mathrm{PP} = \left\{ \mathrm{ADF}\sqrt{\widehat{\gamma}_0} - n(\widehat{\lambda}^2 - \widehat{\gamma}_0)\widehat{\mathrm{se}}(\widehat{\beta}_1)/(2s) \right\} \Big/ \widehat{\lambda}, \tag{9.30}$$

where $\widehat{\gamma}_j = n^{-1}\sum_{t=j+1}^{n} \widehat{\epsilon}_t \widehat{\epsilon}_{t-j}$, $\widehat{\lambda} = \widehat{\gamma}_0 + 2\sum_{j=1}^{q}[1 - j/(q+1)]\widehat{\gamma}_j$, s^2 is the OLS estimate of $\mathrm{Var}(\epsilon_t)$, and n is the sample size. Phillips and Perron have shown that PP has the same limiting distribution as ADF under $H_0 : \beta_1 = 1$. The Phillips-Perron test can be implemented by the `PP.test` in `R`.

9.4.5 Cointegrated VAR

To begin, rewrite (9.28), in which μ is generalized to $\mu_0 + \mu t$, in the form

$$\Delta y_t = \mu + (\beta_1 - 1)y_{t-1} - \sum_{j=1}^{p-1} \beta_{j+1} \Delta y_{t-j} + \epsilon_t. \tag{9.31}$$

This allows the AR(p) model to have a deterministic linear trend with slope μ; note that differencing the linear trend $\mu_0 + \mu t$ yields μ. A k-variate generalization of this yields the following representation of the VAR(p) model, which replaces $\boldsymbol{\mu}$ in (9.27) by $\boldsymbol{\mu}_0 + t\boldsymbol{\mu}$:

$$\Delta \mathbf{y}_t = \boldsymbol{\mu} + (\mathbf{B}_1 - \mathbf{I})\mathbf{y}_{t-1} - \sum_{j=1}^{p-1} \mathbf{B}_{j+1} \Delta \mathbf{y}_{t-j} + \boldsymbol{\epsilon}_t, \tag{9.32}$$

where $\mathbf{B}_j = \sum_{i=j}^{p} \boldsymbol{\Phi}_i$, as in (9.28). Letting $\boldsymbol{\Phi}(B) = \mathbf{I} - \boldsymbol{\Phi}_1 B - \cdots - \boldsymbol{\Phi}_p B^p$, assume that

$$\det(\boldsymbol{\Phi}(z)) \text{ has all zeros at 1 or outside the unit circle.} \tag{9.33}$$

Let $\boldsymbol{\Pi} = \mathbf{B}_1 - \mathbf{I} = -\boldsymbol{\Phi}(1)$. If $\text{rank}(\boldsymbol{\Pi}) = k$ and $\boldsymbol{\mu} = \mathbf{0}$, then $\boldsymbol{\Phi}(1)$ is nonsingular and the VAR(p) process $\boldsymbol{\Phi}(B)\mathbf{y}_t = \boldsymbol{\mu}_0 + \boldsymbol{\epsilon}_t$ is stationary in view of (9.33). If $\text{rank}(\boldsymbol{\Pi}) = 0$, then $\sum_{i=1}^{p} \boldsymbol{\Phi}_i = \mathbf{I}$, implying that

$$\Delta \mathbf{y}_t = \boldsymbol{\mu} - \sum_{j=1}^{p-1} \mathbf{B}_{j+1} \Delta \mathbf{y}_{t-j} + \boldsymbol{\epsilon}_t, \tag{9.34}$$

and therefore $\Delta \mathbf{y}_t$ is a VAR($p-1$) process. The remaining case $0 < \text{rank}(\boldsymbol{\Pi}) = r < k$ implies that there exist $k \times r$ matrices $\boldsymbol{\alpha}$ and $\boldsymbol{\beta}$ such that

$$\boldsymbol{\Pi} = \boldsymbol{\alpha}\boldsymbol{\beta}^T, \quad \text{rank}(\boldsymbol{\alpha}) = \text{rank}(\boldsymbol{\beta}) = r, \tag{9.35}$$

and therefore the VAR(p) model (9.32) can be written as

$$\Delta \mathbf{y}_t = \boldsymbol{\mu} + \boldsymbol{\alpha}\boldsymbol{\beta}^T \mathbf{y}_{t-1} - \sum_{j=1}^{p-1} \mathbf{B}_{j+1} \Delta \mathbf{y}_{t-j} + \boldsymbol{\epsilon}_t, \tag{9.36}$$

noting that $\mathbf{B}_1 = \boldsymbol{\Pi} - \mathbf{I}$; see (9.28). Comparison of (9.36) with (9.34) shows that (9.36) has an extra term $\boldsymbol{\alpha}\boldsymbol{\beta}^T \mathbf{y}_{t-1}$, which is called the *error correction term*. Accordingly, (9.36) is called an *error correction model* (ECM). Since $\boldsymbol{\alpha}$ has rank r, there exists a $k \times (k-r)$ matrix $\mathbf{A}$ such that $\mathbf{A}^T\boldsymbol{\alpha} = \mathbf{0}$. Premultiplying (9.32) by $\mathbf{A}^T$ shows that the $(k-r)$-dimensional series $\mathbf{A}^T\mathbf{y}_t$ has no error correction term (or equivalently $\text{rank}(\mathbf{A}^T\boldsymbol{\Phi}) = 0$). In fact, $\det(\boldsymbol{\Phi}(z))$ has $k-r$ zeros equal to 1 and r zeros outside the unit circle by (9.33).

Definition 9.3. (i) A multivariate time series $\mathbf{y}_t$ is said to be *unit-root nonstationary* if it is nonstationary but $\Delta \mathbf{y}_t$ is stationary in the weak sense.

(ii) A nonzero $k \times 1$ vector $\mathbf{b}$ is called a *cointegration vector* of a unit-root nonstationary time series $\mathbf{y}_t$ if $\mathbf{b}^T\mathbf{y}_t$ is weakly stationary.

(iii) A multivariate time series is said to be *cointegrated* if all its components are unit-root nonstationary and there exists a cointegration vector. If the linear space of cointegrating vectors (with $\mathbf{0}$ adjoined) has dimension $r > 0$, then the time series is said to be *cointegrated with order* r.

The column vectors of $\boldsymbol{\beta} = (\boldsymbol{\beta}_1, \dots, \boldsymbol{\beta}_r)$ have the property that $\boldsymbol{\beta}_i^T \mathbf{y}_t$ is weakly stationary and are therefore cointegrating vectors. An economic interpretation of a cointegrated multivariate time series $\mathbf{y}_t$ is that its components have some common trends that result in $\boldsymbol{\beta}_i^T \mathbf{y}_t$ having long-run equilibrium for $1 \le i \le r$ even though the individual components y_{ti} are nonstationary and have variances diverging to ∞. In particular, if $\beta_{ji} \neq 0$, then linear regression of y_{tj} on the other components of $\mathbf{y}_t$ would not be spurious even though $\mathbf{y}_t$ is unit-root nonstationary.

Extension of reduced-rank regression

Estimation of $\boldsymbol{\alpha}$ and $\boldsymbol{\beta}$ involves extending the reduced-rank regression model $\mathbf{y}_i = \mathbf{A}\mathbf{C}\mathbf{x}_i + \boldsymbol{\epsilon}_i$ considered in Section 9.2.2 to include additional regressors that have unconstrained coefficient matrices. Such an extension was studied by Anderson (1951), who considered the regression model

$$\mathbf{y}_i = \mathbf{A}\mathbf{C}\mathbf{x}_i + \boldsymbol{\Lambda}\mathbf{z}_i + \boldsymbol{\epsilon}_i, \tag{9.37}$$

in which $\mathbf{z}_i \in \mathbb{R}^m$ is an additional regressor and $\boldsymbol{\Lambda}$ is its associated $q \times m$ coefficient matrix. The minimization problem (9.14) is now extended to minimizing

$$\sum_{t=1}^{n} \operatorname{tr}\Big[\boldsymbol{\Gamma}^{1/2}(\mathbf{y}_t - \mathbf{A}\mathbf{C}\mathbf{x}_t - \boldsymbol{\Lambda}\mathbf{z}_t)(\mathbf{y}_t - \mathbf{A}\mathbf{C}\mathbf{x}_t - \boldsymbol{\Lambda}\mathbf{z}_t)^T \boldsymbol{\Gamma}^{1/2}\Big]. \tag{9.38}$$

The solution of this problem yields $\mathbf{A}$ and $\mathbf{C}$ that have the same form as (9.15) except that the covariance and cross-covariance matrices are replaced by partial covariance and partial cross-covariance matrices. Specifically, with $\mathbf{S_{xx}}$, $\mathbf{S_{yy}}$, $\mathbf{S_{xy}}$, and $\mathbf{S_{yx}}$ defined in Section 9.2.2, let

$$\mathbf{S_{xz}} = \frac{1}{n}\sum_{t=1}^{n} \mathbf{x}_t\mathbf{z}_t^T = \mathbf{S}_{\mathbf{zx}}^T, \quad \mathbf{S_{yz}} = \frac{1}{n}\sum_{t=1}^{n} \mathbf{y}_t\mathbf{z}_t^T = \mathbf{S}_{\mathbf{yz}}^T, \quad \mathbf{S_{zz}} = \frac{1}{n}\sum_{t=1}^{n} \mathbf{z}_t\mathbf{z}_t^T,$$

$$\mathbf{S_{xx.z}} = \mathbf{S_{xx}} - \mathbf{S_{xz}}\mathbf{S_{zz}}^{-1}\mathbf{S_{xz}}^T, \qquad \mathbf{S_{yy.z}} = \mathbf{S_{yy}} - \mathbf{S_{yz}}\mathbf{S_{zz}}^{-1}\mathbf{S_{yz}}^T,$$
$$\mathbf{S_{xy.z}} = \mathbf{S_{xy}} - \mathbf{S_{xz}}\mathbf{S_{zz}}^{-1}\mathbf{S_{yz}}^T, \qquad \mathbf{S_{yx.z}} = \mathbf{S_{xy.z}}^T,$$
$$\mathbf{H} = \boldsymbol{\Gamma}^{1/2}\mathbf{S_{yx.z}}\mathbf{S_{xx.z}}^{-1}\mathbf{S_{xy.z}}\boldsymbol{\Gamma}^{1/2}.$$

Note that $\mathbf{S_{xx.z}}$ is the sum of squares of residuals when the $\mathbf{x}_i$ are regressed on the $\mathbf{z}_i$; see Section 1.3.2. Let $\mathbf{b}_1, \dots, \mathbf{b}_r$ be the normalized eigenvectors of $\mathbf{H}$, with $\mathbf{b}_j$ corresponding to the jth-largest eigenvalue $\widehat{\lambda}_j$ of $\mathbf{H}$. Then the $\widehat{\mathbf{A}}$ and $\widehat{\mathbf{C}}$ that minimize (9.38) are given by (9.15), with $\mathbf{S_{yx}}$ and $\mathbf{S_{xx}}$ replaced by $\mathbf{S_{yx.z}}$ and $\mathbf{S_{xx.z}}$, respectively. In the case $\boldsymbol{\Gamma} = \mathbf{S}_{\mathbf{yy.z}}^{-1}$, $\sqrt{\widehat{\lambda}_j}$ gives the jth *partial canonical correlation* between $\mathbf{y}_i$ and $\mathbf{x}_i$, adjusted for $\mathbf{z}_i$.

Maximum likelihood estimation in Gaussian ECM

Suppose the $\boldsymbol{\epsilon}_t$ in the error correction model (9.36) are i.i.d. $N(\mathbf{0}, \boldsymbol{\Sigma})$. Johansen (1988, 1991) has developed the following procedure, consisting of three steps, to compute the MLE of the parameters of (9.36).

Step 1. Perform two auxiliary linear regressions:

$$\Delta\mathbf{y}_t = \widehat{\boldsymbol{\theta}} + \widehat{\boldsymbol{\Phi}}_1 \Delta\mathbf{y}_{t-1} + \cdots + \widehat{\boldsymbol{\Phi}}_{p-1}\Delta\mathbf{y}_{t-p+1} + \mathbf{e}_t, \tag{9.39}$$

$$\mathbf{y}_{t-1} = \widehat{\boldsymbol{\nu}} + \widehat{\boldsymbol{\Psi}}_1 \Delta\mathbf{y}_{t-1} + \cdots + \widehat{\boldsymbol{\Psi}}_{p-1}\Delta\mathbf{y}_{t-p+1} + \mathbf{u}_t, \tag{9.40}$$

in which $\mathbf{e}_t$ and $\mathbf{u}_t$ are the corresponding residuals after using OLS to estimate the regression coefficients, denoted by $\widehat{\boldsymbol{\theta}}$, $\widehat{\boldsymbol{\Phi}}_i$, $\widehat{\boldsymbol{\nu}}$, and $\widehat{\boldsymbol{\Psi}}_i$.

Step 2. Perform canonical correlation analysis on the sample cross-covariance matrix $\widehat{\boldsymbol{\Sigma}}_{\mathbf{eu}} = n^{-1}\sum_{t=1}^n \mathbf{e}_t\mathbf{u}_t^T$. Specifically, let $\widehat{\boldsymbol{\Sigma}}_{\mathbf{ee}} = n^{-1}\sum_{t=1}^n \mathbf{e}_t\mathbf{e}_t^T$, $\widehat{\boldsymbol{\Sigma}}_{\mathbf{uu}} = n^{-1}\sum_{t=1}^n \mathbf{u}_t\mathbf{u}_t^T$, and $\widehat{\boldsymbol{\Sigma}}_{\mathbf{ue}} = \widehat{\boldsymbol{\Sigma}}_{\mathbf{eu}}^T$. Let $\widehat{\lambda}_1 \geq \cdots \geq \widehat{\lambda}_r > 0$ be the ordered positive eigenvalues of $\widehat{\boldsymbol{\Sigma}}_{\mathbf{ee}}^{-1}\widehat{\boldsymbol{\Sigma}}_{\mathbf{eu}}\widehat{\boldsymbol{\Sigma}}_{\mathbf{uu}}^{-1}\widehat{\boldsymbol{\Sigma}}_{\mathbf{ue}}$, and let $\widehat{\mathbf{v}}_1, \ldots, \widehat{\mathbf{v}}_r$ be the corresponding normalized eigenvectors.

The preceding two steps basically perform a partial canonical correlation analysis between $\Delta\mathbf{y}_t$ and $\mathbf{y}_{t-1}$, adjusted for $\Delta\mathbf{y}_{t-1}, \ldots, \Delta\mathbf{y}_{t-p+1}$. This is not surprising since the error correction model (9.36) has the same form as (9.37), with $\Delta\mathbf{y}_t$ taking the place of $\mathbf{y}_i$, $\mathbf{y}_{t-1}$ taking the place of $\mathbf{x}_i$ and the components of $\Delta\mathbf{y}_{t-1}, \ldots, \Delta\mathbf{y}_{t-p+1}$ forming the vector $\mathbf{z}_i$.

Step 3. Estimate $\boldsymbol{\alpha}$ and $\boldsymbol{\beta}$ by

$$\widehat{\boldsymbol{\alpha}} = (\mathbf{v}_1, \ldots, \mathbf{v}_r), \qquad \widehat{\boldsymbol{\beta}}^T = (\mathbf{v}_1, \ldots, \mathbf{v}_r)^T \widehat{\boldsymbol{\Sigma}}_{\mathbf{ee}}^{-1}\widehat{\boldsymbol{\Sigma}}_{\mathbf{eu}}\widehat{\boldsymbol{\Sigma}}_{\mathbf{uu}}^{-1}. \tag{9.41}$$

The parameters $\boldsymbol{\mu}$, $\mathbf{B}_2, \ldots, \mathbf{B}_p$ can then be estimated by performing least squares regression (with intercept term) of $\Delta\mathbf{y}_t - \widehat{\boldsymbol{\alpha}}\widehat{\boldsymbol{\beta}}^T\mathbf{y}_{t-1}$ on $\Delta\mathbf{y}_{t-1}, \ldots, \Delta\mathbf{y}_{t-p+1}$. The residuals $\Delta\mathbf{y}_t - \widehat{\Delta\mathbf{y}_t}$ can be used to provide the MLE of $\boldsymbol{\Sigma}$ by

$$\widehat{\boldsymbol{\Sigma}} = n^{-1}\sum_{t=1}^n \big(\Delta\mathbf{y}_t - \widehat{\Delta\mathbf{y}_t}\big)\big(\Delta\mathbf{y}_t - \widehat{\Delta\mathbf{y}_t}\big)^T.$$

The maximum likelihood estimate of the reduced-rank regression parameter matrix $\mathbf{B}$ when the $\boldsymbol{\epsilon}_k$ are independent $N(\mathbf{0}, \boldsymbol{\Sigma})$ is given by (9.16) with $\boldsymbol{\Gamma} = \mathbf{S}_{\mathbf{yy}}^{-1}$. This turns out to be equivalent to putting the GLS choice $\boldsymbol{\Gamma} = \boldsymbol{\Sigma}^{-1}$ in (9.16), where $\boldsymbol{\Sigma}$ is the MLE of $\boldsymbol{\Sigma}$ in the full-rank regression model that does not impose rank constraints on $\mathbf{B}$; see Reinsel and Velu (1998, pp. 31, 40).

Johansen's test for the number of cointegration vectors

Putting the MLE into the log-likelihood function yields the maximized log-likelihood $\widehat{l}_r$ when there are r cointegrating vectors. Johansen (1988) has shown that $2\widehat{l}_r$ is given by $-n\sum_{j=1}^{r}\log(1-\widehat{\lambda}_j)$ plus an additive constant that does not depend on r. Recall that $\sqrt{\widehat{\lambda}_j}$ is the jth partial canonical correlation between $\Delta\mathbf{y}_t$ and $\mathbf{y}_{t-1}$, adjusted for $\Delta\mathbf{y}_{t-1}, \ldots, \Delta\mathbf{y}_{t-p+1}$, and is defined as $1 \le j \le k$. This leads to Johansen's (1988) *likelihood ratio statistics*

$$\mathrm{LR}_r = -n \sum_{j=r+1}^{k} \log(1-\widehat{\lambda}_j) \tag{9.42}$$

for testing $H_0 : \mathrm{rank}(\boldsymbol{\Pi}) \le r$ versus $H_1 : \mathrm{rank}(\boldsymbol{\Pi}) > r$, and

$$\mathrm{LR}_{r,r+1} = -n\log(1-\widehat{\lambda}_{r+1})$$

for testing $H_0 : \mathrm{rank}(\boldsymbol{\Pi}) = r$ versus $H_1 : \mathrm{rank}(\boldsymbol{\Pi}) = r+1$. Under the null hypothesis $\mathrm{rank}(\boldsymbol{\Pi}) = r$, LR_r and $\mathrm{LR}_{r,r+1}$ do not have limiting χ^2-distributions (unlike the situations in Section 2.4.2 because of unit-root nonstationarity), and their limiting distributions involve integrals of Brownian motion; see Appendix C.

The `R` function `ca.jo` can be used to calculate the test statistic and critical values of Johansen's likelihood ratio (LR) test and to compute maximum likelihood estimates, as illustrated in the following example.

An example: U.S. monthly interest rates

Consider the U.S. monthly average Treasury constant maturity rates $r_{m,t}$ with maturities $m = 1, 5$, and 10 years from April 1953 to July 2005 in Section 9.4.4. Let $\mathbf{y}_t = (r_{1,t}, r_{5,t}, r_{10,t})$. Fitting a VAR(2) model to this multivariate time series, we perform Johansen's cointegration test for the number of cointegrating vectors using the `R` function `ca.jo`. The results are summarized in Table 9.2, in which $*$ indicates that the LR statistic shows significant departure from H_0, which is therefore rejected by Johansen's test.

From the results in Table 9.2, we conclude that there are two cointegration vectors. Table 9.3 gives the maximum likelihood estimates of the 3×2 matrices $\boldsymbol{\beta} = (\boldsymbol{\beta}_1, \boldsymbol{\beta}_2)$ and $\boldsymbol{\alpha} = (\boldsymbol{\alpha}_1, \boldsymbol{\alpha}_2)$ in the ECM with two cointegrating vectors. Figure 9.5 plots $\widehat{\boldsymbol{\beta}}_i^T \mathbf{y}_t$ for $i = 1, 2$.

9.5 Long-memory models and regime switching/structural change

9.5.1 Long memory in integrated models

The "long memory" of certain time series models has already been noted in Sections 6.3.2 and 6.3.3 in connection with volatility persistence; i.e., the

Table 9.2. Test statistic and critical values of the LR test.

H_0	LR	10%	5%	1%
$r = 0$	50.61*	19.80	21.89	26.41
$r \leq 1$	31.28*	13.78	15.75	19.83
$r \leq 2$	3.57	7.56	9.09	12.74

Table 9.3. MLEs of $\boldsymbol{\alpha}$ and $\boldsymbol{\beta}$.

	$\widehat{\boldsymbol{\beta}}_1$	$\widehat{\boldsymbol{\beta}}_2$	$\widehat{\boldsymbol{\alpha}}_1$	$\widehat{\boldsymbol{\alpha}}_2$
$r_{1,t}$	1.000	1.000	0.075	−0.004
$r_{5,t}$	−3.388	1.305	0.126	−0.001
$r_{10,t}$	2.397	−2.278	0.090	0.000

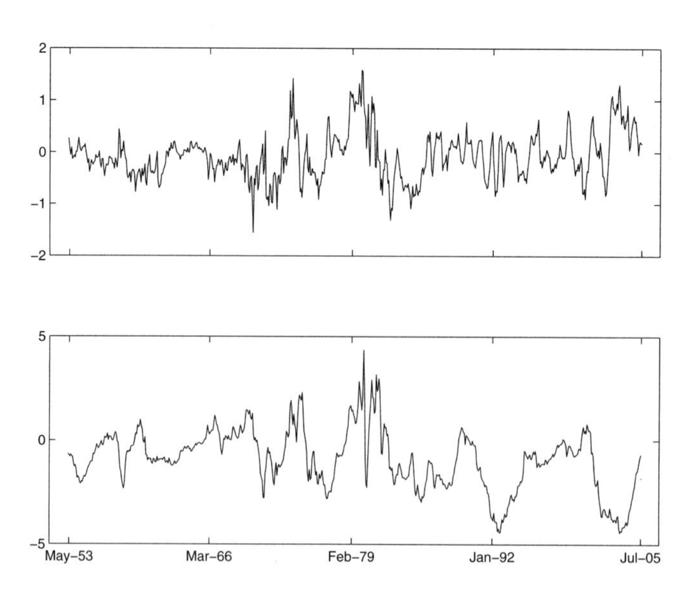

Fig. 9.5. The time series $\widehat{\boldsymbol{\beta}}_i^T \mathbf{y}_t$. Top panel: $i = 1$; bottom panel: $i = 2$.

persistent effect of events that occurred long ago on the current volatility. In particular, estimates of the parameters ω, α, and β in the GARCH(1, 1) model reveal high volatility persistence in many empirical studies of stock returns, with the maximum likelihood estimate $\widehat{\lambda}$ of $\lambda := \alpha + \beta$ close to 1. As shown in (6.21), the conditional variance σ_t^2 of u_t given the past observations can be expressed as

$$\sigma_t^2 = \sigma^2 + \alpha[\eta_{t-1} + \lambda\eta_{t-2} + \lambda^2\eta_{t-3} + \cdots].$$

For $\lambda = 1$, the contributions of the past innovations η_{t-i} to the conditional variance do not decay over time but are "integrated" instead, yielding the IGARCH model in Section 6.3.3. Similarly, integrated ARMA models (ARIMA(p, d, q) with integer $d \geq 1$) exhibit long memory; for example, the current and past disturbances in the random walk $y_t = \sum_{i=1}^t \epsilon_i$, which is ARIMA(0, 1, 0), have the same weight 1 in the sum that gives the current value of y_t.

We conclude this section with a brief introduction to a concept called factional differencing that has led to *stationary long-memory* models. To begin, note that taking positive integers d to define ARIMA(p, d, q) models has resulted in their nonstationarity. By using $-\frac{1}{2} < d < \frac{1}{2}$ to perform "fractional integration," it is possible to obtain covariance stationary processes with long memory. For $d < 1/2$, a *fractionally integrated* ARMA model, denoted by ARFIMA(p, d, q), is of the form $(1 - B)^d y_t = x_t$, where x_t is a stationary ARMA(p, q) model. The inverse of the operator $(1 - B)^d$ is

$$(1 - B)^{-d} = \sum_{j=0}^{\infty} h_j B^j, \qquad \text{where } h_j = \frac{d(d+1)\dots(d+j-1)}{j!}. \tag{9.43}$$

Since $\sum_{j=1}^{\infty} h_j^2 < \infty$ if $d < \frac{1}{2}$, the operator $(1-B)^{-d}x_t$ is covariance stationary. If $-\frac{1}{2} < d < \frac{1}{2}$, then the autocorrelation coefficient at lag k of y_t decays at a hyperbolic rate of k^{2d-1} instead of an exponential rate, and therefore y_t is considered to have long memory. Maximum likelihood estimation of the parameters in an ARFIMA(p, d, q) model when the innovations are normal, with d also regarded as an unspecified parameter, can be performed by using the `R` function `fracdiff`.

9.5.2 Change-point AR-GARCH models

Although estimation of the differencing parameter d has revealed long memory in many financial time series, Perron (1989) and Rappoport and Reichlin (1989) have argued that it is relatively rare for economic events to have an actual permanent effect. They have pointed out that if parameter changes are ignored in fitting time series models whose parameters undergo occasional changes, then the fitted models tend to exhibit long memory. As noted by

Lai and Xing (2006), this can be explained by two timescales in the underlying time series model. One is the "long" timescale for parameter changes, and the other is the "short" timescale of the modeled dynamics when the parameters stay constant. To incorporate possible parameter changes in fitting AR-GARCH models to financial time series, one approach is to use regime-switching models, which are introduced in Section 9.5.3. Besides the inherent computational complexity, another issue with regime-switching models is the number of regimes to be chosen and their interpretation. Lai and Xing (2006, 2007) have recently developed an alternative approach that models parameter changes more directly than the regime-switching approach, resulting in a more flexible model that is also easier to implement. In fact, some key formulas were already provided in Section 4.3.4 for segments of the data for which the parameters remain constant. We therefore provide details of the change-point approach in this section and summarize the regime-switching approach in the next section.

Note that, unlike historic volatility, the volatility in the AR-GARCH model is not directly observable and is inferred from the model parameters that are estimated from the observed data. In particular, consider the AR(2)-GARCH(1,1) model

$$\begin{aligned} r_t &= \mu + \rho_1 r_{t-1} + \rho_2 r_{t-2} + u_t, \quad u_t = \sigma_t \epsilon_t, \\ \sigma_t^2 &= \omega + \alpha u_{t-1}^2 + \beta \sigma_{t-1}^2, \quad \epsilon \sim N(0,1). \end{aligned} \tag{9.44}$$

Lai and Xing (2006) have used the `garch` toolbox in `MATLAB` to fit (9.44) to the weekly log returns (multiplied by 100) $r_t = \log(P_t/P_{t-1}) \times 100$ of the NASDAQ index during the period November 11, 1984 to September 15, 2003; see the top panel of Figure 6.2. The estimated parameters in the fitted model are $\widehat{\mu} = 0.320$, $\widehat{\rho}_1 = 0.092$, $\widehat{\rho}_2 = 0.077$, $\widehat{\omega} = 0.234$, $\widehat{\alpha} = 0.224$, $\widehat{\beta} = 0.772$, so $\widehat{\alpha} + \widehat{\beta} = 0.996$, which is near 1. In comparison with these estimates, Figure 9.6 plots the sequential estimates $\widehat{\mu}_t$, $\widehat{\rho}_{1,t}$, $\widehat{\rho}_{2,t}$, $\widehat{\omega}_t$, $\widehat{\alpha}_t$, and $\widehat{\beta}_t$, which are based on data up to time t for times t from January 1986 to September 2003, starting with an initial estimate based on the period November 1984 to December 1985. These sequential estimates suggest that the parameters changed over time and that $\widehat{\alpha}_t + \widehat{\beta}_t$ became close to 1 after 2000. The most conspicuous change is exhibited by the time course of $\widehat{\omega}_t$, which shows a sharp rise around Black Monday in October 1987.

Note that the roles of the parameters ω and (α, β) in (9.44) are quite different in describing the volatility dynamics of a GARCH(1,1) model. Whereas α and β are related to local (short-run) variance oscillations, $\omega/(1-\alpha-\beta)$ is the long-run variance level (which is the unconditional variance) of u_t; see (6.16). This suggests a tractable change-point model that allows occasional jumps in ω while keeping α and β time-invariant. Such an approach was recently developed by Lai and Xing (2007) in the more general setting of ARX-GARCH

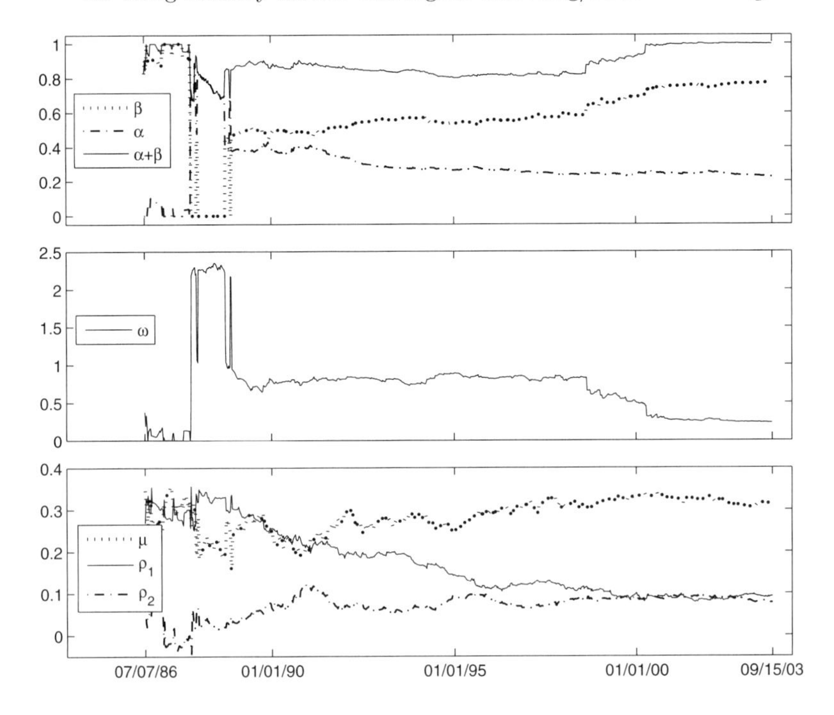

Fig. 9.6. Sequential estimates of AR(2)-GARCH(1,1) parameters.

models (i.e., autoregressive models with exogenous variables and GARCH errors). Earlier, Lai, Liu, and Xing (2005) and Lai and Xing (2006) considered the simpler model in which the GARCH parameters α and β are absent so that $\sigma_t^2 = \omega_t$ with piecewise constant ω_t. The remainder of this section focuses on this simpler model, which can be formulated more generally as a dynamic regression model of the form

$$y_t = \boldsymbol{\beta}_t^T \mathbf{x}_t + \sigma_t \epsilon_t, \qquad t > k, \tag{9.45}$$

in which ϵ_t are i.i.d. standard normal, $\mathbf{x}_t$ is an observed regressor that is determined by the events up to time $t-1$, and $\boldsymbol{\beta}_t$ and σ_t are piecewise constant parameters, with $\boldsymbol{\beta}_t$ of dimension k. Note that this dynamic regression model differs from the DLM in Section 5.3.2 in two important ways. First, the DLM assumes that $\boldsymbol{\beta}_t$ changes continually via the linear difference equation in (5.39). Second, the DLM assumes constant error variance, whereas (9.45) also allows jumps in the error variance.

A Bayesian model of occasional parameter jumps

Let $\boldsymbol{\theta}_t = (\boldsymbol{\beta}_t^T, \sigma_t)$ and $I_t = \mathbf{1}_{\{\boldsymbol{\theta}_t \neq \boldsymbol{\theta}_{t-1}\}}$, which is the indicator variable (assuming the values 0 and 1) that a parameter change occurs at time t. The prior distribution of the parameter sequence $\boldsymbol{\theta}_t$ assumes that $I_t, t > k+1$ are independent Bernoulli random variables with $P(I_t = 1) = p$, which means that the parameter jumps in (9.45) occur with probability p independently over time. Setting $I_{k+1} = 1$, sequential estimation of the k-dimensional regression parameter $\boldsymbol{\beta}_t$ and the error variance σ_t^2 begins at time $t \geq k+2$. The prior distribution of $\boldsymbol{\theta}_t$ at $t = k+1$ and at other jump times t is the same as that in Section 4.3.4. Specifically, letting $\tau_t = (2\sigma_t^2)^{-1}$, the dynamics of $\boldsymbol{\theta}_t$ can be described by

$$\boldsymbol{\theta}_t = (1 - I_t)\boldsymbol{\theta}_{t-1} + I_t(\mathbf{z}_t^T, \gamma_t),$$

where $(\mathbf{z}_1^T, \gamma_1), (\mathbf{z}_2^T, \gamma_2), \cdots$ are i.i.d. random vectors such that

$$\gamma_t \sim \text{gamma}(g, \lambda), \quad \mathbf{z}_t | \gamma_t \sim N(\mathbf{z}, \mathbf{V}/(2\gamma_t)). \tag{9.46}$$

The Bayesian model therefore involves hyperparameters $p, \mathbf{z}, \mathbf{V}, \lambda$, and g that have to be specified a priori or estimated from the observed data.

Explicit formulas for $E(\boldsymbol{\theta}_t | \mathbf{x}_1, y_1, \ldots, \mathbf{x}_t, y_t)$

We first derive the posterior distribution of $\boldsymbol{\theta}_t$ based on the observations $\mathcal{Y}_t := (\mathbf{x}_1, y_1, \ldots, \mathbf{x}_t, y_t)$ up to time t. An important ingredient in the derivation is $J_t = \max\{j \leq t : I_j = 1\}$, the most recent change time before or at t. Conditional on $\mathcal{Y}_t$ and $J_t = j$, the posterior distribution of $\boldsymbol{\theta}_t$ $(= \boldsymbol{\theta}_{t-1} = \cdots = \boldsymbol{\theta}_j)$ is the same as that given in Section 4.3.4 based on $(\mathbf{x}_j, y_j, \ldots, \mathbf{x}_t, y_t)$; see (4.37):

$$\tau_t \sim \text{gamma}\left(g_{j,t}, \frac{1}{a_{j,t}}\right), \quad \boldsymbol{\beta}_t | \tau_t \sim N\left(\mathbf{z}_{j,t}, \frac{1}{2\tau_t}\mathbf{V}_{j,t}\right), \tag{9.47}$$

where $g_{i,j} = g + (j - i + 1)/2$ for $i \leq j \leq n$ and

$$\mathbf{V}_{i,j} = \left(\mathbf{V}^{-1} + \sum_{l=i}^{j} \mathbf{x}_l \mathbf{x}_l^T\right)^{-1}, \quad \mathbf{z}_{i,j} = \mathbf{V}_{i,j}\left(\mathbf{V}^{-1}\mathbf{z} + \sum_{l=i}^{j} \mathbf{x}_l y_l\right),$$

$$a_{i,j} = \lambda^{-1} + \mathbf{z}^T \mathbf{V}^{-1} \mathbf{z} + \sum_{l=i}^{j} y_l^2 - \mathbf{z}_{i,j}^T \mathbf{V}_{i,j}^{-1} \mathbf{z}_{i,j}. \tag{9.48}$$

It remains to evaluate $p_{j,t} := P\{J_t = j | \mathcal{Y}_t\}$. Letting

$$f_{00} = \lambda^g (\det(\mathbf{V}))^{1/2} \Gamma(g), \qquad f_{ij} = a_{i,j}^{-g_{i,j}} (\det(\mathbf{V}_{i,j}))^{1/2} \Gamma(g_{i,j}), \tag{9.49}$$

Lai and Xing (2006, 2007) have derived the following recursive formula for $p_{j,t} = p^*_{j,t} / \sum_{i=k+1}^{t} p^*_{i,t}$:

$$p^*_{j,t} = \begin{cases} p f_{tt}/f_{00} & \text{if } j = t, \\ (1-p) p_{j,t-1} f_{jt}/f_{j,t-1} & \text{if } j \le t-1. \end{cases} \tag{9.50}$$

From (9.47) and (4.38), it follows that

$$E(\boldsymbol{\beta}_t | \mathcal{Y}_t) = \sum_{j=1}^{t} p_{j,t} \mathbf{z}_{j,t}, \quad E(\sigma_t^2 | \mathcal{Y}_t) = \sum_{j=1}^{t} p_{j,t} \frac{a_{j,t}}{2(g_{j,t}-1)}, \tag{9.51}$$

in which the weights $p_{j,t}$ can be evaluated by (9.50).

Explicit formulas for $E(\boldsymbol{\theta}_t | \mathcal{Y}_n)$, $1 \le t \le n$

Let $\widetilde{J}_t = \min\{j > t : I_{j+1} = 1\}$ be the future change-time closest to t. Conditional on $\mathcal{Y}_n$ and $(J_t, \widetilde{J}_t) = (i, j)$ with $i \le t < j$, the posterior distribution of $\boldsymbol{\theta}_t (= \boldsymbol{\theta}_i = \cdots = \boldsymbol{\theta}_j)$ is

$$\tau_t \sim \text{gamma}\left(g_{i,j}, \frac{1}{a_{i,j}} \right), \quad \boldsymbol{\beta}_t | \tau_t \sim N\left(\mathbf{z}_{i,j}, \frac{1}{2\tau_t} \mathbf{V}_{i,j} \right); \tag{9.52}$$

see Section 4.3.5. Lai and Xing (2006, 2007) have extended the recursive formulas for $p_{i,t} = P\{J_t = i | \mathcal{Y}_t\}$ given above to similar formulas for

$$\xi_{ijt} = \begin{cases} P\{J_t = i, \widetilde{J}_t = j | \mathcal{Y}_n\} & \text{if } i \le t < j, \\ P\{J_t = i, I_{t+1} = 1 | \mathcal{Y}_n\} & \text{if } i \le t = j. \end{cases}$$

Specifically, let $q_{t+1,j} = q^*_{t+1,j} / \sum_{i=t+1}^{n} q^*_{t+1,i}$, where

$$q^*_{t+1,j} = \begin{cases} p f_{t+1,t+1}/f_{00} & \text{if } j = t+1, \\ (1-p) q_{t+2,j} f_{t+1,j}/f_{t+2,j} & \text{if } j > t+1. \end{cases}$$

Then $\xi_{ijt} = \xi^*_{ijt} / \{p + \sum_{1 \le i' \le t < j' \le n} \xi^*_{ijt}\}$, where

$$\xi^*_{ijt} = \begin{cases} p p_{it} & i \le t = j, \\ (1-p) p_{it} q_{t+1,j} f_{ij} f_{00} / (f_{it} f_{t+1,j}) & i \le t \le j \le n. \end{cases} \tag{9.53}$$

Hence, analogous to (9.51), the Bayes estimates of $\boldsymbol{\beta}_t$ and σ_t^2 based on $\mathcal{Y}_n$ are

$$E(\boldsymbol{\beta}_t|\mathcal{Y}_n) = \sum_{1\le i\le t\le j\le n} \xi_{ijt}\mathbf{z}_{i,j}, \quad E(\sigma_t^2|\mathcal{Y}_n) = \sum_{1\le i\le t\le j\le n} \xi_{ijt}\frac{a_{i,j}}{2(g_{i,j}-1)}. \tag{9.54}$$

Estimation of hyperparameters

Lai and Xing (2006) propose to estimate first the hyperparameters $g, \lambda, \mathbf{z}$, and $\mathbf{V}$ in (9.46) by using the method of moments and then estimate the hyperparameter p by maximum likelihood, noting that the conditional density $f(y_t|\mathcal{Y}_{t-1})$ of y_t given $\mathcal{Y}_{t-1}$ can be expressed as $f(y_t|\mathcal{Y}_{t-1}) = \sum_{j=k+1}^{t} p^*_{j,t}$, where $p^*_{j,t}$ is defined in (9.50) as a function of p. Therefore the log-likelihood function is given by

$$l(p) = \log\left(\prod_{t=k+1}^{n} f(y_t|\mathbf{X}_{1,t-1})\right) = \sum_{t=k+1}^{n} \log\left(\sum_{j=k+1}^{t} p^*_{j,t}\right). \tag{9.55}$$

in which the other parameters $g, \lambda, \mathbf{z}$, and $\mathbf{V}$ are replaced by their estimated values.

9.5.3 Regime-switching models

The change-point AR-GARCH models of the preceding section have been developed to provide a more flexible alternative to regime-switching models, for which there is extensive literature in econometrics, following Hamilton's (1989) seminal paper on autoregressive models with piecewise constant regression and volatility parameters whose unknown values are generated by a finite-state Markov chain. The regime-switching model is therefore of the form

$$y_t = \beta_{s_t}^T \mathbf{x}_t + \sigma_{s_t}\epsilon_t, \tag{9.56}$$

in which ϵ_t are i.i.d. standard normal, $\mathbf{x}_t$ is an observed regressor that is determined by the events up to time $t-1$, and the value of $(\beta_{s_t}, \sigma_{s_t})$ depends on an unobserved ("hidden") state s_t of a finite-state Markov chain.

Regime-switching ARCH models

Diebold (1986) noted that in fitting GARCH models to interest rate data the choice of a constant term ω, which does not accommodate shifts in monetary policy regimes, might have led to the IGARCH model in Section 6.3.3. Regime-switching volatility models have been introduced to allow parameter jumps over a small number of regimes. A seminal paper on regime-switching volatility models is by Hamilton and Susmel (1994), who consider the case in which the parameter vector of an ARCH process is assumed to depend on an unobservable finite-state Markov chain. They modify the usual ARCH model (6.9) by introducing a scale factor κ_t that varies with an unobservable Markovian regime:

$$u_t = \kappa_t \sigma_t \epsilon_t, \quad \sigma_t^2 = \omega + \sum_{i=1}^{k} \alpha_i (u_{t-i}/\kappa_{t-i})^2 + \gamma (u_{t-1}/\kappa_{t-1})^2 \mathbf{1}_{\{u_{t-1} \geq 0\}}, \quad (9.57)$$

where $\mathbf{1}_{\{u_{t-1} \geq 0\}}$ is an indicator variable that assumes the value 1 or 0 and is used to incorporate the leverage effect. Suppose that there are K possible values of κ_t and that the transition probabilities of the Markov chain are given. Because of the finite number of hidden states assumed, the likelihood function is still tractable and can be maximized numerically. Hamilton and Susmel have found that (9.57) provides a better fit and better forecasts for the weekly returns of the value-weighted portfolio of stocks from the week ending July 3, 1962 to the week ending December 29, 1987. They attribute most of the persistence in stock price volatility to Markovian switching among low-, moderate-, and high-volatility regimes. Subsequent empirical work on regime-switching volatility models of interest rates is summarized in Section 10.6.2.

9.6 Stochastic volatility and multivariate volatility models

9.6.1 Stochastic volatility models

As noted in (6.14), the GARCH(1,1) model $\sigma_t^2 = \omega + \beta \sigma_{t-1}^2 + \alpha u_{t-1}^2$ is actually an ARMA model for u_t^2. Although it resembles an AR(1) model for σ_t^2, it involves the observed u_{t-1}^2 instead of an unobservable innovation in the usual AR(1) model. Replacing αu_{t-1}^2 by a zero-mean random disturbance has the disadvantage that σ_t^2 may become negative. A simple way to get around this difficulty is to consider $\log \sigma_t^2$ instead of σ_t^2, leading to the *stochastic volatility* (SV) model

$$u_t = \sigma_t \epsilon_t, \quad \sigma_t^2 = e^{h_t}, \quad h_t = \phi_0 + \phi_1 h_{t-1} + \cdots + \phi_p h_{t-p} + \eta_t, \quad (9.58)$$

which has AR(p) dynamics for $\log \sigma_t^2$. The ϵ_t and η_t in (9.58) are assumed to be independent normal random variables with $\epsilon_t \sim N(0,1)$ and $\eta_t \sim N(0, \sigma^2)$. A complication of the SV model is that unlike in usual AR(p) models, the h_t in (9.58) is an unobserved state undergoing AR(p) dynamics, while the observations are u_t such that $u_t | h_t \sim N(0, e^{h_t})$. The likelihood function of $\boldsymbol{\theta} = (\sigma, \phi_0, \dots, \phi_p)^T$, based on a sample of n observations $u_1, \dots, u_n$, involves n-fold integrals, making it prohibitively difficult to compute the MLE by numerical integration for usual sample sizes.

A quasi-maximum likelihood (QML) estimator

To fix the ideas, we consider in the following the case $p = 1$ and let $\phi_0 = \omega$, $\phi_1 = \phi$. Letting $y_t = \log u_t^2$, note that $y_t = h_t + \log \epsilon_t^2$, where $\log \epsilon_t^2$ is distributed like $\log \chi_1^2$, which has mean -1.27. Let $\xi_t = \log \epsilon_t^2 - E(\log \chi_1^2)$. Note that this is a linear state-space model with unobserved states h_t and observations y_t satisfying

$$h_t = \omega + \phi h_{t-1} + \eta_t, \qquad y_t = h_t + E(\log \chi_1^2) + \xi_t; \tag{9.59}$$

see Section 5.3. Let $h_{t|t-1}$ denote the minimum-variance linear predictor of h_t and $h_{t-1|t-1}$ denote the best linear estimate of h_{t-1} based on observations up to time $t-1$. Then the Kalman filter gives the recursion

$$\begin{aligned}
&h_{t|t-1} = \omega + \phi h_{t-1|t-1}, \quad h_{t|t} = h_{t|t-1} + V_{t|t-1}\big[y_t - E(\log \chi_1^2) - h_{t|t-1}\big]/v_t,\\
&V_{t|t-1} = \phi^2 V_{t-1|t-1} + \sigma^2, \quad V_{t|t} = V_{t|t-1}(1 - V_{t|t-1}/v_t),
\end{aligned}$$

where $v_t = V_{t|t-1} + \mathrm{Var}(\log \chi_1^2)$ is the variance of $e_t := y_t - E(\log \chi_1^2) - h_{t|t-1}$ and $V_{t|t} = \mathrm{Var}(h_{t|t})$; see Section 5.3.1. If ξ_t were normal (with mean 0 and variance $\mathrm{Var}(\log \chi_1^2)$), then e_t would be $N(0, v_t)$ and therefore the log-likelihood function would be given by

$$l(\omega, \phi, \sigma^2) = -\frac{1}{2}\sum_{t=1}^{n} \log v_t - \frac{1}{2}\sum_{t=1}^{n} e_t^2/v_t \tag{9.60}$$

up to additive constants that do not depend on (ω, ϕ, σ^2). By a result of Dunsmuir (1979), the estimator that maximizes (9.60) is still consistent and asymptotically normal. However, (9.60) is a quasi-likelihood rather than an actual likelihood function that requires the ξ_t to be nonnormal, and the QML estimator that maximizes (9.60) has been shown to be less efficient (having a larger mean squared error) than the Bayes estimator described below; see Jacquier, Polson, and Rossi (1994).

Bayes estimates and an introduction to Gibbs sampling

The technique used to compute the posterior distribution in Example 4.5, which ignores the normalizing constant, depends heavily on the fact that the posterior distribution belongs to some known parametric family (e.g., normal in Example 4.5). For more complicated problems, this is not the case and one does not have an explicit formula for the normalizing constant. Calculating the posterior mean or other functionals of the posterior distribution by direct numerical integration is difficult unless $\boldsymbol{\theta}$ is low-dimensional. *Gibbs sampling* is a Monte Carlo method that circumvents this difficulty. Let $\mathbf{X}$

denote the vector of observations. Suppose $\boldsymbol{\theta}$ can be partitioned into J subvectors $\boldsymbol{\theta}_1, \ldots, \boldsymbol{\theta}_J$ such that the conditional distribution of $\boldsymbol{\theta}_j$, given $\mathbf{X}$ and $\boldsymbol{\theta}_{\neq j} := \{\boldsymbol{\theta}_i : i \neq j\}$, has a known form, denoted by $H(\boldsymbol{\theta}_j | \boldsymbol{\theta}_{\neq j}, \mathbf{X})$. The Gibbs sampler is an iterative scheme that proceeds from the current state $\boldsymbol{\theta}^{(t)} = (\boldsymbol{\theta}_1^{(t)}, \ldots, \boldsymbol{\theta}_J^{(t)})$ as follows: Draw $\boldsymbol{\theta}_1^{(t+1)}$ from $H(\boldsymbol{\theta}_1 | \boldsymbol{\theta}_{\neq 1}^{(t)}, \mathbf{X})$. Then draw $\boldsymbol{\theta}_2^{(t+1)}$ from $H(\boldsymbol{\theta}_2 | \boldsymbol{\theta}_1^{(t+1)}, \boldsymbol{\theta}_3^{(t)}, \ldots, \boldsymbol{\theta}_J^{(t)}, \mathbf{X})$, ..., and finally draw $\boldsymbol{\theta}_J^{(t+1)}$ from $H(\boldsymbol{\theta}_J | \boldsymbol{\theta}_{\neq J}^{(t+1)}, \mathbf{X})$. The state is then updated as $\boldsymbol{\theta}^{(t+1)} = (\boldsymbol{\theta}_1^{(t+1)}, \ldots, \boldsymbol{\theta}_J^{(t+1)})$. Under certain conditions this "Monte Carlo Markov chain" converges in distribution to the posterior distribution of $\boldsymbol{\theta}$ given $\mathbf{X}$.

Consider the SV model $u_t = \sigma_t \epsilon_t$, $\sigma_t^2 = e^{h_t}$, $h_t = \omega + \phi h_{t-1} + \eta_t$, in which $\epsilon_t \sim N(0,1)$ and $\eta_t \sim N(0, \sigma^2)$ are independent. The prior distribution of (σ^2, ω, ϕ) is

$$\frac{m\lambda}{\sigma^2} \sim \chi_m^2, \qquad (\omega, \phi) | \sigma^2 \sim N((\omega_0, \phi_0), \sigma^2 \mathbf{V}_0)\big|_{|\phi|<1}, \tag{9.61}$$

where $|\phi_0| < 1$ and $N(\cdot,\cdot)\big|_{|\phi|<1}$ denotes the bivariate normal distribution restricted to the region $\{(\omega, \phi) : |\phi| < 1\}$ so that the corresponding AR(1) model for h_t is stationary; see (5.11). Note that (9.61) says that σ^2 has an inverted χ_m^2 distribution; see Section 4.3.4. Therefore the conditional distribution of $\boldsymbol{\theta} := (\omega, \phi, \sigma^2)$ given $\mathbf{h} := (h_1, \ldots, h_n)$ again has the same form as (9.61) but with $(m, \lambda, \omega_0, \phi_0, \mathbf{V}_0)$ replaced by their posterior counterparts, which are given in Section 4.3.4. The difficulty with posterior estimation in SV models is that $\mathbf{h}$ is actually unobservable and the observations are $u_1, \ldots, u_n$. Gibbs sampling can be used to obtain a Monte Carlo approximation to the posterior distribution of $(\boldsymbol{\theta}, \mathbf{h})$. Let $\mathbf{h}_{-t} = (h_1, \ldots, h_{t-1}, h_{t+1}, \ldots, h_n)$, which removes h_t from $\mathbf{h}$. Using $f(\cdot|\cdot)$ to denote conditional densities, we can use the AR(1) dynamics for h_t to obtain

$$f(h_t | \mathbf{u}, \boldsymbol{\theta}, \mathbf{h}_{-t}) \propto \frac{1}{\sigma_t} \exp\left(-\frac{u_t^2}{2\sigma_t^2} \right) \frac{1}{\sigma_t^2} \exp\left(-\frac{(h_t - u_t)^2}{2\nu^2} \right), \tag{9.62}$$

where $\nu^2 = \sigma^2/(1+\phi^2)$ and $\mu_t = \big[\omega(1-\phi) + \phi(h_{t-1} + h_{t+1})\big]/(1+\phi^2)$. Moreover, the conditional distribution of $\boldsymbol{\theta}$ given $\mathbf{u}$ and $\mathbf{h}$ is a combined normal and inverted gamma:

$$\frac{m\lambda + \sum_{t=2}^n v_t^2}{\sigma^2} \sim \chi_{m+n-1}^2, \quad (\omega, \phi)\big|\sigma^2 \sim N((\omega_*, \phi_*), \sigma^2 \mathbf{V}_*)\big|_{|\phi|<1}, \tag{9.63}$$

where given $\mathbf{h}$ and (ω, ϕ), we calculate $v_t = h_t - \omega - \phi \log h_{t-1}$ for $2 \le t \le n$ and generate σ^2 by the inverted Gamma distribution in (9.63). With σ^2 thus generated, let $\mathbf{z}_t = (1, \log h_{t-1})^T$,

$$V_*^{-1} = \left(\sum_{t=2}^{n} \mathbf{z}_t \mathbf{z}_t^T \right) \Big/ \sigma^2 + \mathbf{V}_0^{-1},$$

$$(\omega_*, \phi_*)^T = V_* \left\{ \mathbf{V}_0^{-1} (\omega_0, \phi_0)^T + \sum_{t=2}^{n} h_t z_t / \sigma^2 \right\},$$

and generate (ω, ϕ) from the truncated bivariate normal distribution in (9.63). The Gibbs sampler iterates these steps in generating $h_1, \dots, h_n$, σ^2, ω, and ϕ until convergence; see Jacquier, Polson and Rossi (1994) for details.

In practice, we discard the first m random draws of the Gibbs iterations to form a Gibbs sample $(\mathbf{h}_{m+1}, \boldsymbol{\theta}_{m+1}), \dots, (\mathbf{h}_B, \boldsymbol{\theta}_B)$ for Bayesian inference, regarding m as the "burn-in" period. Implementation details and software can be found on the `WinBUGS` Website (`http://www.mrc-bsu.cam.ac.uk/bugs/`).

9.6.2 Multivariate volatility models

For a multivariate time series $\mathbf{u}_t \in \mathbb{R}^p$, let $\boldsymbol{\Sigma}_t = \mathrm{Cov}(\mathbf{u}_t | \mathcal{F}_{t-1})$ be the conditional covariance matrix of $\mathbf{u}_t$ given the history of events up to time $t-1$. There is extensive literature on dynamic models of $\boldsymbol{\Sigma}_t$. A straightforward extension of the univariate GARCH(1, 1) model $\sigma_t^2 = \omega + \beta \sigma_{t-1}^2 + \alpha u_{t-1}^2$ to the multivariate setting is to use the *vech* (half-vectorization) operator, which transforms a $p \times p$ symmetric matrix $\mathbf{M}$ into a vector. The vector vech($\mathbf{M}$) consists of the $p(p+1)/2$ lower-diagonal (including diagonal) elements of $\mathbf{M}$. This leads to a multivariate GARCH model of the form

$$\mathrm{vech}(\boldsymbol{\Sigma}_t) = \boldsymbol{\omega} + \mathbf{B}\mathrm{vech}(\boldsymbol{\Sigma}_{t-1}) + \mathbf{A}\mathrm{vech}(\mathbf{u}_t \mathbf{u}_t^T), \tag{9.64}$$

where $\mathbf{A}$ and $\mathbf{B}$ are $p(p+1)/2 \times p(p+1)/2$ matrices. Besides the large number (of order p^4) of parameters involved, $\mathbf{A}$ and $\mathbf{B}$ also have to satisfy certain constraints for $\boldsymbol{\Sigma}_t$ to be positive definite, making it difficult to fit such models.

Instead of using the vech operator, an alternative approach is to replace σ_s^2 and u_s^2 in the univariate GARCH model (6.13) by $\boldsymbol{\Sigma}_s$ and $\mathbf{u}_s \mathbf{u}_s^T$ and the scalar parameters $\omega, \alpha_j, \beta_i$ by matrices, leading to

$$\boldsymbol{\Sigma}_t = \mathbf{A}\mathbf{A}^T + \sum_{i=1}^{k} \mathbf{A}_i (\mathbf{u}_{t-i} \mathbf{u}_{t-i}^T) \mathbf{A}_i^T + \sum_{j=1}^{h} \mathbf{B}_j \boldsymbol{\Sigma}_{t-j} \mathbf{B}_j^T; \tag{9.65}$$

see Engle and Kroner (1995), who refer to their earlier work with Baba and Kraft. The model, named BEKK after the four authors, involves a lower-triangular matrix $\mathbf{A}$ and always gives nonnegative definite $\boldsymbol{\Sigma}_t$. The number of parameters is of order p^2.

Many other multivariate generalizations of the GARCH(h, k) model have been proposed in the literature; see the recent survey by Bauwens, Laurent, and Rombouts (2006). An important statistical issue is the curse of dimensionality in parameter estimation unless p is small. Another issue that arises if one models the GARCH dynamics separately for the volatilities of the components of $\mathbf{u}_t$ and for their correlations as in (9.64) is that the resultant matrix may not be nonnegative definite. A much simpler approach is to generalize the exponentially weighted moving average model in Section 6.2 to covariance matrices, yielding

$$\widehat{\boldsymbol{\Sigma}}_t = \lambda \widehat{\boldsymbol{\Sigma}}_{t-1} + (1-\lambda)\mathbf{u}_t \mathbf{u}_t^T. \tag{9.66}$$

9.7 Generalized method of moments (GMM)

9.7.1 Instrumental variables for linear relationships

Consider the linear equation $y_t = \boldsymbol{\beta}^T \mathbf{x}_t + \epsilon_t$, in which $\mathbf{x}_t$ is a random vector that is correlated with the unobserved random disturbance ϵ_t such that $E(\epsilon_t) = 0$ and $\mathrm{Var}(\epsilon_t) = \sigma^2$. This is in contrast to the regression model in Chapter 1, where ϵ_t and $\mathbf{x}_t$ are assumed to be uncorrelated. Economic theory, however, often precludes this assumption. For example, consider the demand curve $y_t = \beta x_t + \epsilon_t$, in which y_t represents the logarithm of the quantity in demand and x_t is the log-price of a certain commodity. The log-price also affects the supply of the commodity via the equation $y_t = \widetilde{\beta} x_t + \widetilde{\epsilon}_t$, with $E(\widetilde{\epsilon}_t) = 0$ and $\mathrm{Var}(\widetilde{\epsilon}_t) = \widetilde{\sigma}^2$. The two simultaneous equations imply $(\beta - \widetilde{\beta})x_t = \widetilde{\epsilon}_t - \epsilon_t$. Hence, if ϵ_t and $\widetilde{\epsilon}_t$ are independent, then

$$\begin{aligned} \mathrm{Cov}(x_t, \epsilon_t) &= E\{(\widetilde{\epsilon}_t - \epsilon_t)\epsilon_t\}/(\beta - \widetilde{\beta}) = -\sigma^2/(\beta - \widetilde{\beta}), \\ \mathrm{Var}(x_t) &= E(\widetilde{\epsilon}_t - \epsilon_t)^2/(\beta - \widetilde{\beta})^2 = (\sigma^2 + \widetilde{\sigma}^2)/(\beta - \widetilde{\beta})^2. \end{aligned} \tag{9.67}$$

If we apply OLS to estimate β, then it follows from (9.67) and the consistency of sample variances and covariances that

$$\widehat{\beta} = \beta + \frac{\sum_{i=1}^n (x_i - \bar{x})\epsilon_i}{\sum_{i=1}^n (x_i - \bar{x})^2} \longrightarrow \beta - \frac{(\beta - \widetilde{\beta})\sigma^2}{(\sigma^2 + \widetilde{\sigma}^2)} \tag{9.68}$$

in probability. Hence the OLS estimate of the demand elasticity β is inconsistent. The case where the regressor $\mathbf{x}_t$ is determined endogenously in the linear model $y_t = \boldsymbol{\beta}^T \mathbf{x}_t + \epsilon_t$ therefore requires careful subject-matter considerations to determine if $\mathbf{x}_t$ and ϵ_t can be reasonably assumed to be uncorrelated before applying the OLS estimate of Chapter 1.

The method of instrumental variables circumvents the difficulty of regressors being correlated with the random disturbance by finding *instruments* $\mathbf{z}_t$ such that (i) $\mathbf{z}_t$ is uncorrelated with ϵ_t and (ii) $(\sum_{t=1}^n \mathbf{z}_t \mathbf{x}_t^T)/n$ converges in probability to a nonsingular matrix. In this case, we can estimate $\boldsymbol{\beta}$ by

$$\widehat{\boldsymbol{\beta}}_{IV} = \left(\sum_{t=1}^{n} \mathbf{z}_t \mathbf{x}_t^T\right)^{-1} \sum_{t=1}^{n} \mathbf{z}_t^T y_t. \tag{9.69}$$

Putting $y_t = \mathbf{x}_t^T \boldsymbol{\beta} + \epsilon_t$ in (9.69) yields

$$\widehat{\boldsymbol{\beta}}_{IV} = \boldsymbol{\beta} + \left(\sum_{t=1}^{n} \mathbf{z}_t \mathbf{x}_t^T\right)^{-1} \sum_{t=1}^{n} \mathbf{z}_t^T \epsilon_t. \tag{9.70}$$

Under the weak regularity condition that $(\mathbf{z}_t, \epsilon_t)$ satisfies the law of large numbers so that $n^{-1} \sum_{t=1}^{n} \mathbf{z}_t \epsilon_t$ converges in probability to $E(\mathbf{z}_t \epsilon_t) = 0$ by (i), it follows from (9.70) that $\widehat{\boldsymbol{\beta}}_{IV}$ is consistent. For example, consider in the preceding paragraph another product whose price z_t depends on the supply curve of the commodity but not on the demand curve, so that $\mathrm{Cov}(z_t, x_t) \neq 0$ but $\mathrm{Cov}(z_t, \epsilon_t) = 0$. Using z_t as an instrument, the estimate (9.69) has the form

$$\widehat{\beta}_{IV} = \sum_{t=1}^{n} z_t y_t \Big/ \sum_{t=1}^{n} z_t x_t = \beta + \sum_{t=1}^{n} z_t \epsilon_t \Big/ \sum_{t=1}^{n} z_t x_t$$

and $n^{-1} \sum_{t=1}^{n} z_t \epsilon_t$ converges to $E(z_t \epsilon_t) = 0$, while $n^{-1} \sum_{t=1}^{n} z_t x_t$ converges to a nonzero limit in probability, under conditions that guarantee the law of large numbers given in Appendices A and B.

We have assumed so far that the number of instruments (i.e., the dimensionality of $\mathbf{z}_t$) is the same as the number of parameters (i.e., the dimensionality of $\boldsymbol{\beta}$). Since the instruments are introduced via the moment restrictions

$$E\{(y_t - \boldsymbol{\beta}^T \mathbf{x}_t)\mathbf{z}_t\} = 0, \tag{9.71}$$

one can often come up with an *overidentified* system of moment restrictions with the dimensionality of $\mathbf{z}_t$ exceeding that of $\boldsymbol{\beta}$. For example, in the preceding example, there may be several products whose prices are correlated with the supply curve but uncorrelated with the demand curve of the commodity. In this case, the idea is to choose $\boldsymbol{\beta}$ to minimize

$$\left\{\sum_{t=1}^{n} (y_t - \boldsymbol{\beta}^T \mathbf{x}_t)\mathbf{z}_t\right\}^T \mathbf{W} \left\{\sum_{t=1}^{n} (y_t - \boldsymbol{\beta}^T \mathbf{x}_t)\mathbf{z}_t\right\}, \tag{9.72}$$

where $\mathbf{W}$ is a positive definite matrix. Note that $n^{-1} \sum_{t=1}^{n} (y_t - \boldsymbol{\beta}^T \mathbf{x}_t)\mathbf{z}_t$ corresponds to the sample analog of the left-hand side of (9.71). The optimal choice of $\mathbf{W}$ involves minimization of (9.72) over $\mathbf{W}$, which turns out to be equivalent to the following *two-stage least squares* (2SLS) procedure; see Campbell, Lo, and MacKinlay (1997, pp. 530–531). Let

$$\mathbf{Y} = \begin{pmatrix} y_1 \\ \vdots \\ y_n \end{pmatrix}, \quad \mathbf{X} = \begin{pmatrix} x_1 \\ \vdots \\ x_n \end{pmatrix}, \quad \mathbf{Z} = \begin{pmatrix} \mathbf{z}_1^T \\ \vdots \\ \mathbf{z}_n^T \end{pmatrix}.$$

Stage 1. Perform multivariate regression of $\mathbf{x}_t$ on $\mathbf{z}_t$ as in Section 9.2.1, yielding the fitted matrix $\widehat{\mathbf{X}} = \mathbf{Z}(\mathbf{Z}^T\mathbf{Z})^{-1}\mathbf{Z}^T\mathbf{X}$.

Stage 2. Regress $\mathbf{y}_t$ on $\widehat{\mathbf{x}}_t$, and estimate $\boldsymbol{\beta}$ by $\widehat{\boldsymbol{\beta}} = (\widehat{\mathbf{X}}^T\widehat{\mathbf{X}})^{-1}\widehat{\mathbf{X}}^T\mathbf{Y}$.

9.7.2 Generalized moment restrictions and GMM estimation

The orthogonality condition (9.71) for instrumental variables can be generalized to a moment condition of the form

$$E\big[\mathbf{g}(\boldsymbol{\theta}_0, \mathbf{y}_t)\big] = \mathbf{0}, \tag{9.73}$$

where $\mathbf{g}: \mathbb{R}^d \times \mathbb{R}^p \to \mathbb{R}^k$, and $\boldsymbol{\theta}_0$ is the true value of an unknown parameter vector $\boldsymbol{\theta}$ to be estimated from a multivariate sample $\mathbf{y}_1, \dots, \mathbf{y}_n$. Replacing the left-hand side of (9.73) by its sample counterpart, GMM estimates $\boldsymbol{\theta}_0$ using the minimizer of

$$Q(\boldsymbol{\theta}) := \left\{\sum_{t=1}^n \mathbf{g}(\boldsymbol{\theta}, \mathbf{y}_t)\right\}^T \mathbf{W}(\boldsymbol{\theta}) \left\{\sum_{t=1}^n \mathbf{g}(\boldsymbol{\theta}, \mathbf{y}_t)\right\}, \tag{9.74}$$

where $\mathbf{W}(\boldsymbol{\theta})$ is a $k \times k$ positive definite weighting matrix.

As in the case of instrumental variables, an important issue is how $\mathbf{W}(\boldsymbol{\theta})$ can be chosen optimally. This issue has been addressed by Hansen (1982) via an asymptotic argument that linearizes $n^{-1}\sum_{t=1}^n g(\boldsymbol{\theta}, \mathbf{y}_t)$ around the true parameter $\boldsymbol{\theta}_0$. Hansen assumes that $\mathbf{y}_t$ is strictly stationary and uses the ergodic theorem (see Appendix B) to conclude that $n^{-1}\sum_{t=1}^n \mathbf{g}(\boldsymbol{\theta}, \mathbf{y}_t)$ converges with probability 1 to $E\{\mathbf{g}(\boldsymbol{\theta}, \mathbf{y}_t)\}$, which is assumed to be finite for every $\boldsymbol{\theta}$. Under additional assumptions on $\mathbf{g}$, he also proves that the minimizer $\widehat{\boldsymbol{\theta}}_n$ of (9.74) converges to $\boldsymbol{\theta}_0$ with probability 1 and that

$$\mathbf{S} = \lim_{n\to\infty} n^{-1}\mathrm{Cov}\left(\sum_{t=1}^n \mathbf{g}(\boldsymbol{\theta}_0, \mathbf{y}_t)\right) \tag{9.75}$$

exists. Let $\mathbf{D}(\boldsymbol{\theta}) = E\big\{(\partial/\partial\boldsymbol{\theta})\mathbf{g}(\boldsymbol{\theta}, \mathbf{y}_t)\big\}$ denote the Jacobian matrix under the assumption that $\mathbf{g}$ is continuously differentiable in $\boldsymbol{\theta}$ belonging to some neighborhood of $\boldsymbol{\theta}_0$. Let $\mathbf{D} = \mathbf{D}(\boldsymbol{\theta}_0)$ and $\mathbf{W} = \mathbf{W}(\boldsymbol{\theta}_0)$. Under additional regularity conditions, Hansen establishes the asymptotic normality of the GMM estimator $\widehat{\boldsymbol{\theta}}_n$, i.e., $\sqrt{n}(\widehat{\boldsymbol{\theta}}_n - \boldsymbol{\theta}_0)$ has a limiting $N(\mathbf{0}, \mathbf{V})$ distribution, where

$$\mathbf{V} = (\mathbf{D}^T\mathbf{W}\mathbf{D})^{-1}\mathbf{D}^T\mathbf{W}\mathbf{S}\mathbf{W}\mathbf{D}(\mathbf{D}^T\mathbf{W}\mathbf{D})^{-1}. \tag{9.76}$$

Such $\mathbf{V}$ is minimized at $\mathbf{W} = \mathbf{S}^{-1}$, giving $\mathbf{V}_{\min} = (\mathbf{D}^T\mathbf{S}^{-1}\mathbf{D})^{-1}$ (in the sense that $\mathbf{V} - \mathbf{V}_{\min}$ is nonnegative definite). Since the optimal weighting matrix $\mathbf{S}^{-1}$ involves the unknown $\boldsymbol{\theta}_0$, Hansen proposes to start with a simple positive definite matrix $\mathbf{W}$ (e.g., the identity matrix) to obtain a consistent estimate $\widehat{\boldsymbol{\theta}}_n$, which is then used to replace the unknown $\boldsymbol{\theta}_0$ in (9.75), thereby obtaining a consistent estimate of $\mathbf{S}$.

In particular, if $\mathbf{g}(\boldsymbol{\theta}_0, \mathbf{y}_t)$ and $\mathbf{g}(\boldsymbol{\theta}_0, \mathbf{y}_s)$ are uncorrelated for $t > s$, then a consistent estimate of $\mathbf{S}$ is

$$\widehat{\mathbf{S}} = n^{-1}\sum_{t=1}^{n} \mathbf{g}(\widehat{\boldsymbol{\theta}}_n, \mathbf{y}_t)\mathbf{g}^T(\widehat{\boldsymbol{\theta}}_n, \mathbf{y}_t). \tag{9.77}$$

To handle the serially correlated case, let $\widehat{\boldsymbol{\Gamma}}_0$ denote the right-hand side of (9.77) and define for $\nu \geq 1$

$$\widehat{\boldsymbol{\Gamma}}_\nu = n^{-1}\sum_{t=\nu+1}^{n} \left\{\mathbf{g}(\widehat{\boldsymbol{\theta}}_n, \mathbf{y}_t)\mathbf{g}^T(\widehat{\boldsymbol{\theta}}_n, \mathbf{y}_{t-\nu}) + \mathbf{g}(\widehat{\boldsymbol{\theta}}_n, \mathbf{y}_{t-\nu})\mathbf{g}^T(\widehat{\boldsymbol{\theta}}_n, \mathbf{y}_t)\right\}.$$

A consistent estimator of $\mathbf{S}$ in this case is the Newey-West estimator

$$\widehat{\mathbf{S}} = \widehat{\boldsymbol{\Gamma}}_0 + \sum_{\nu=1}^{q}\left(1 - \frac{\nu}{q+1}\right)\widehat{\boldsymbol{\Gamma}}_\nu, \tag{9.78}$$

in which $q \to \infty$ but $q/n^{1/4} \to 0$ as $n \to \infty$; see Newey and West (1987).

Asymptotic theory and inference

Using arguments similar to those in Section 2.4.1 for the MLE, the limiting $N(\mathbf{0}, \mathbf{V})$ distribution of $\sqrt{n}(\widehat{\boldsymbol{\theta}}_n - \boldsymbol{\theta}_0)$ for the GMM estimator can be derived from the asymptotic normality of $\sum_{t=1}^n \mathbf{g}(\boldsymbol{\theta}_0, \mathbf{y}_t)$; i.e., that $n^{-1/2}\sum_{t=1}^n \mathbf{g}(\boldsymbol{\theta}_0, \mathbf{y}_t)$ has a limiting normal distribution with mean $\mathbf{0}$ and covariance matrix $\mathbf{S}$. For the special case $\mathbf{W}(\boldsymbol{\theta}_0) = \mathbf{S}^{-1}$, this implies that $n^{-1}Q(\boldsymbol{\theta}_0)$ has a limiting χ^2_k-distribution, where $Q(\boldsymbol{\theta})$ is the quadratic objective function defined in (9.74). Although $\widehat{\boldsymbol{\theta}}_n$ converges to $\boldsymbol{\theta}_0$ and $\widehat{\mathbf{S}}$ converges to $\mathbf{S}$ with probability 1, it does not follow that $n^{-1}\left[\sum_{t=1}^n \mathbf{g}(\widehat{\boldsymbol{\theta}}_n, \mathbf{y}_t)\right]^T \widehat{\mathbf{S}}^{-1}\sum_{t=1}^n \mathbf{g}(\widehat{\boldsymbol{\theta}}_n, \mathbf{y}_t)$ also has a limiting χ^2_k-distribution. In fact, since $\widehat{\boldsymbol{\theta}}_n$ is the minimizer of (9.74) in which $\mathbf{W}(\boldsymbol{\theta})$ is replaced by $\widehat{\mathbf{S}}$,

$$\left[\sum_{t=1}^{n} \frac{\partial}{\partial\boldsymbol{\theta}}\mathbf{g}(\boldsymbol{\theta}, \mathbf{y}_t)\right]^T_{\boldsymbol{\theta}=\widehat{\boldsymbol{\theta}}_n} \widehat{\mathbf{S}}^{-1}\sum_{t=1}^{n} \mathbf{g}(\widehat{\boldsymbol{\theta}}_n, \mathbf{y}_t) = \mathbf{0},$$

implying that d linear combinations of the components of $\sum_{t=1}^n \mathbf{g}(\widehat{\boldsymbol{\theta}}_n, \mathbf{y}_t)$ are equal to 0, where d is the dimensionality of $\boldsymbol{\theta}$. However, Hansen (1982) has shown that

$$n^{-1}\Big[\sum_{t=1}^{n}\mathbf{g}(\widehat{\boldsymbol{\theta}}_n,\mathbf{y}_t)\Big]^T\widehat{\mathbf{S}}^{-1}\sum_{t=1}^{n}\mathbf{g}(\widehat{\boldsymbol{\theta}}_n,\mathbf{y}_t) \text{ has a limiting } \chi^2_{k-d}\text{-distribution} \tag{9.79}$$

in the case of overidentifying moment restrictions (i.e., $k > d$), which is analogous to (2.48) for the GLR statistic. This can be used to test hypotheses about $\boldsymbol{\theta}_0$, as illustrated below.

9.7.3 An example: Comparison of different short-term interest rate models

Chan et al. (1992) have applied (9.79) to construct tests of various short-term interest rate models whose dynamics are described by the stochastic differential equation

$$dr_t = (\alpha + \beta r_t)dt + \sigma r_t^{\gamma} dw_t, \tag{9.80}$$

where w_t is standard Brownian motion; see Section 10.4.1 for details. They estimate the parameters of the continuous-time model by the discrete-time approximation

$$r_{t+1} - r_t = \alpha + \beta r_t + \epsilon_{t+1}, \quad E_t(\epsilon_{t+1}) = 0, \quad E_t(\epsilon_{t+1}^2) = \sigma^2 r_t^{2\gamma}, \tag{9.81}$$

where $E_t(\cdot)$ denotes the conditional expectation given the observation up to time t, or more precisely $E_t(\cdot) = E(\cdot|\mathcal{F}_t)$, in which $\mathcal{F}_t = \{r_1, \ldots, r_t\}$. Chan et al. (1992) consider (9.81) as a set of overidentifying moment restrictions and point out that using GMM has the following advantage over MLE. Whereas the discrete-time version of (9.80) assumes that the $\epsilon_t (= \Delta w_t)$ are i.i.d. normal so that MLE can indeed be used, GMM does not require this distributional assumption and still gives consistent and asymptotically normal parameter estimates if ϵ_t is a martingale difference sequence satisfying certain conditional moment assumptions; see Appendix A. Table 9.4 summarizes several well-known models of the short-term interest rate, which are special cases of (9.80) corresponding to certain values of α, β, and γ.

Let $\boldsymbol{\theta} = (\alpha, \beta, \gamma, \sigma^2)^T$ and $\mathbf{g}(\boldsymbol{\theta}, \mathbf{y}_t) = \mathbf{g}_t(\boldsymbol{\theta})$, where $\mathbf{y}_t = (r_t, r_{t-1})^T$ and

$$\mathbf{g}_t(\boldsymbol{\theta}) = \left(\epsilon_t,\ \epsilon_t r_{t-1},\ \epsilon_t^2 - \sigma^2 r_{t-1}^{2\gamma},\ (\epsilon_t - \sigma^2 r_{t-1}^{2\gamma}) r_{t-1}\right)^T, \tag{9.82}$$

in which $\epsilon_t = r_t - (1+\beta)r_{t-1} - \alpha$ by (9.81). Since $\mathbf{g}_t(\boldsymbol{\theta}_0)$ is a martingale difference sequence, we can use (9.77) to estimate the optimal weighting matrix for the GMM estimator. The parameter constraints for various models in Table 9.4 can be tested by a GMM that makes use of (9.79), with $k = 4$, $d = 2$ for Brownian motion (BM) and geometric Brownian motion (GBM), and $d = 3$

Table 9.4. Stochastic models of the short-term interest rate.

	Stochastic Process	Parameter Constraints
BM	$dr_t = \alpha dt + \sigma dz_t$	$\beta = \gamma = 0$
GBM	$dr_t = \beta r_t dt + \sigma r_t dz_t$	$\alpha = 0,\ \gamma = 1$
Vasicek	$dr_t = (\alpha + \beta r_t)dt + \sigma dz_t$	$\gamma = 0$
CIR	$dr_t = (\alpha + \beta r_t)dt + \sigma\sqrt{r_t}dz_t$	$\gamma = 1/2$

for the Vasicek and CIR (Cox, Ingersoll, and Ross, 1985) models. Chan et al. (1992) carried out such tests for these models using 307 1-month Treasury bill yields from June 1964 to December 1989. They found that the p-values of the χ^2-tests for goodness of fit for the Brownian motion, Vasicek, and CIR models are less than 5%, indicating that these models are misspecified, whereas the GBM model has a p-value larger than 0.2 and is not rejected.

Exercises

9.1. Let $\mathbf{x}$ and $\mathbf{y}$ be $p \times 1$ and $q \times 1$ random vectors, respectively, and let

$$\boldsymbol{\Sigma} = \text{Cov}\begin{pmatrix} \mathbf{x} \\ \mathbf{y} \end{pmatrix} = \begin{pmatrix} \boldsymbol{\Sigma}_{\mathbf{xx}} & \boldsymbol{\Sigma}_{\mathbf{xy}} \\ \boldsymbol{\Sigma}_{\mathbf{yx}} & \boldsymbol{\Sigma}_{\mathbf{yy}} \end{pmatrix},$$

as in Section 9.1.2. Consider the problem of maximizing $\mathbf{a}^T \boldsymbol{\Sigma}_{\mathbf{xy}} \mathbf{b}$ subject to $\mathbf{a}^T \boldsymbol{\Sigma}_{\mathbf{xx}} \mathbf{a} = \mathbf{b}^T \boldsymbol{\Sigma}_{\mathbf{xx}} \mathbf{b} = 1$.

(a) Using the Lagrange multipliers θ and μ for the two constraints, show that $\theta = \mu$ and that θ is a solution of the equation

$$\det\begin{pmatrix} -\theta\boldsymbol{\Sigma}_{\mathbf{xx}} & \boldsymbol{\Sigma}_{\mathbf{xy}} \\ \boldsymbol{\Sigma}_{\mathbf{yx}} & -\theta\boldsymbol{\Sigma}_{\mathbf{yy}} \end{pmatrix} = 0. \tag{9.83}$$

(b) Show that for any nonsingular square matrices $\mathbf{C}$ and $\mathbf{D}$,

$$\det\begin{pmatrix} -\theta\mathbf{C}\boldsymbol{\Sigma}_{\mathbf{xx}}\mathbf{C}^T & \mathbf{C}\boldsymbol{\Sigma}_{\mathbf{xy}}\mathbf{D}^T \\ \mathbf{D}\boldsymbol{\Sigma}_{\mathbf{yx}}\mathbf{C}^T & -\theta\mathbf{D}\boldsymbol{\Sigma}_{\mathbf{yy}}\mathbf{D}^T \end{pmatrix} = 0,$$

and therefore the solutions of (9.83) are invariant with respect to the transformation $\widetilde{\mathbf{x}} = \mathbf{C}\mathbf{x}$, $\widetilde{\mathbf{y}} = \mathbf{D}\mathbf{y}$.

(c) Show that if θ is a solution of (9.83), then θ^2 is an eigenvalue of $\boldsymbol{\Sigma}_{\mathbf{xx}}^{-1}\boldsymbol{\Sigma}_{\mathbf{xy}}\boldsymbol{\Sigma}_{\mathbf{yy}}^{-1}\boldsymbol{\Sigma}_{\mathbf{yx}}$ and is also an eigenvalue of $\boldsymbol{\Sigma}_{\mathbf{yy}}^{-1}\boldsymbol{\Sigma}_{\mathbf{yx}}\boldsymbol{\Sigma}_{\mathbf{xx}}^{-1}\boldsymbol{\Sigma}_{\mathbf{xy}}$.

9.2. Suppose a random vector $\mathbf{y} = (y_1, \ldots, y_p)^T$ has mean $\boldsymbol{\mu}$ and a positive definite covariance matrix $\mathbf{V}$.

(a) Making use of Section 1.5.1, show that the minimum-variance linear predictor $\widehat{y}_i$ of y_i based on $y_1, \ldots, y_{i-1}$ is of the form (9.19), where $\boldsymbol{\beta}_i := (\beta_1^{(i)}, \ldots, \beta_{i-1}^{(i)})^T$ is given by $\boldsymbol{\beta}_i = \mathbf{V}_{i-1}^{-1}(V_{1i}, \ldots, V_{i-1,i})^T$ and $\mathbf{V}_{i-1} = (V_{kl})_{1 \le k,l \le i-1}$. Moreover, prove (9.20).

(b) Let $\widehat{\mathbf{y}} = (\widehat{y}_1, \ldots, \widehat{y}_p)^T$ with $\widehat{y}_1 = \mu_1$, $\mathbf{e} = \mathbf{y} - \widehat{\mathbf{y}}$. Suppose $\mu_1 = \cdots = \mu_p = \theta$. Then $\mathbf{e}$ can be written as a linear transformation of $\mathbf{y}$, having the form $\mathbf{e} = \mathbf{A}(\mathbf{y} - \theta\mathbf{1})$. Making use of (a), show that $\mathbf{A}$ is a lower-diagonal matrix, with unit diagonal elements and below-diagonal entries $-\beta_j^{(i)}$, and that $\mathrm{Cov}(\mathbf{e}) = \mathrm{diag}(d_1, \ldots, d_p)$.

(c) Show that $\mathbf{A} = \mathbf{L}^{-1}$, where L_{ij} and d_i are defined recursively by (9.17) and (9.18).

9.3. The file `m_logret_4auto.txt` contains the monthly log returns of four automobile manufacturers (General Motors Corp., Toyota Motor Corp., Ford Motor Co., and Honda Motor Co.) from January 1994 to June 2007. The file `m_logret_4soft.txt` contains the monthly log returns of four application software companies (Adobe Systems Inc., Microsoft Corp., Oracle Corp., and SPSS Inc.) from January 1994 to June 2007.

(a) Perform a canonical correlation analysis for these two sets of returns. Give the first two estimated canonical variate pairs and the corresponding canonical correlations.

(b) Perform reduced-rank regression of the log returns of automobile stocks on those of software company stocks, taking $\mathrm{rank}(\mathbf{B}) = 2$ in (9.10).

9.4. The file `impvol_sp500_atm_tom.txt` contains at-the-money implied volatilities (i.e., $I_t(1,\tau)$) with different times to maturity ($\tau = T - t$) of European calls on the S&P 500 index for the period from January 3, 2005 to April 10, 2006.

(a) Plot the implied volatility surface versus different dates and different times to maturity. You can use `wireframe` in the `R` package `lattice` or `surf` in `MATLAB` to plot functions of two variables as surfaces.

(b) Perform PCA for the differenced series $\Delta_t(1,\tau) = \log I_t(1,\tau) - \log I_{t-1}(1,\tau)$. Plot the first three eigenvectors versus τ.

9.5. The file `m_cofi_4rates.txt` contains the monthly rates of the 11th District Cost of Funds Index (COFI), the prime rate of U.S. banks, 1-year and 5-year U.S. Treasury constant maturity rates, and U.S. Treasury 3-month secondary market rates from September 1989 to June 2007. The COFI rates are obtained from the Federal Home Loan Bank of San Francisco, and the other rates are obtained from the Federal Reserve Bank of St. Louis. COFI is a weighted-average interest rate paid by savings institutions headquartered in Arizona, California, and Nevada and is one of the most popular adjustable-rate mortgage (ARM) indices. The prime rate is the interest rate at which banks lend to their most creditworthy customers.

(a) Perform the augmented Dickey-Fuller test of the unit-root hypothesis for each of these rates.

(b) Assuming the VAR(2) model (9.27) for the multivariate time series of these five rates, perform Johansen's test for the number of cointegration vectors.

(c) Estimate the cointegration vectors and use them to describe the equilibrium relationship between the five rates.

(d) Regress COFI on the four other rates. Discuss the economic meaning of this regression relationship and whether the regression is spurious.

9.6. A dynamic regression model that differs from the dynamic linear model in Section 5.3.2 assumes piecewise constant regression parameter vectors in

$$y_t = \boldsymbol{\beta}_t^T \mathbf{x}_t + \sigma \epsilon_t, \qquad t = 1, 2, \ldots, n, \tag{9.84}$$

in which the ϵ_t are assumed to be i.i.d. standard normal and $\mathbf{x}_t$ is an observed regressor that is determined by the events up to time $t-1$. This is the same as the Bayesian change-point model (9.45) except that σ_t in (9.45) is replaced by the constant σ, which is tautamount to letting the variance of the gamma prior distribution in (9.46) approach 0. The posterior distribution of $\boldsymbol{\beta}_t$ given $\mathbf{x}_1, y_1, \ldots, \mathbf{x}_t, y_t$ in this case is the normal mixture $\sum_{j=1}^t p_{j,t} N(\mathbf{z}_{j,t}, \sigma^2 \mathbf{V}_{j,t})$, in which $p_{j,t} = p_{j,t}^* / \sum_{i=1}^t p_{i,t}^*$ and $p_{j,t}^*$ is given by (9.50) with

$$\begin{aligned} f_{00} &= (\det(\mathbf{V}))^{1/2} \exp\left\{\mathbf{z}^T \mathbf{V}^{-1} \mathbf{z} / (2\sigma^2)\right\}, \\ f_{ij} &= (\det(\mathbf{V}_{i,j}))^{1/2} \exp\left\{\mathbf{z}_{i,j}^T \mathbf{V}_{i,j}^{-1} \mathbf{z}_{i,j} / (2\sigma^2)\right\}. \end{aligned}$$

Apply the change-point regression model (9.84), with univariate β_t and x_t, to re-analyze the monthly log returns of the Apple Computer stock (the first column in the file `m_logret_10stocks.txt`) in Exercise 5.10, which also uses related information on CAPM contained in the the file `m_sp500ret_3mtcm.txt`.

(a) As in Exercise 5.10, we use January 1994 to June 1998 as the training period in which β_t is assumed to remain constant. Estimate z, σ^2, and V by $\widehat{\beta}$, $\widehat{\sigma}^2$, and $\widehat{V}$ which is the estimated variance of $\widehat{\beta}$, respectively; see Section 1.1.4.

(b) For the period July 1998 to December 2006, we use the posterior mean in the Bayesian change-point model (9.84) to estimate β_t based on observations up to time t. The posterior mean, however, requires specification of the hyperparameter $\Phi := (p, z, \sigma^2, V)$. With the estimates $\widehat{z}$, $\widehat{\sigma}^2$, and $\widehat{V}$ obtained from (a), we can calculate the log-likelihood function of p based on the training period from January 1994 to June 1998. Maximize this log-likelihood function over $p = 10^{-4}i$, $1 \leq i \leq 500$, to obtain the estimated hyperparameter $\widehat{\Phi}$.

(c) Substituting Φ by $\widehat{\Phi}$ in the posterior mean of β_t given the observations up to time t, obtain the estimated $\widehat{\beta}_t$ as t varies from July 1998 to December 2006. Plot $\widehat{\beta}_t$ versus t and compare these sequential estimates with the beta obtained by fitting CAPM to the period July 1998 to December 2006.

10

Interest Rate Markets

Chapters 3 and 8 treat the interest rate, which appears in the fundamental formulas therein, as fixed and observable from the current yield of a short-maturity Treasury bill. However, similar to volatilities of asset returns, which are regarded as fixed (but unknown) parameters in the theory of single-period investments in Chapter 3 but are treated via time series models in Chapter 6 for other applications, interest rates also vary over time, and time series models of interest rates have been developed and used in interest rate markets to forecast future interest rates and to price and hedge interest rate derivatives. An additional complication of interest rate markets is that there are actually many interest rates at a given time: interest rates for different maturities, fixed versus floating rates, short rates versus forward rates, etc.; see Section 10.1. An overview of interest rate markets and basic concepts, such as present value, LIBOR, caps and floors, interest rate swaps, forward rates and short rates, and the zero-coupon yield curve, is given in Section 10.1.

Section 10.2 describes statistical methods to estimate the zero-coupon yield curve from the current prices of default-free bonds, some of which have coupon payments. It points out that imposing smoothness constraints on the estimated yield curve has advantages when this yield curve is used as the initial term structure. Section 10.3 studies multivariate time series of yields for different maturities over time and applies cointegration and principal component analysis to analyze and model them. Whereas Section 10.3 focuses on statistical (empirical) analysis of real-world interest rates for different maturities, Sections 10.4 and 10.5 give an overview of subject-matter models in the finance literature for the valuation of interest rate derivatives. It is shown briefly at the beginning of Section 10.4 and then more completely in Section 10.5.2 that arbitrage-free pricing entails a risk-neutral measure (or more precisely an equivalent martingale measure), as in the Black-Scholes theory for equity options. The interest rate models are therefore specified under the risk-neutral (instead of the real-world) measure. This causes some difficulties in relating the financial time series models in Section 10.3 to those considered in Sections 10.4 and 10.5. The current practice in the financial industry is to choose a model from those listed in Sections 10.4 and 10.5 and to "calibrate" it to daily (or somewhat longer time windows of) market data of actively traded options (e.g., quotes on caps and swap options). Section 10.6 gives an overview of the financial engineering (calibration) and the financial econometrics (empirical time series) approaches to interest rate models. It also summarizes the statistical methods and different kinds of data used by the two approaches.

10.1 Elements of interest rate markets

For a "risk-free" asset (e.g., Treasury bills, bonds), the rate of return is called an *interest rate*. If the asset has cash value P_0 at time 0 and the interest rate is fixed at r and compounded once per unit time, then the cash value of the asset becomes $P_t = P_0(1+r)^t$ at time t; see Section 3.1.2. The unit of time is usually taken as years. If interest is compounded m times a year, we replace $1+r$ above by $(1 + r/m)^m$. Letting $m \to \infty$ corresponds to continuous compounding, with $P_t = P_0 e^{rt}$; i.e., $dP_t/P_t = rdt$. For more general situations in which the interest rate r_t varies with time, the differential equation $dP_t/P_t = r_t dt$ has the solution

$$P_t = P_0 \exp\left(\int_0^t r_s ds\right). \tag{10.1}$$

The presence of interest rates implies that a dollar received at time T in the future is worth $\pi(T)$ today, which is less than a dollar and is called the *present value* of a dollar at time T. Thus, $\pi(T) = 1/P_T$. In the case of continuous compounding with constant interest rate r, $\pi(T) = e^{-rT}$. The present value of an investment that generates cash flows $C(t_i)$ at times t_i in the future ($i = 1, \ldots, n$) is $\sum_{i=1}^n \pi(t_i)C(t_i)$. The relationship between interest rates and their maturities is called the *term structure*.

Fixed-income investments are investments that are in the form of contracts, also called *securities*, which promise to give the holder certain cash flows at specified times in the future. They include (i) savings accounts and certificates of deposit (CDs) at banks, (ii) money market accounts at financial institutions, (iii) U.S. Treasury bills, Treasury notes, and Treasury bonds, and (iv) bonds offered outside the country of the borrower (e.g., Eurobonds).

Interest rate markets include not only fixed-income securities but also interest rate derivatives such as bond options, interest rate caps and floors, interest rate swaps and swaptions, and interest rate futures contracts.

10.1.1 Bank account (money market account) and short rates

The P_t in (10.1) can be regarded as the value of a bank account at time $t \geq 0$, and r_s in (10.1) is commonly called the *short rate* (or *instantaneous spot rate*); the short rate can be changed on a daily basis by the bank. The *bank account numeraire* refers to the integral $\exp(\int_0^t r_s ds)$ in (10.1) and is denoted by $B(t)$. The value at time t of one unit of currency payable at time $T \geq t$ is $B(t)/B(T) = \exp(-\int_t^T r_s ds)$, which is the *discount factor* between time t and time T. For $t = 0$, this is the present value of one unit of currency at time T.

10.1.2 Zero-coupon bonds and spot rates

A *zero-coupon bond* pays a specific amount, called the *face* (or *par*) *value*, at maturity without intermediate coupon payments. U.S. Treasury bills (T-bills) are zero-coupon bonds with fixed terms to maturity of 13, 26, and 52 weeks. U.S. Treasury notes (T-notes) are semiannual coupon bonds with maturities of 1 to 10 years; U.S. Treasury bonds (T-bonds) are semiannual coupon bonds with maturities of more than 10 years. Examples are given in Figures 10.1 and 10.2.

The *n-year zero rate* (also called *zero-coupon yield*) is the *yield* (i.e., interest rate) of a zero-coupon bond that matures in n years. For a coupon-bearing bond, the yield is the interest rate implied by its payment structure. Let A = face value of bond, B = bond price, C_j = coupon payment at time t_j ($j = 1, \ldots, J$), and n denote the number of years to maturity. Under continuous compounding, the bond's *yield to maturity* (or simply yield) is given by the equation

$$B = Ae^{-yn} + \sum_{j=1}^{J} C_j e^{-yt_j}. \tag{10.2}$$

When interest is compounded m times per year, (10.2) can be modified by replacing e^{-y} by $(1 + y/m)^{-m}$. Bond prices are often quoted in two different forms in the market. The *dirty* price is the actual amount paid in return for the full amount of all future coupon payments and the principal. It is the sum of the *clean* price and the *accrued interest*.

The price at time $t \leq T$ of a zero-coupon bond with face value 1 and maturity date T is denoted by $P(t, T)$. Clearly $P(T, T) = 1$. The *spot rate* $R(t, T)$ at time t of the bond is the yield (under continuous compounding) given by

$$R(t, T) = -\frac{\log P(t, T)}{T - t} \tag{10.3}$$

or, equivalently

$$P(t, T) = \exp\big\{ -(T - t)R(t, T)\big\}. \tag{10.4}$$

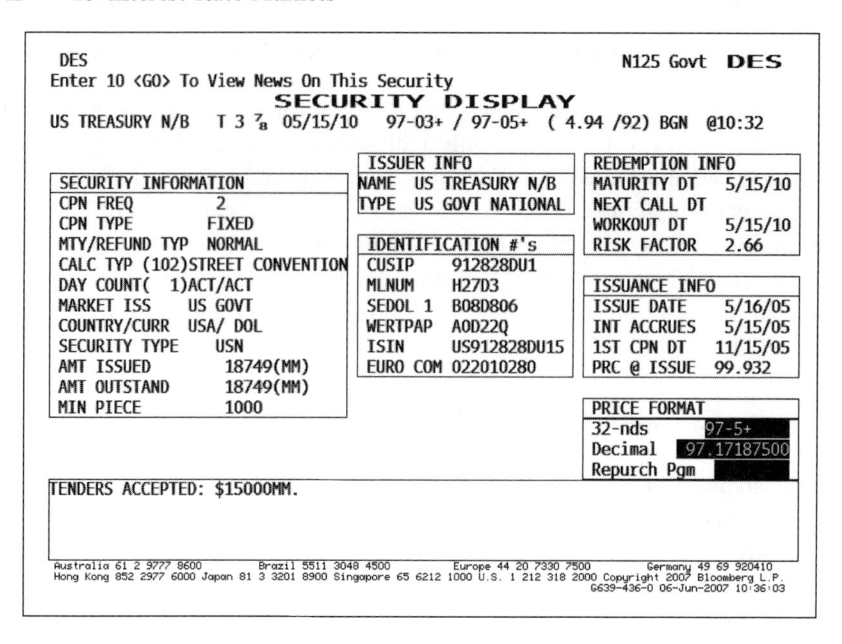

DES N125 Govt DES
Enter 10 <GO> To View News On This Security
SECURITY DISPLAY
US TREASURY N/B T 3 7/8 05/15/10 97-03+ / 97-05+ (4.94 /92) BGN @10:32

SECURITY INFORMATION
CPN FREQ 2
CPN TYPE FIXED
MTY/REFUND TYP NORMAL
CALC TYP (102)STREET CONVENTION
DAY COUNT(1)ACT/ACT
MARKET ISS US GOVT
COUNTRY/CURR USA/ DOL
SECURITY TYPE USN
AMT ISSUED 18749(MM)
AMT OUTSTAND 18749(MM)
MIN PIECE 1000

ISSUER INFO
NAME US TREASURY N/B
TYPE US GOVT NATIONAL

IDENTIFICATION #'s
CUSIP 912828DU1
MLNUM H27D3
SEDOL 1 B08D806
WERTPAP A0D22Q
ISIN US912828DU15
EURO COM 022010280

REDEMPTION INFO
MATURITY DT 5/15/10
NEXT CALL DT
WORKOUT DT 5/15/10
RISK FACTOR 2.66

ISSUANCE INFO
ISSUE DATE 5/16/05
INT ACCRUES 5/15/05
1ST CPN DT 11/15/05
PRC @ ISSUE 99.932

PRICE FORMAT
32-nds 97-5+
Decimal 97.17187500
Repurch Pgm

TENDERS ACCEPTED: $15000MM.

Australia 61 2 9777 8600 Brazil 5511 3048 4500 Europe 44 20 7330 7500 Germany 49 69 920410
Hong Kong 852 2977 6000 Japan 81 3 3201 8900 Singapore 65 6212 1000 U.S. 1 212 318 2000 Copyright 2007 Bloomberg L.P.
G639-436-0 06-Jun-2007 10:36:03

Fig. 10.1. A T-note on June 6, 2007. Used with permission of Bloomberg LP.

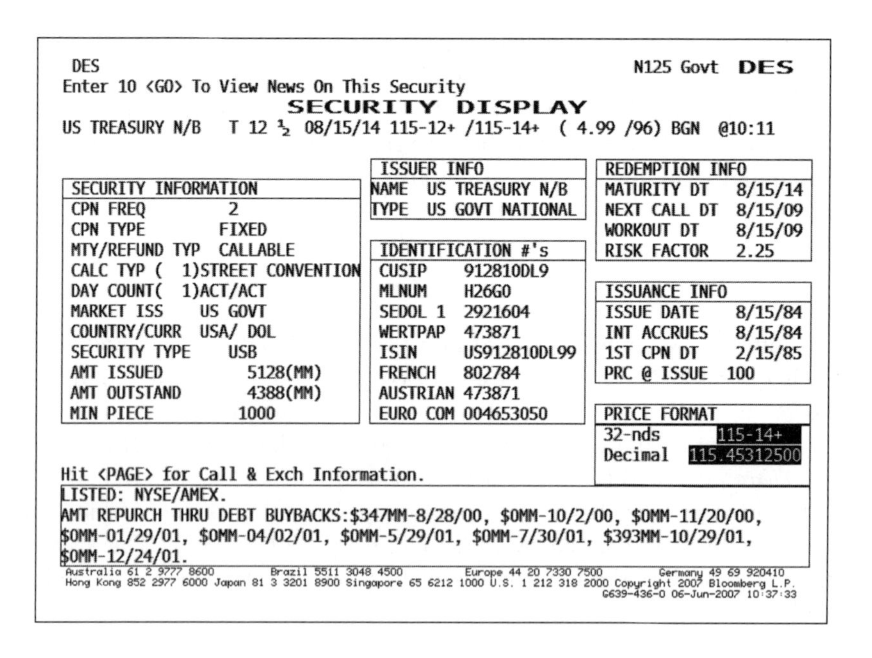

DES N125 Govt DES
Enter 10 <GO> To View News On This Security
SECURITY DISPLAY
US TREASURY N/B T 12 1/2 08/15/14 115-12+ /115-14+ (4.99 /96) BGN @10:11

SECURITY INFORMATION
CPN FREQ 2
CPN TYPE FIXED
MTY/REFUND TYP CALLABLE
CALC TYP (1)STREET CONVENTION
DAY COUNT(1)ACT/ACT
MARKET ISS US GOVT
COUNTRY/CURR USA/ DOL
SECURITY TYPE USB
AMT ISSUED 5128(MM)
AMT OUTSTAND 4388(MM)
MIN PIECE 1000

ISSUER INFO
NAME US TREASURY N/B
TYPE US GOVT NATIONAL

IDENTIFICATION #'s
CUSIP 912810DL9
MLNUM H26G0
SEDOL 1 2921604
WERTPAP 473871
ISIN US912810DL99
FRENCH 802784
AUSTRIAN 473871
EURO COM 004653050

REDEMPTION INFO
MATURITY DT 8/15/14
NEXT CALL DT 8/15/09
WORKOUT DT 8/15/09
RISK FACTOR 2.25

ISSUANCE INFO
ISSUE DATE 8/15/84
INT ACCRUES 8/15/84
1ST CPN DT 2/15/85
PRC @ ISSUE 100

PRICE FORMAT
32-nds 115-14+
Decimal 115.45312500

Hit <PAGE> for Call & Exch Information.
LISTED: NYSE/AMEX.
AMT REPURCH THRU DEBT BUYBACKS:$347MM-8/28/00, $0MM-10/2/00, $0MM-11/20/00,
$0MM-01/29/01, $0MM-04/02/01, $0MM-5/29/01, $0MM-7/30/01, $393MM-10/29/01,
$0MM-12/24/01.

Australia 61 2 9777 8600 Brazil 5511 3048 4500 Europe 44 20 7330 7500 Germany 49 69 920410
Hong Kong 852 2977 6000 Japan 81 3 3201 8900 Singapore 65 6212 1000 U.S. 1 212 318 2000 Copyright 2007 Bloomberg L.P.
G639-436-0 06-Jun-2007 10:37:33

Fig. 10.2. A T-bond on June 6, 2007. Used with permission of Bloomberg LP.

For a coupon-bearing bond, the *par yield* at time t specifies the coupon rate $\rho(t,T)$ that causes the price of the bond, issued at t and maturing at T, to equal its par value. For example, in the case of a bond that pays coupons annually and matures in T years, the par yield $\rho(t,T)$ in year $t<T$ is given by

$$\rho(t,T)\sum_{s=t+1}^{T} P(t,s)+P(t,T)=1;$$

i.e., $\rho(t,T)=\{1-P(t,T)\}/\sum_{s=t+1}^{T} P(t,s)$.

Duration and convexity

Duration, a measure of interest rate sensitivity, is a weighted average of the maturities of all the individual payments. Specifically, if payments are received at times $t_0,t_1,\ldots,t_K$, the duration of this cash flow is

$$D=\frac{\sum_{i=0}^{K} \mathrm{PV}(t_i)t_i}{\sum_{i=0}^{K} \mathrm{PV}(t_i)}, \tag{10.5}$$

in which $\mathrm{PV}(t_i)$ is the present value of the payment that occurs at time t_i. From (10.5), it follows that the duration of a zero-coupon bond is equal to its maturity and that the duration of a coupon-bearing bond is strictly less than its maturity. Moreover, it follows from (10.2) that

$$D=-\frac{1}{B}\frac{dB}{dy}, \tag{10.6}$$

where B is the bond price and y is the yield to maturity. The right-hand side of (10.6) is interpreted as the percentage change in the bond's price for a unit change in the yield. When interest is compounded m times a year so that e^{-y} in (10.2) is replaced by $(1+y/m)^{-m}$,

$$\frac{dB}{dy}=-\frac{DB}{1+y/m}=-D_M B, \tag{10.7}$$

where $D_M=D/(1+y/m)$ is called the *modified duration* of the bond, and (10.6) still holds with D replaced by D_M.

Duration is an important concept in hedging interest rate risk. Let P be the value of the portfolio. Assuming that changes in bond yields are the same for all bonds included in the portfolio, a small change Δy in yield leads to an approximate change $-DP\Delta y$, where D is the portfolio's duration (or modified duration), given by $D=\sum_{i=1}^{k} w_i D_i$ in which D_i is the duration and w_i is the portfolio weight of the ith bond ($\sum_{i=1}^{k} w_i=1$). The portfolio is called *duration hedged* if $\sum_{i=1}^{k} w_i D_i=0$. Financial institutions often attempt to hedge against interest rate risk by matching the (weighted) average duration of their assets to that of their liabilities. A key assumption in duration hedging is that all interest rates change by the same amount, which is often violated in practice. A more refined approach that dispenses with this assumption is given in Section 12.3.1.

Duration hedging is based on linear approximation to the price-yield curve. A better approximation is to include a second-order term in the Taylor expansion. The *convexity* of a portfolio is

$$C = \frac{1}{P}\frac{d^2P}{dy^2} \qquad \left(\text{so } \Delta P \approx -DP\Delta y + \frac{1}{2}CP(\Delta y)^2\right). \tag{10.8}$$

10.1.3 Forward rates

Forward rate agreement (FRA)

A *forward rate agreement* is a contract at the current time t for a loan between the expiration date T_1 of the contract and the maturity date T_2 of the loan. The contract gives its holder a loan at time T_1 with a fixed rate of simple interest for the period $T_2 - T_1$, to be paid at time T_2 besides the principal; the holder also receives at time T_2 an interest payment based on the forward rate $F(t, T_1, T_2)$, defined below.

Forward rate

Under continuous compounding, the *forward rate* is defined as

$$F(t, T_1, T_2) = \frac{1}{T_2 - T_1} \log \frac{P(t, T_1)}{P(t, T_2)}. \tag{10.9}$$

In the case $T_1 = t$, the forward rate becomes the spot rate $R(t, T_2)$ defined by (10.3). The *instantaneous forward rate* at time t is defined as

$$f(t, T) = \lim_{\delta \to 0} F(t, T, T + \delta) = -\frac{1}{P(t, T)} \frac{\partial P(t, T)}{\partial T}. \tag{10.10}$$

Hence the price $P(t, T)$ of a zero-coupon bond can be expressed as

$$P(t, T) = \exp\left[-\int_t^T f(t, u)du\right]. \tag{10.11}$$

The short rate introduced in Section 10.1.1 is related to the instantaneous forward rate by

$$r(t) = \lim_{T \to t} f(t, T) = f(t, t). \tag{10.12}$$

Note that the relation (7.35) between bond price and the instantaneous forward rate in Section 7.5.1 is a consequence of (10.2), (10.4), and (10.11).

LIBOR (London InterBank Offered Rate)

The LIBOR is the interest rate that banks charge each other for loans. These rates apply to loans with various maturities, such as 1 day, 1 month, 3 months, 6 months, 1 year, and up to 5 years. It is an annualized, simple rate of interest that will be delivered at the end of a specified period. The LIBOR *zero curve* up to 1 year is determined by the 1-month, 3-month, 6-month, and 12-month LIBOR rates. Eurodollar futures contracts can be used to extend the zero curve to longer maturities; see Hull (2006, pp. 141–142).

There are therefore different types of forward rates, and a major distinction can be made between interbank rates (LIBOR) and government (U.S. Treasury, Japanese Treasury, etc.) rates. The same notations $P(t,T)$, $R(t,T)$, $f(t,T)$, however, are used to refer to the rates in different sectors in what follows. In the bond options market, the forward rate is often associated with LIBOR, for which simple, instead of continuously compounded, interest rates are used. In particular, the *forward LIBOR rate* is defined as

$$F(t, T_1, T_2) = \frac{1}{T_2 - T_1}\left[\frac{P(t,T_1)}{P(t,T_2)} - 1\right], \tag{10.13}$$

in which $P(t,T)$ is given by the LIBOR term structure. Note that the last equality in (10.10) holds under either (10.9) or (10.13).

10.1.4 Swap rates and interest rate swaps

The *interest rate swap* is an extension of the FRA involving two financial institutions. In an interest rate swap, financial institution A agrees to pay B cash flows equal to the interest at a predetermined fixed rate while B agrees to pay A cash flows equal to the interest at a floating rate on a *notional principal* (which is not exchanged between A and B) for a prespecified period of time. The floating rate is usually LIBOR.

Under a swap contract that is initialized at time $T = T_0$, there are swap payments at times $T_1, \dots, T_M$ with $T_i - T_{i-1} = \tau_i$. The contract specifies that one party pays a fixed rate $\tau_i K$ at time T_i and receives from the other party the floating rate $\tau_i L(T_{i-1}, T_{i-1}, T_i)$, where $L(t, T_{i-1}, T_i) = \{P(t, T_{i-1})/P(t, T_i) - 1\}/\tau_i$ is used to denote the forward LIBOR rate (10.13). The party that pays the fixed rate is called the *payer* and the other party, paying the floating rate, is called the *receiver* of the swap. To ensure no arbitrage, K has to be chosen at time $t \le T$ so that the swap has value 0, i.e.,

$$K = \frac{P(t, T_0) - P(t, T_M)}{\sum_{i=1}^{M} P(t, T_i)\tau_i}, \tag{10.14}$$

noting that $\sum_{i=0}^{M-1} \{P(t,T_i)/P(t,T_{i+1}) - 1\}P(t,T_{i+1}) = P(t,T_0) - P(t,T_M)$. The value of K given by (10.14) is called the *forward swap rate* and is denoted by $s(t,T_0,T_M)$. For the particular case $t = T_0$ and $\tau_i = \tau$, it is called the *swap rate* with maturity $M\tau$ and is denoted by s_M; see Figure 10.3. Combining (10.14) with (10.13) yields an alternative expression for the forward swap rate that links it to the underlying forward rates:

$$s(t, T_0, T_M) = \sum_{i=0}^{M-1} w_i F(t, T_i, T_{i+1}) \tag{10.15}$$

with

$$w_i = P(t, T_{i+1})\tau_{i+1} \Bigg/ \left\{ \sum_{j=0}^{M-1} P(t, T_{j+1})\tau_{j+1} \right\}.$$

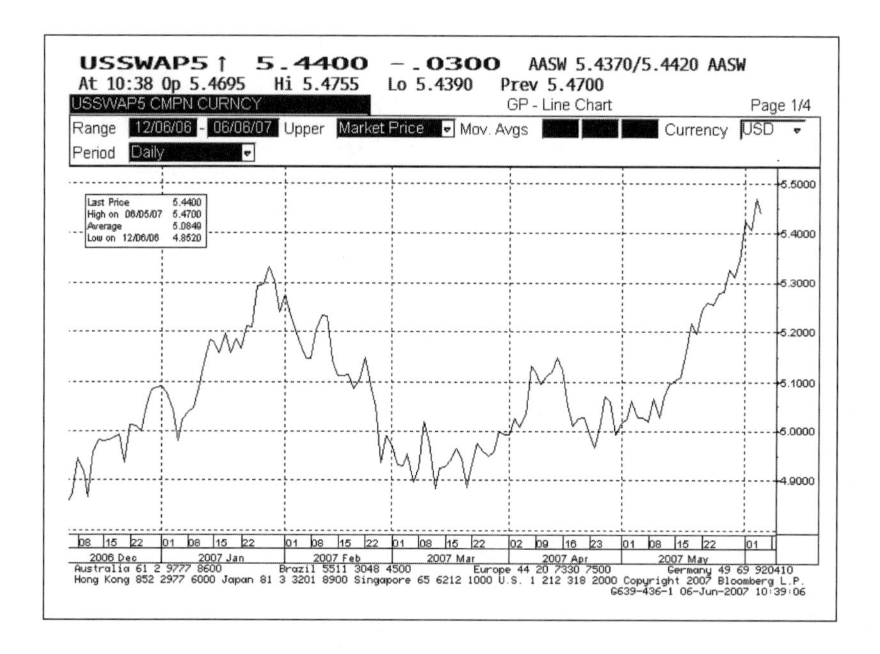

Fig. 10.3. Time series U.S. 5-year swap rates from December 6, 2006 to June 6, 2007. Used with permission of Bloomberg LP.

10.1.5 Caps, floors, and swaptions

In addition to interest rate swaps, swap options (also called *swaptions*), which will be described in Section 10.5.3, and interest rate caps and floors are the most popular interest rate derivatives. An interest rate *cap* is designed to provide insurance against the rate of interest on a floating-rate loan rising above a specified level R^*, called the *cap rate*, when the floating rate of the loan is periodically reset to equal LIBOR at dates $T_1, \dots, T_M$. At each reset date T_i during the life of the cap with expiration date $\widetilde{T}(> T_M)$, if LIBOR exceeds R^*, the cap's payoff at date T_{i+1} is LIBOR minus R^*, and there is no payoff if LIBOR falls below R^*. Note that the payment dates (when payoffs from the cap are assessed) are $T_1, \dots, T_M$. In a similar way (but with R^* minus LIBOR as the payoff), an interest rate *floor* with *floor rate* R^* provides insurance against LIBOR falling below R^*.

10.2 Yield curve estimation

The *term structure* of interest rates at time t can be described by $P(t,T)$, the price of a zero-coupon bond with face value 1, as a function of the maturity date $T \ge t$. Equivalently, it can be described by the relationship, called the *yield curve*, between the yield of a zero-coupon bond and its maturity; the yield for maturity $T-t$ can be expressed in terms of $P(t,T)$ via (10.3) under continuous compounding.

To estimate the yield curve at initial time 0, one uses a set of n reference default-free bonds (e.g., U.S. Treasury bonds). The data consist of the face value A_j, price B_j, maturity T_j, and coupon payments $C_{t_i,j}$ at dates $t_1, t_2, \dots, t_{i_j}$ for the jth bond ($1 \le j \le n$). For most bonds, the time between coupon payments is 6 months. In the absence of arbitrage, $P(0,\cdot)$ satisfies the system of linear equations

$$B_j = \sum_{i=1}^{i_j} C_{t_i,j} P(0,t_i) + A_j P(0,T_j), \qquad 1 \le j \le n. \tag{10.16}$$

Note that the relation (7.35) between the bond price and the instantaneous forward rate in Section 7.5.1 is a corollary of (10.16) and (10.11). The system (10.16), however, may have no solution if there are arbitrage opportunities or infinitely many solutions if there are not enough bonds. Moreover, even if this system yields a unique solution, it gives $P(0,T)$ only for T belonging to the set of maturities and coupon times in the sample. To circumvent this difficulty, least squares regression is used to estimate either (i) the $P(0,t_i)$ and $P(0,T_j)$ as parameters (with $\widehat{P}(0,t_i)$ and $\widehat{P}(0,T_j)$ as the least squares estimates) or (ii) the parameters of a nonparametric or parametric regression model for $P(0,T)$. Either case yields an estimate $\widehat{P}(0,\cdot)$ of $P(0,\cdot)$, and the method of least squares estimates the parameters in (i) or (ii) by minimizing $\sum_{j=1}^{n}(B_j - \widehat{B}_j)^2$, where $\widehat{B}_j$ is defined by (10.16) with P replaced by $\widehat{P}$. For case (i), the yield curve cannot be estimated beyond the longest maturity and interpolation is needed to estimate $P(0,T)$ when T does not belong to the set of maturities and coupon times in the sample; see Figure 10.4 (for $T = 1, 4, 6, 7, 8, 9, 15, 20$ years). Sections 10.2.1 and 10.2.2 consider case (ii) and, in particular, the choice of parametric models or basis functions in the nonparametric approach.

In view of (10.10), the initial forward rate $f(0,T)$ can be estimated from $\widehat{P}(0,T)$ if $\widehat{P}(0,T)$ is differentiable. Because of this and other applications, smoothness constraints are often imposed on $\widehat{P}(0,T)$. The forward rate curve is useful for pricing and hedging interest rate derivatives such as caps and swap options, and interbank data such as deposit rates and swap rates are often used to estimate $f(0,T)$; see Figure 10.5.

10.2.1 Nonparametric regression using spline basis functions

In the nonparametric regression approach to yield curve estimation, commonly used basis functions for $P(0,s)$ are cubic splines or exponential cubic splines (i.e., cubic splines for $\log P(0,s)$), so the regression parameters can be estimated in closed form by OLS; see Section 7.2.3. A variant of the latter, proposed by Vasicek and Fong (1982), is to approximate $P(0,s)$ by a piecewise cubic polynomial in $e^{-\beta s}$ in which β is a nonlinear parameter and can be estimated by nonlinear least squares. How knots should be chosen for the regression splines is discussed in Section 7.2.3. A simple choice that is often used for yield curve estimation consists of cardinal splines with evenly spaced knots in the range of observed maturities. Another choice that has sometimes been used is smoothing cubic splines or more general penalized spline models, with the smoothing parameter chosen by cross-validation or generalized cross-validation; see Sections 7.2.4 and 7.5.1.

10.2.2 Parametric models

A disadvantage of the nonparametric regression approach is that it may give negative forward rate estimates even when care is taken to make sure that the estimate of $P(0,s)$ is always positive. Another issue with the use of spline basis functions is that they imply that $P(0,T)$ diverges as $T \to \infty$ instead of converging to a limit as required by economic theory. To circumvent this difficulty, Vasicek and Fong (1982) proposed to use splines in $e^{-\beta s}$ instead. An alternative approach is to use parametric models for $P(0,T)$ that always yield positive forward rates and that are flexible enough to give the following

<HELP> for explanation. N299 Govt **IYC**
Hit <PAGE> for graph or <MENU> for list of curves.
YIELD CURVE – US TREASURY ACTIVES Page 2/2
DATE 6/ 6/07

	DESCRIPTION	PRICE	SRC	UPDATE	YIELD	HEDGED YIELD
3MO	1) B 0 09/06/07	B 4.6600	BGN	10:37	4.7941	4.7941
6MO	2) B 0 12/06/07	B 4.7500	BGN	10:00	4.9480	4.9480
1YR	3)					
2YR	4) T 4 $\frac{7}{8}$ 05/31/09	B 99.8281	BGN	10:36	4.9666	4.9666
3YR	5) T 4 $\frac{1}{2}$ 05/15/10	B 98.7969	BGN	10:36	4.9440	4.9440
4YR	6)					
5YR	7) T 4 $\frac{3}{4}$ 05/31/12	B 99.1563	BGN	10:36	4.9429	4.9429
6YR	8)					
7YR	9)					
8YR	10)					
9YR	11)					
10YR	12) T 4 $\frac{1}{2}$ 05/15/17	B 96.3125	BGN	10:41	4.9744	4.9744
15YR	13)					
20YR	14)					
30YR	15) T 4 $\frac{3}{4}$ 02/15/37	B 94.9688	BGN	10:41	5.0795	5.0795
1MO	16) B 0 07/05/07	B 4.6600	BGN	10:36	4.7419	4.7419

To change price source for securities, use <FMPS>.
To change price source for swaps, use <XDF>.
Yields are based on STANDARD settlement and are Conventional

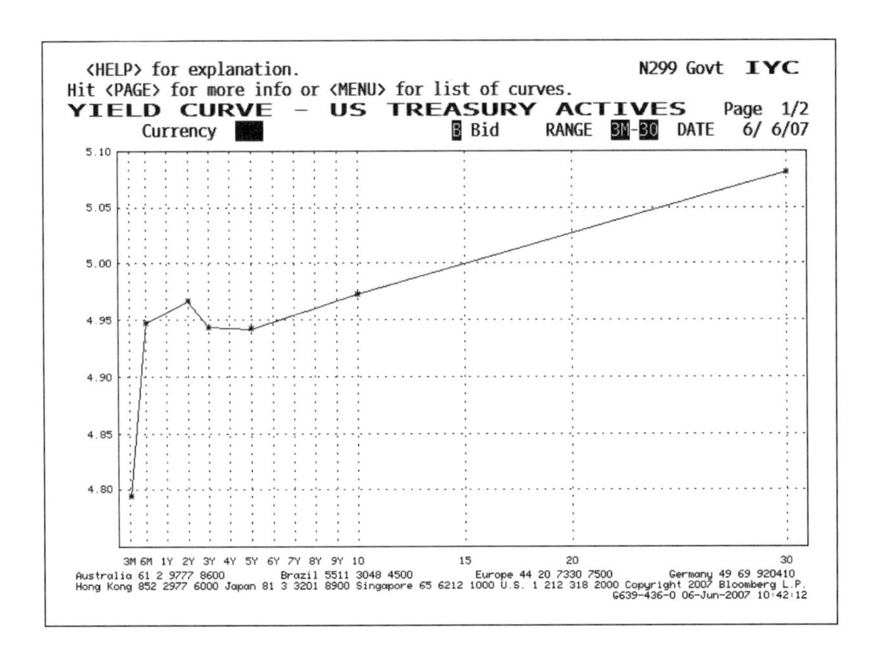

Fig. 10.4. Bond data (top panel) and yield curve (bottom panel) of U.S. Treasuries on June 6, 2006. Used with permission of Bloomberg LP.

<HELP> for explanation. N125 Govt **FWCV**

FORWARD CURVE ANALYSIS
US Dollar

BASE CURVE DEFAULTS - BGN		TERM	YIELD	9/10/07 P	12/10/07 P	6/ 9/08 P
Curve Dated: 6/ 6/07		1 Wk	5.3175	5.3268 R	5.3309 R	5.2352 R
Settlement Date: 6/ 8/07	D	1 Mo	5.3200	5.3359 O	5.3400 O	5.2393 O
Coupon/Spot: C	E R	2 Mo	5.3400	5.3465 J	5.3451 J	5.2451 J
Bid/Ask/Mid: B	P A	3 Mo	5.3600	5.3581 E	5.3509 E	5.2503 E
FMC # or SWDF # 23	O T	4 Mo	5.3719	5.3719 C	5.3590 C	5.2556 C
	S E	5 Mo	5.3828	5.3824 T	5.3655 T	5.2613 T
	I S	6 Mo	5.3959	5.3907 E	5.3687 E	5.2665 E
1 GO Graph	T	9 Mo	5.4300	5.4138 D	5.3769 D	5.2825 D
		1 Yr	5.4563	5.4269	5.3893	5.2992
2 GO Update Curve		2 Yr	5.3900	5.3768	5.3643	5.3363
	S R	3 Yr	5.3840	5.3813	5.3800	5.3775
3 GO Forwards	W A	4 Yr	5.4030	5.4051	5.4082	5.4183
Analysis	A T	5 Yr	5.4300	5.4348	5.4406	5.4522
	P E	7 Yr	5.4790	5.4853	5.4924	5.5062
4 GO FWCM <GO>	S	10Yr	5.5420	5.5452	5.5491	5.5562
		15Yr	5.6310	5.6400	5.6494	5.6676
		20Yr	5.6810	5.6874	5.6945	5.7086
		30Yr	5.7000	n/a	n/a	n/a

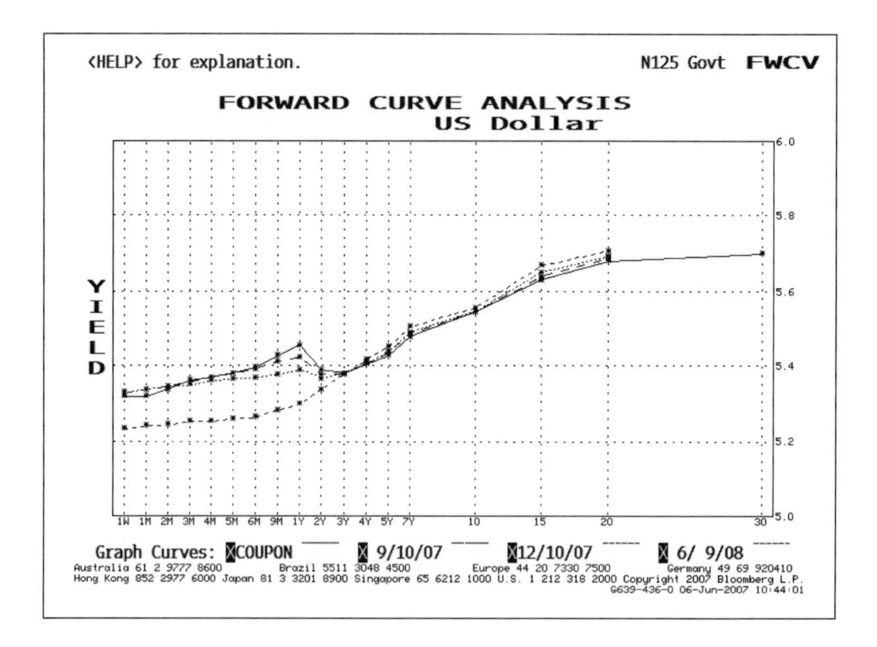

Fig. 10.5. Forward rate data (top panel) and curve (bottom panel) on June 6, 2006. Used with permission of Bloomberg LP.

shapes, which have been observed in government bonds, of the yield to maturity (see Section 10.1.2) versus maturity by varying the model parameters: (a) increasing, (b) decreasing, (c) flat (i.e., approximately the same for all maturities), (d) humped, and (e) inverted; see Figure 10.6.

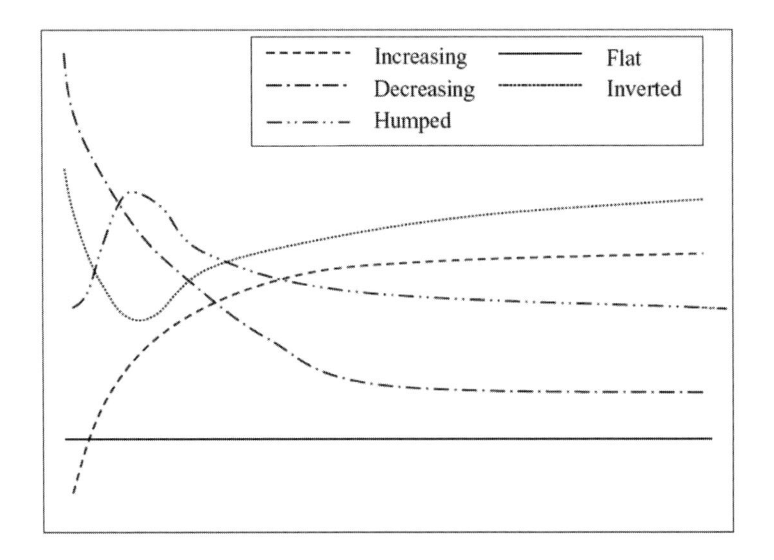

Fig. 10.6. Five common shapes of yields to maturity.

Nelson and Siegel's model

Nelson and Siegel (1987) assume the parametric model

$$f(0,s) = \beta_0 + \beta_1 \exp\Big(-\frac{s}{\tau}\Big) + \beta_2 \frac{s}{\tau} \exp\Big(-\frac{s}{\tau}\Big) \tag{10.17}$$

for the instantaneous forward rate, in which $\beta_0, \beta_1, \beta_2$, and τ are unknown parameters with the following economic interpretation: β_0 is the long-maturity limiting forward rate, τ is a time constant measuring how fast the forward rate tends to change with maturity, and β_1 and β_2 are coefficients used to accommodate different shapes for $P(0,t)$. Combining (10.17) with (10.11) yields

$$P(0,t) = \exp\Big\{-\beta_0 t - (\beta_1+\beta_2)\tau(1-e^{-t/\tau}) + t\beta_2 e^{-t/\tau}\Big\}. \tag{10.18}$$

Svensson's model

Svensson (1994) generalized the Nelson-Siegel model (10.17) by including an additional component and two more parameters:

$$f(0,s) = \beta_0 + \beta_1 \exp\left(-\frac{s}{\tau_1}\right) + \beta_2 \frac{s}{\tau_1} \exp\left(-\frac{s}{\tau_1}\right) + \beta_3 \frac{s}{\tau_2} \exp\left(-\frac{s}{\tau_2}\right). \quad (10.19)$$

This gives additional U and humped shapes for $P(0,t)$:

$$P(0,t) = \exp\Big\{ -\beta_0 t - (\beta_1+\beta_2)\tau_1(1-e^{-t/\tau_1}) + t\beta_2 e^{-t/\tau_1} \\ -\beta_3\tau_2(1-e^{-t/\tau_2}) + t\beta_3 e^{-t/\tau_2}\Big\}. \quad (10.20)$$

An example

As an illustration, we use Svensson's model to estimate the yield curve on December 20, 2006 from a set of $n = 31$ T-notes and T-bonds; see Exercise 10.1. Each note or bond has a specified coupon rate paid semiannually. Table 10.1 gives the estimated parameters in Svensson's model and the root mean squared error (RMSE) between the estimated and quoted prices $\{\sum_{i=1}^n (\widehat{B}_i - B_i)^2/n\}^{1/2}$. The estimated yield curve and the yields to maturity of the 31 notes and bonds are plotted in Figure 10.7.

Table 10.1. Estimated parameters and RMSE of the Svensson model.

τ_1	τ_2	β_0	β_1	β_2	β_3	RMSE
0.0245	2.4464	0.0513	8.4362	−8.4245	−0.0228	0.1813

10.3 Multivariate time series of bond yields and other interest rates

Whereas Section 10.2 considers estimation of the zero-coupon yield curve at current time t, going back in time gives a time series of yield curves. To simplify matters, we consider in this section the multivariate time series $\mathbf{y}_t = (y_{t,1}, \dots, y_{t,k})^T$ of yields for a specified set of maturities $t+\tau_1, \dots, t+\tau_k$, instead of the time series of yield curves $y_t(T), T \geq t$. Thus, we consider at each time t the yield curve only for maturities $\tau_1, \dots, \tau_k$ (e.g., 3 months, 6 months, and 1, 3, 5, 7, 10, and 20 years).

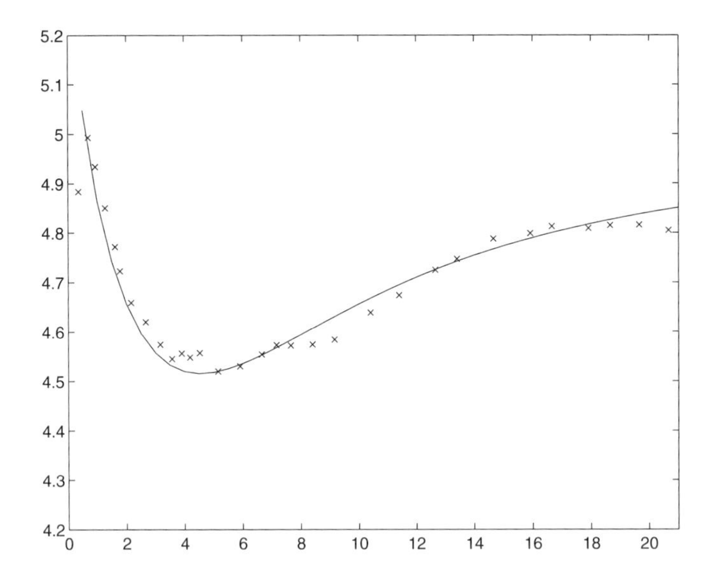

Fig. 10.7. Yields to maturity and the estimated yield curve (solid line).

Unit-root nonstationarity and cointegration

Figure 10.8 plots the time series of 3-month and 6-month daily U.S. Treasury bill rates in the secondary market from December 9, 1958 to March 27, 2007, obtained from the Federal Reserve Bank of St. Louis. The plots exhibit nonstationarity of both series that appears to be of the unit-root type. Performing an augmented Dickey-Fuller test for unit root (see Section 9.4.4) for the 3-month and 6-month rates gives the p-values 0.198 and 0.244, respectively, showing no significant departures from the unit-root hypothesis. Moreover, Figure 10.8 shows that the nonstationary movements of the 3-month and 6-month rates track each other. Performing Johansen's cointegration test confirms that these two series are cointegrated (see Section 9.4.5).

Principal component analysis (PCA)

We next perform a similar cointegration analysis on the time series of U.S. Treasury 1-year, 2-year, 3-year, 5-year, 7-year, 10-year, and 20-year weekly constant maturity rates from October 1, 1993 to March 23, 2007. The data are obtained from the Federal Reserve Bank of St. Louis. We apply the `R` function `ca.jo` to these data, and the results are given in Table 10.2, in which $*$ indicates that the likelihood ratio (LR) statistic (9.42) shows significant departure from the null hypothesis H_0 on the cointegration rank r.

The preceding cointegration analysis suggests that the short-, intermediate- and long-term Treasury rates are highly correlated, as is also revealed by the

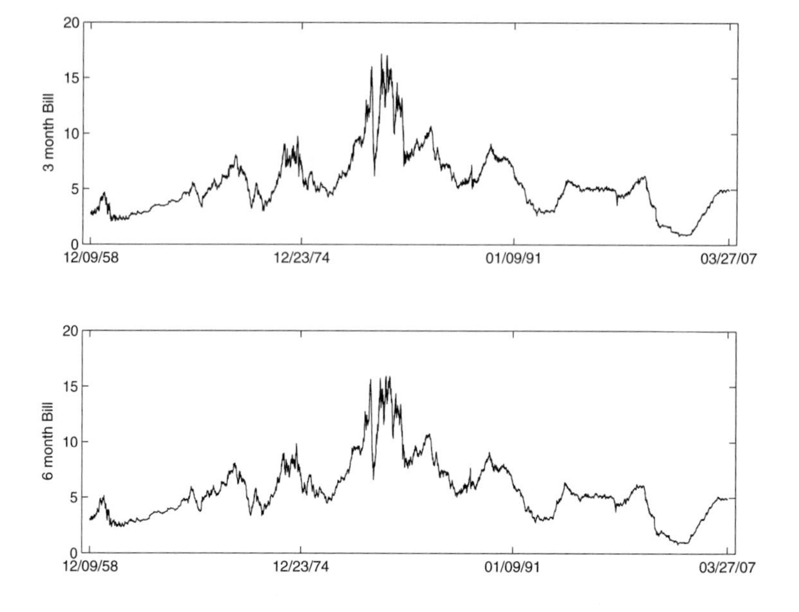

Fig. 10.8. Time series of daily 3-month (top panel) and 6-month (bottom panel) U.S. Treasury bill rates.

Table 10.2. Values of test statistic and critical values of the cointegration test.

H_0	LR	10%	5%
$r = 0$	69.55*	30.77	33.18
$r \leq 1$	31.48*	24.71	27.17
$r \leq 2$	13.90	18.70	20.78

plots of these time series. Figure 10.9 plots the time series of U.S. Treasury 1-year, 7-year, and 20-year weekly constant maturity rates to provide an illustration. To identify the important factors by using PCA, let r_{kt} denote the weekly Treasury rate for the seven maturities $k = 1, 2, 3, 5, 7, 10$, and 20 years, and let $d_{kt} = r_{k,t} - r_{k,t-1}$, $1 \leq t \leq 704$ (the number of weeks in the period). Let $\widehat{\mu}_k$ and $\widehat{\sigma}_k^2$ be the sample mean and sample variance of the series $\{d_{kt}\}$, and let $x_{tk} := (d_{kt} - \widehat{\mu}_k)/\widehat{\sigma}_k$. We can use (2.16) to represent $\mathbf{X}_k = (x_{1k}, \ldots, x_{nk})^T$ in terms of the principal components. Table 10.3 gives the PCA results for the covariance and correlation matrices of $\{d_{kt}\}$. The first three principal components account for 99% of the total variance in both cases.

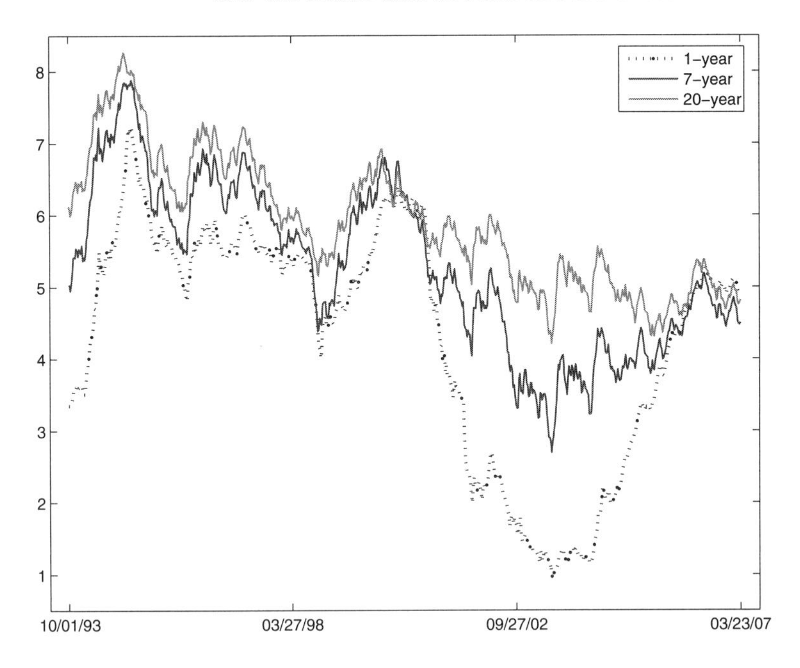

Fig. 10.9. Time series of weekly 1-year, 7-year, and 20-year U.S. Treasury constant maturity rates from October 1, 1993 to March 23, 2007.

Time series of other interest rates

Note the similarities in the time series plots of the U.S. Treasury constant maturity rates in Figure 10.9 and the swap rates in Figure 2.1. The results of PCA in Tables 10.3 and 2.1 are also quite similar, with the first three principal components accounting for most of the variability. In fact, the Treasury rate movements indicated by the first three principal components are similar to those for the swap rates in Section 2.2.3, consisting of a parallel shift component, a tilt component, and a curvature component. A similar statistical analysis can be performed on the time series of LIBOR; see Exercise 10.2. Figure 10.10 plots the daily LIBOR of 1 month, 3 months, and 6 months in U.S. dollars from January 2, 1987 to March 22, 2007, obtained from `http://www.Economagic.com`.

10.4 Stochastic interest rates and short-rate models

To price interest rate derivatives that are contingent on future interest rates, stochastic models of interest rate dynamics are prescribed and pricing is based on arbitrage-free arguments similar to those in the Black-Scholes theory described in Section 8.1.2. A major class of stochastic models in the literature

Table 10.3. PCA of seven U.S. Treasury rates.

	PC1	PC2	PC3	PC4	PC5	PC6	PC7
	(a) Using sample covariance matrix						
Standard dev. ($\times 10^2$)	25.71	6.91	2.87	1.83	1.33	1.16	1.07
Proportion	0.912	0.066	0.011	0.005	0.002	0.002	0.002
Factor loadings	0.279	0.590	−0.696	0.290	0.072	−0.018	0.030
	0.384	0.398	0.221	−0.514	−0.286	0.216	−0.502
	0.416	0.239	0.318	−0.238	0.099	−0.195	0.751
	0.423	−0.044	0.341	0.434	0.311	−0.508	−0.400
	0.412	−0.240	0.087	0.269	0.286	0.780	0.060
	0.384	−0.363	−0.129	0.239	−0.790	−0.102	0.111
	0.324	−0.495	−0.475	−0.527	0.317	−0.199	−0.081
	(b) Using sample correlation matrix						
Standard dev.	2.507	0.741	0.307	0.183	0.130	0.108	0.100
Proportion	0.898	0.078	0.013	0.005	0.002	0.002	0.001
Factor loadings	0.340	−0.653	0.633	−0.228	−0.064	−0.013	−0.017
	0.382	−0.337	−0.314	0.524	0.321	0.240	0.455
	0.391	−0.182	−0.354	0.219	−0.106	−0.250	−0.753
	0.394	0.054	−0.331	−0.389	−0.373	−0.490	0.449
	0.391	0.219	−0.105	−0.292	−0.299	0.777	−0.101
	0.383	0.338	0.093	−0.340	0.769	−0.120	−0.094
	0.362	0.511	0.495	0.523	−0.249	−0.150	0.069

consists of diffusion processes for the short rate r_t defined in Section 10.1.1:

$$dr_t = m(t, r_t)dt + s(t, r_t)dw_t. \tag{10.21}$$

Let $\mu(t,r) = m(t,r)/r$, $\sigma(t,r) = s(t,r)/r$, so $dr_t/r_t = \mu(t, r_t)dt + \sigma(t, r_t)dw$. Using Ito's formula and hedging arguments similar to those of Section 8.1.2 but applied to a portfolio consisting of two bonds with different maturities, it

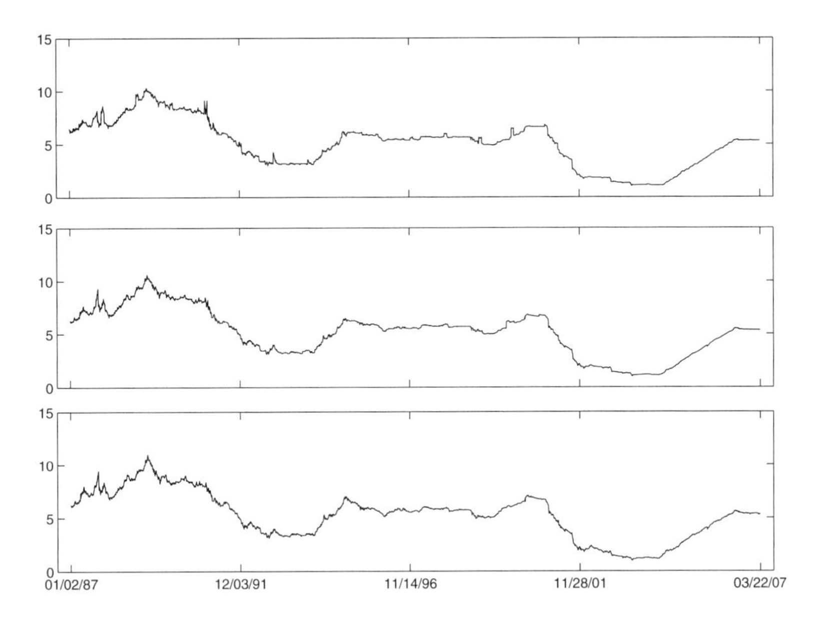

Fig. 10.10. Time series of daily LIBOR of 1 month, 3 months, and 6 months in U.S. dollars from January 2, 1987 to March 22, 2007.

can be shown that in the absence of arbitrage,

$$\lambda(t,r) = \frac{\mu(t,r) - r}{\sigma(t,r)} \tag{10.22}$$

(which is analogous to the Sharpe ratio $(\mu_i - r)/\sigma_i$ in Section 3.3.2) is the same for all derivatives dependent on r_t (e.g., bonds with the same maturities); $\lambda(t,r)$ is called the *market price of risk*. Moreover, the price $g(t,r)$ of an interest rate derivative satisfies the Black-Scholes-type PDE

$$\frac{\partial g}{\partial t} + (m - \lambda s)\frac{\partial g}{\partial r} + \frac{s^2}{2}\frac{\partial^2 g}{\partial r^2} = rg, \quad 0 \le t < T. \tag{10.23}$$

In the case of a bond with par value 1 ($g = P$), we have the terminal condition $g(T,r) = 1$. The solution of (10.23) with this terminal condition can be expressed as the expectation

$$\widehat{E}\left\{\exp\left(-\int_t^T r_s ds\right)\bigg| r_t = r\right\}; \tag{10.24}$$

see Appendix A. In (10.24), $\widehat{E}$ denotes the expectation under which r_t is the diffusion process

$$dr_t = (m - \lambda s)dt + s dw_t$$

(corresponding to the "risk-neutral" measure or, more precisely, the "equivalent martingale measure," which will be described in Section 10.4.2). Unlike the Black-Scholes model and pricing theory, which are commonly used in the equity options market, there is no single dominant model with a specific choice of $m(t, r)$ and $s(t, r)$ in (10.23) that is widely accepted for pricing various derivative securities in the interest rate market. Instead, one encounters a plethora of models and risk-neutral measures that are different from the physical (real-world) measure. Rebonato (2004) gives an overview of the historical development of these models. Another modeling issue for interest rate derivatives is that unlike the relatively short maturity of equity options, the interest rate market considers bonds with maturities as long as 10–30 years. Because of anticipated changes in the economy over such a long period, there are no convincing models that can actually capture the real-world interest rate movements 10–30 years in the future. What one can model at best are the perceived movements that are reflected in the prices of interest rate derivatives.

10.4.1 Vasicek, Cox-Ingersoll-Ross, and Hull-White models

When r_t is a Gaussian process under the risk-neutral measure, the expectation (10.24) has a closed-form expression in view of the formula for moment generating functions of normal random variables. Interest rates are supposed to fluctuate around some level, and the following classical models (under the risk-neutral measure) have been motivated by this *mean-reverting* property:

- Vasicek's (1977) model: $dr_t = \kappa(\theta - r_t)dt + \sigma dw_t,$ (10.25)
- Hull and White's (1990) model: $dr_t = (\theta_t - \kappa r_t)dt + \sigma dw_t,$ (10.26)

in which θ_t is a nonrandom function of t such that $P(0, T)$ agrees with the term structure at time 0. Note that (10.25) is a special case of (10.26) with $\theta_t = \kappa\theta$, but this choice often fails to match the initial term structure exactly. A conceptual difficulty with Gaussian process models for r_t is that $\{r_t < 0\}$ has positive probability. The following (non-Gaussian) modification of the Vasicek model proposed by Cox, Ingersoll, and Ross (1985, abbreviated by CIR) has positive r_t when $2\kappa\theta > \sigma^2$:

- CIR model: $dr_t = \kappa(\theta - r_t)dt + \sigma\sqrt{r_t}dw_t.$ (10.27)

The price $P(t, T)$ of a zero-coupon bond with par value 1 given by the expectation (10.24) has a closed-form expression for the models (10.25)–(10.27):

$$P(t, T) = \alpha(t, T)e^{-\beta(t,T)r_t}. \tag{10.28}$$

The functions α and β in (10.28) are given below.

(i) For the Vasicek model, $\beta(t,T) = (1 - e^{-\kappa(T-t)})/\kappa$,

$$\alpha(t,T) = \exp\left\{\left(\theta - \frac{\sigma^2}{2\kappa^2}\right)[\beta(t,T) - (T-t)] - \frac{\sigma^2}{4\kappa}\beta^2(t,T)\right\}.$$

(ii) For the CIR model, letting $h = \sqrt{\kappa^2 + 2\sigma^2}$,

$$\alpha(t,T) = \left[\frac{2he^{(\kappa+h)(T-t)/2}}{2h + (\kappa+h)[e^{(T-t)h} - 1]}\right]^{2\kappa\theta/\sigma^2},$$

$$\beta(t,T) = \frac{2(e^{(T-t)h} - 1)}{2h + (\kappa+h)[e^{(T-t)h} - 1]}.$$

(iii) For the Hull-White model, $\beta(t,T) = (1 - e^{-\kappa(T-t)})/\kappa$,

$$\log\alpha(t,T) = \log\frac{P(0,T)}{P(0,t)} - \beta(t,T)\frac{\partial}{\partial t}\log P(0,t) - \frac{\sigma^2(1-e^{-2\kappa t})}{4\kappa}\beta^2(t,T).$$

From (10.28), it follows that the spot rate $R(t,T)$ defined in (10.3) is an affine function of the short rate r_t:

$$R(t,T) = -\frac{\log\alpha(t,T)}{T-t} + r_t\frac{\beta(t,T)}{T-t}. \tag{10.29}$$

Moreover, the instantaneous forward rate (10.10) is also an affine function of r_t:

$$f(t,T) = -\frac{\partial}{\partial T}\log P(t,T) = -\frac{\partial}{\partial T}\log\alpha(t,T) + r_t\frac{\partial}{\partial T}\beta(t,T). \tag{10.30}$$

10.4.2 Bond option prices

Jamshidian (1989) has derived explicit formulas for the prices of European options on bonds for the preceding short-rate models. In the Vasicek model, the price $Z(t,T,\widetilde{T},K)$ at time t of a European option with strike K, maturity T, and written on a zero-coupon bond that matures at time $\widetilde{T}$ and has par value 1 is

$$Z(t,T,\widetilde{T},K) = \omega\Big[P(t,\widetilde{T})\Phi(\omega h) - KP(t,T)\Phi(\omega(h-\sigma_P))\Big], \tag{10.31}$$

where $\omega = 1$ for a call, $\omega = -1$ for a put, and

$$\sigma_P = \sigma\sqrt{\frac{1-e^{-2\kappa(T-t)}}{2\kappa}}\beta(T,\widetilde{T}), \quad h = \frac{1}{\sigma_P}\log\frac{P(t,\widetilde{T})}{P(t,T)K} + \frac{\sigma_P}{2}.$$

The formula (10.31) still holds for the Hull-White model, with the same definition of σ_P but with h redefined as

$$h = \frac{1}{\sigma_P} \log \frac{P(0, \widetilde{T})}{P(0, T)K} + \frac{\sigma_P}{2}.$$

In the CIR model, the price at time t of a European call option with strike K, maturity T, and written on a zero-coupon bond that matures at time $\widetilde{T}$ and has par value 1 is

$$\begin{aligned} Z(t, T, \widetilde{T}, K) &= P(t, \widetilde{T})\chi^2\left(2\mu\left[\rho + \psi + \beta(T, \widetilde{T})\right]; \frac{4\kappa\theta}{\sigma^2}, \frac{2\rho^2 r_t e^{h(T-t)}}{\rho + \psi + \beta(T, \widetilde{T})}\right) \\ &\quad - KP(t, T)\chi^2\left(2\mu[\rho + \psi]; \frac{4\kappa\theta}{\sigma^2}, \frac{2\rho^2 r_t e^{h(T-t)}}{\rho + \psi}\right) \end{aligned} \tag{10.32}$$

when the short rate at time t is r_t, where

$$\rho = \frac{2h}{\sigma^2 e^{h(T-t)} - 1}, \qquad \psi = \frac{\kappa + h}{\sigma^2}, \qquad \mu = \frac{\ln(\alpha(T, \widetilde{T})/K)}{\beta(T, \widetilde{T})},$$

and $\chi^2(\cdot; \nu, \lambda)$ is the distribution function of the noncentral chi-square distribution with ν degrees of freedom and noncentrality parameter λ; see Section 8.3.2. The `R` function `pchisq` or the `MATLAB` function `ncx2cdf` can be used to evaluate $\chi^2(\cdot; \nu, \lambda)$. The price of a European put option can be found by the *put–call parity* relation

$$Z_{\text{call}} + KP(t, T) = Z_{\text{put}} + P(t, \widetilde{T}). \tag{10.33}$$

10.4.3 Black-Karasinski model

By considering the dynamics of $\log r_t$ with $r_t > 0$, similar to that of r_t in the Hull-White model, Black and Karasinski (1991) assumed that under the risk-neutral measure

$$d \log r_t = (\theta_t - a_t \log r_t)dt + \sigma_t dw_t, \tag{10.34}$$

where θ_t, a_t, and σ_t are nonrandom functions of t chosen to match the initial term structure of interest rates and some market volatility curves. The model, however, does not yield closed-form expressions for bond prices (10.24) and bond option prices. A trinomial tree can be used to approximate the continuous-time process r_t and thereby compute the bond prices and bond option prices by backward induction; see Hull (2006, pp. 660–672).

10.4.4 Multifactor affine yield models

The models (10.25), (10.26), and (10.27) are *one-factor models*, which only rely on one source of randomness dw_t in the short rate r_t. In these models, the yields for different maturities are perfectly correlated. A simple way to introduce additional sources of randomness is to replace dw_t by $d\mathbf{w}_t$, where $\mathbf{w}_t$ is d-dimensional Brownian motion (see Appendix A), and to introduce a d-dimensional diffusion process of the form $d\mathbf{x}_t = \mathbf{B}\mathbf{x}_t dt + \boldsymbol{\Sigma}^{1/2} d\mathbf{w}_t$ so that $r_t = \mu + \boldsymbol{\theta}^T d\mathbf{x}_t$. For example, consider the two-factor model

$$r_t = x_t + y_t,$$
$$dx_t = -ax_t dt + \sigma dw_t^{(1)}, \qquad dy_t = -by_t dt + \widetilde{\sigma} dw_t^{(2)},$$

where $a, b, \sigma, \widetilde{\sigma}$ are positive constants and $(w_t^{(1)}, w_t^{(2)})$ is a two-dimensional Brownian motion. Then the price of a zero-coupon bond maturing at time T can be expressed as

$$P(t,T) = \exp\left\{ -\frac{1-e^{-a(T-t)}}{a} x_t - \frac{1-e^{-b(T-t)}}{b} y_t + \frac{1}{2}\alpha(t,T) \right\},$$

where

$$\begin{aligned} \alpha(t,T) = & \frac{\sigma^2}{a^2}\left[T - t + \frac{2}{a}e^{-a(T-t)} - \frac{1}{2a}e^{-2a(T-t)} - \frac{3}{2a} \right] \\ & + \frac{\widetilde{\sigma}^2}{b^2}\left[T - t + \frac{2}{b}e^{-b(T-t)} - \frac{1}{2b}e^{-2b(T-t)} - \frac{3}{2b} \right]. \end{aligned}$$

Empirical results on PCA of yield curves suggest that usually $d = 2$ or 3 suffices for the number of factors; see the examples in Section 2.2.3 and Section 10.3.

10.5 Stochastic forward rate dynamics and pricing of LIBOR and swap rate derivatives

Instead of modeling the stochastic dynamics of the short rate process r_t, an alternative approach is to model the instantaneous forward rate $f(t,T)$, noting that $P(t,T)$ can be retrieved from $f(t,\cdot)$ via (10.11). This is the Heath-Jarrow-Morton (HJM) framework, which will be described in Section 10.5.4. An even more fundamental entity is the forward rate (10.9) that involves a three-dimensional time index (t, T_1, T_2), and therefore a stochastic model for (10.9) is a random field rather than a stochastic process such as r_t that involves one-dimensional time t. However, because the relevant dates in interest rate derivatives are typically fixed at $T_1, \ldots, T_M$, one only has to work with a vector-valued process $(F_1(t), \ldots, F_M(t))$, where $F_i(t) = F(t, T_i, T_{i+1})$. In Section 10.5.1, we describe Black's model for the dynamics of $F_i(t)$ and the associated formulas, of Black-Scholes type, that are commonly used in the interest rate market for pricing caps/floors and swaptions. Sections 10.5.2 and 10.5.3 describe developments in the last decade, called the LIBOR and swap market models, that resolve certain inconsistencies in Black's model and provide theoretical justifications and extensions of the standard market formulas for pricing caps/floors and swaptions.

10.5.1 Standard market formulas based on Black's model of forward prices

The market convention of pricing commonly traded interest rate derivatives is based on Black's (1976) formula, whose futures option counterpart has been given in Section 8.1.2. First consider a European option with strike price K and expiration date T on a zero-coupon bond maturing at time $\widetilde{T} > T$. Black's formula yields the option price

$$Z(t, T, \widetilde{T}, K) = \omega P(t, \widetilde{T})\big[F_t \Phi(\omega d_1) - K\Phi(\omega d_2)\big] \tag{10.35}$$

at time t, where $\omega = 1$ for a call, $\omega = -1$ for a put, and

$$d_1 = \frac{\log(F_t/K) + \sigma^2(T-t)/2}{\sigma\sqrt{T-t}}, \quad d_2 = d_1 - \sigma\sqrt{T-t},$$

under the assumption that F_t follows the geometric Brownian motion $dF_t/F_t = \mu dt + \sigma dw_t$; compare this with (8.6) and (8.7), in which $q = r$. Besides this GBM assumption on the forward price F_t of the bond (with maturity date $\widetilde{T}$) at the expiration date T of the option, (10.35) also assumes that the interest rate is a deterministic function so that the price of a forward contract is the same as that of a futures contract with the same delivery date; see Section 5.8 of Hull (2006), where it is also pointed out that the forward and futures prices need no longer be the same in the case of stochastic interest rates.

We next consider an interest rate cap described in Section 10.1.5. Let $F_i(t)$ denote the forward rate $F(t, T_i, T_{i+1})$, where the T_i $(i = 1, \ldots, M)$ denote the reset dates as in Section 10.1.5. Let $\tau_i = T_{i+1} - T_i$, which is called a *tenor*. The cap can be viewed as a portfolio of M *caplets* that are call options with payoff $N\tau_i(F_i - R^*)_+$ at time T_{i+1}, in which $F_i = F(T_i, T_i, T_{i+1})$, R^* is the cap rate, and N the notional principal of the cap; see Hull (2006, p. 620). Note that F_i is the interest rate observed at time T_i for the period between T_i and T_{i+1}; i.e., $F_i = F(T_i, T_i, T_{i+1})$. Assuming GBM dynamics $dF_i/F_i = \mu_i dt + \sigma_i dw_t^{(i)}$, the value at time $t \leq T_i$ of the ith caplet is given by (10.35) as

$$Z_i(t) = N\tau_i P(t, T_{i+1})\{F_i(t)\Phi(d_{1i}) - R^*\Phi(d_{2i})\}, \tag{10.36}$$

where $d_{1i} = \{\log(F_i(t)/R^*) + \sigma_i^2(T_i - t)/2\}/\{\sigma_i\sqrt{T_i - t}\}$, $d_{2i} = d_{1i} - \sigma_i\sqrt{T_i - t}$. The cap price at time t (before the first reset date) is the sum $\sum_{i=1}^M Z_i(t)$ of the prices of these caplets. The standard market price of floors is similar, using put options in place of the call options.

10.5.2 Arbitrage-free pricing: martingales and numeraires

The stochastic interest rate (short rate or forward rate) models used for pricing interest rate derivatives are chosen to be *arbitrage-free.* A fundamental result on arbitrage-free models is the central role of "equivalent martingale measures," and a basic tool for specifying them is "change of numeraires." A *numeraire* is the positive price process of any non-dividend-paying asset. Suppose $\mathbb{P}$ is the probability measure (called the "physical" or "real-world" measure) induced by the stochastic processes of interest rates and N_t is a numeraire. The *fundamental theorem of asset pricing* states that a stochastic model is arbitrage-free if and only if there exists a measure $\mathbb{Q}$ equivalent to $\mathbb{P}$ such that the price (relative to N_t) of any traded asset X_t that does not have any intermediate payments is a martingale under $\mathbb{Q}$. Therefore

$$\frac{X_t}{N_t} = E_{\mathbb{Q}}\left\{\frac{X_T}{N_T}\bigg|\mathcal{F}_t\right\}, \qquad 0 \le t \le T, \tag{10.37}$$

where $\mathcal{F}_t$ is the information set up to time t; see Brigo and Mercurio (2006, pp. 27–28). The measure $\mathbb{Q}$ is called an *equivalent martingale measure.*

Taking the bank account $B(t)$ of Section 10.1.1 as the numeraire N_t, the equivalent martingale measure $\mathbb{Q}$ is the risk-neutral measure used in Section 10.4, for which (10.37) reduces to

$$X_t = E_{\mathbb{Q}}\left\{e^{-\int_t^T r_s ds} X_T \Big| \mathcal{F}_t\right\}. \tag{10.38}$$

Taking the zero-coupon bond price $P(t, T)$ with maturity T as the numeraire, the equivalent martingale measure $\mathbb{Q}$ is often called *forward risk-neutral* with respect to the bond and is denoted by $\mathbb{P}_T$. In this case, (10.37) reduces to

$$X_t = P(t, T)E_{\mathbb{P}_T}\left(X_T \Big| \mathcal{F}_t\right) \tag{10.39}$$

since $P(T, T) = 1$. In particular, let $X = X(P(T, \widetilde{T}))$ be the payoff of a derivative at time T that is dependent on $P(T, \widetilde{T})$ for some $\widetilde{T} > T$. Then the value $V(t)$ of the derivative can be expressed via (10.37), with $X_t = V(t)$, as

$$V(t) = P(t, T)E_{\mathbb{P}_T}\left[X(P(T, \widetilde{T}))\Big|\mathcal{F}_t\right]. \tag{10.40}$$

10.5.3 LIBOR and swap market models

Market models, introduced by Brace, Gaterek, and Musiela (1997) and Jamshidian (1997), assume arbitrage-free stochastic processes for forward rates or swap rates and thereby derive Black's formulas for caps/floors or swaptions.

LIBOR market model

Consider the probability measure $\mathbb{P}_{T_i}$ that is forward risk-neutral with respect to the numeraire $P(t, T_i)$; see Section 10.5.2. Note that the forward rate $F_i(t)$ satisfies $F_i(t)P(t, T_{i+1}) = \{P(t, T_i) - P(t, T_{i+1})\}/\tau_i$; see (10.13). Suppose that, under $\mathbb{P}_{T_i}$, $F_i(t)$ is a driftless GBM defined by

$$dF_i(t) = \nu_i(t)F_i(t)dw_i(t), \qquad t \le T_i, \tag{10.41}$$

in which $\nu_i(t)$ is a deterministic function. Then $\{F_i(t), t \le T_i\}$ is a martingale under $\mathbb{P}_{T_i}$, and it follows from (10.41) that $F_i(T_i)$ is lognormal under P_{T_i}, with

$$\begin{aligned} \mathrm{Var}_{\mathbb{P}_{T_i}}\big[\log F_i(T_i)\big|\mathcal{F}_t\big] &= \int_t^{T_i} \nu_i^2(u)du, \\ E_{\mathbb{P}_{T_i}}\big[\log F_i(T_i)\big|\mathcal{F}_t\big] &= \log F_i(t) - \frac{1}{2}\int_t^{T_i} \nu_i^2(u)du. \end{aligned}$$

Hence we can apply (10.40) to evaluate the price of the ith caplet by

$$N\tau_i P(t, T_{i+1})E_{\mathbb{P}_{T_i}}\big[(F_i(T_i) - R^*)_+\big] = N\tau_i P(t, T_{i+1})\big\{F_i(t)\Phi(d_{1i}) - R^*\Phi(d_{2i})\big\}, \tag{10.42}$$

where N is the notional principal and d_{1i} and d_{2i} are the same as those in Black's formula (10.36) when the ith caplet volatility σ_i satisfies

$$\sigma_i^2 = \frac{1}{T_i - t}\int_t^{T_i} \nu_i^2(u)du, \qquad 1 \le i \le M; \tag{10.43}$$

i.e., σ_i^2 is the average instantaneous variance $\nu_i^2(\cdot)$ over the time interval (t, T_i).

Swap market model and Black's formula for swaption prices

Swaptions (or swap options) are options on interest rate swaps giving the holder the right to enter into an interest rate swap (see Section 10.1.4) at a certain time T in the future. A European payer swaption gives the holder the right to pay fixed rate R^* and receive floating rate (LIBOR) at times $T_1 \le \cdots \le T_M$ with $T_1 > T$ (which is the swaption's expiration date). Let

$T_0 = T$, and denote the forward swap rate $s(t, T_0, T_M)$ in (10.15) simply by $s(t)$ for $t \le T$. The swaption is exercised at its expiration date T only when $s(T) > R^*$. Therefore the payoff of this swaption is $N \sum_{i=1}^{M} \tau_i \big(s(T) - R^*\big)_+ P(T, T_i)$, where N is the notional principal. Consider the probability measure $\mathbb{Q}$ that is forward risk-neutral with respect to the numeraire $N \sum_{i=1}^{M} \tau_i P(t, T_i)$, which is the present value of the interest rate swap on the notional principal N; see Section 10.5.2. Suppose that, under $\mathbb{Q}$, $s(t)$ follows a driftless geometric Brownian motion $ds(t) = \varsigma(t)s(t)dw_t$, where $\varsigma(t)$ is a deterministic function. Then an argument similar to that in (10.42) yields

$$N\sum_{i=1}^{M}\tau_i P(t,T_i)E_{\mathbb{Q}}\left[(s(T)-R^*)_+\right] = \left\{N\sum_{i=1}^{M}\tau_i P(t,T_i)\right\}\{s(t)\Phi(d_1)-R^*\Phi(d_2)\}, \tag{10.44}$$

where

$$d_1 = \frac{\log(s(t)/R^*) + \sigma^2(T-t)/2}{\sigma\sqrt{T-t}},$$

$$d_2 = d_1 - \sigma\sqrt{T-t}, \qquad \sigma^2 = \frac{1}{T-t}\int_t^T \varsigma^2(u)du.$$

The swaption price (10.44) is the same as the widely used Black's formula for swaptions, which was originally derived by heuristic arguments similar to those in Section 10.5.1. Note that the swap market model expresses the swaption volatility σ as the square root of the average instantaneous variance $\varsigma^2(\cdot)$ over the interval (t, T).

Incompatibility of LIBOR and swap market models

The LIBOR market model retrieves Black's formula (10.36) for pricing caplets by assuming that $F_i(\cdot)$ is GBM and hence lognormal, whereas the swap market model retrieves Black's formula (10.44) for pricing swaptions by assuming that $s(T_0)(= s(T_0, T_0, T_M))$ is lognormal. However, these two assumptions preclude each other since, by (10.15), $s(T_0) = \sum_{i=0}^{M-1} w_i(T_0)F_i(T_0)$, in which $w_i(t) = P(t, T_{i+1})\tau_i / \{\sum_{j=0}^{M-1} P(t, T_{j+1})\tau_j\}$ is a nonlinear function of $F_i(T_0)$; see (10.13). One can use Ito's formula and the LIBOR market model to derive the stochastic dynamics of $s(t)$, from which it follows that the instantaneous swap rate volatility $\varsigma(t)$ is related to the instantaneous volatilities $\nu_i(t)$ in (10.41) by

$$\varsigma^2(u) = \sum_{i=1}^{M}\sum_{j=1}^{M} \nu_i(t)\nu_j(t)\rho_{ij}(t)\psi_i(t)\psi_j(t), \tag{10.45}$$

where $\rho_{ij}(t)$ is the instantaneous correlation between the forward rates $F_i(t)$ and $F_j(t)$ and

$$\psi_j(t) = \frac{w_j(t) + \sum_{i=1}^{M} F_i(t) dw_j(t)/dF_j(t)}{\sum_{i=1}^{M} w_i(t) F_i(t)};$$

see Rebonato (2002, p.175).

10.5.4 The HJM models of the instantaneous forward rate

Heath, Jarrow, and Morton (1992) proposed to model the instantaneous forward rate process $f(t,T)$ for every given maturity T by a k-factor model of the form

$$df(t,T) = \sum_{i=1}^{k} \sigma_i(t,T) s_i(t,T) dt + \sum_{i=1}^{k} \sigma_i(t,T) dw_i(t) \tag{10.46}$$

under an equivalent martingale measure, where $\mathbf{w}(t) = (w_1(t), \ldots, w_k(t))^T$ is k-dimensional Brownian motion. They showed that, in the absence of arbitrage,

$$s_i(t,T) = \int_t^T \sigma_i(t,u) du. \tag{10.47}$$

The initial condition $f(0,T)$ in the stochastic differential equation (10.46) is the market forward rate curve described in Section 10.2.1. For the one-factor case $k = 1$, we can rewrite (10.46) as

$$f(t,T) = f(0,T) + \int_0^t \sigma(u,T) s(u,T) du + \int_0^t \sigma(u,T) dw(u), \tag{10.48}$$

which yields the following representation of the short rate $r(t) = f(t,t)$:

$$r(t) = f(0,t) + \int_0^t \sigma(u,t) s(u,t) du + \int_0^t \sigma(u,t) dw(u). \tag{10.49}$$

An important advantage of (10.46) over short-rate models is that it can include multiple factors naturally. Although a number of multifactor short-rate models have been proposed, most of them involve decomposing $r(t)$ into a sum of unobservable state variables $x_i(t)$ that are used to introduce additional Brownian motions $w_i(t)$ as sources of randomness but do not have a physical interpretation. For $k \geq 2$ or for general volatility functions $\sigma(t,T)$ in the case $k = 1$, the short rate $r(t)$ in the Heath-Jarrow-Morton (HJM) models is non-Markovian, making it difficult to use tree methods to compute prices of interest rate derivatives because they lead to non-recombining trees. Monte Carlo simulations are needed to compute these prices in HJM models; see Hull (2006, pp. 681–682).

10.6 Parameter estimation and model selection

10.6.1 Calibrating interest rate models in the financial industry

The current practice in the financial industry is to estimate the parameters of a chosen interest rate model by calibrating it to daily market data; i.e., minimizing the sum of squared differences between the theoretical and observed values of the selected calibrating instrument. The underlying motivation is to enable market makers to price derivatives in a way that is consistent with the market prices of the calibrating instrument, which evolves with interest rates over time. The parametric approach to yield curve estimation in Section 10.2.2 is an example of calibration of the Nelson-Siegel and Svensson forward rate models, using the U.S. Treasury notes and bonds as *calibrating instruments.* The data used to estimate the yield curve in Figure 10.7 are based on 31 notes and bonds on December 20, 2006, and provide an example of daily calibration that allows the parameters of the underlying interest rate to change from one day to the next.

The calibrating instruments in the financial industry are chosen to be as similar as possible to the derivative being valued. Therefore, caps, floors, and swaptions, which are the most popular interest rate derivatives, are often used as calibrating instruments (instead of the bond prices mentioned above). Moreover, instead of actual prices, the market quotes of these derivatives are their implied volatilities, as shown in Figures 10.11 and 10.12. As pointed out in Rebonato's (2004) survey of interest rate models, the fundamental work of Black and Scholes (1973) on option pricing considered in Chapter 8 paved the way for the first phase of interest rate model development. In particular, Black (1976) extended the Black-Scholes theory for equity options to interest rate derivatives, replacing the volatility of the spot price in the Black-Scholes formula with that of the *forward* price. The short-rate models in Section 10.4 constituted the second phase of the development of interest rate models. Since the volatility parameters of these short-rate models are not the same as the implied volatilities from Black's model, the quoted implied volatilities based on Black's model are first converted to prices, which are then compared with the corresponding prices of the derivative computed from the short-rate model. The next generation of term structure models consists of the HJM models in Section 10.5.4 followed by the market models in Section 10.5.3. Instead of modeling the forward price of a bond in Black's approach, the HJM models use the instantaneous forward rate $f(t, T)$ under an equivalent martingale measure, whereas the LIBOR market model assumes that the forward rate $F(t, T_i, T_{i+1})$ follows a geometric Brownian motion under the forward risk-neutral measure $\mathbb{P}_{T_i}$ and recovers the prices of caplets produced by Black's approach. Unlike HJM models, which consider the dynamics of the entire curve $f(t, \cdot)$ as time t changes, the LIBOR market model treats each caplet in isolation using its own equivalent martingale measure $\mathbb{P}_{T_i}$. Similarly, the swap market model also prices a given swaption using its own equivalent martingale measure, and there is no internal consistency between the LIBOR and swap market models. Rebonato (2004) remarks that "the joint facts that traders carried on using the Black formula for caplets and swaptions despite its then-perceived lack of sound theoretical standing, and that this approach could be later theoretically justified, contributed to turning the approach into a market standard." He points out that to understand the calibration of these models, "it is essential to grasp how they are put to use" by (a) relative-value bond traders and plain vanilla option traders, who use models not only to describe prices but also to prescribe trading strategies, and (b) complex-derivative traders, who do not have access to readily accessible market prices for their products and require models to price an exotic product, using the observable market data of bond pricing and the implied volatilities of plain vanilla hedging instruments (caplets and swaptions) as inputs to the models.

Curncy TK N125 CurncyTK
<MENU> to return
Enter your search terms
Page 1/55 World Currency Value
USD (United States Dollar) Caps and Floors
Priced within Forward Term Type Quoted as Struck as
All All All All All
986 of 1851 Tickers

		Symbol	Value
1) USD CAP STRIKE	6 MO	USCPSTF	5.342
2) USD CAP STRIKE	1 YR	USCPST1	5.309
3) USD CAP STRIKE	18 M	USCPST1F	5.280
4) USD CAP STRIKE	2 YR	USCPST2	5.260
5) USD CAP STRIKE	2.5YR	USCPST2F	5.262
6) USD CAP STRIKE	3 YR	USCPST3	5.260
7) USD CAP STRIKE	3.5YR	USCPST3F	5.278
8) USD CAP STRIKE	4 YR	USCPST4	5.291
9) USD CAP STRIKE	4.5YR	USCPST4F	5.309
10) USD CAP STRIKE	5 YR	USCPST5	5.320
11) USD CAP STRIKE	6 YR	USCPST6	5.349
12) USD CAP STRIKE	7 YR	USCPST7	5.369
13) USD CAP STRIKE	8 YR	USCPST8	5.396
14) USD CAP STRIKE	9 YR	USCPST9	5.418
15) USD CAP STRIKE	10YR	USCPST10	5.437
16) USD CAP STRIKE	12YR	USCPST12	5.470
17) USD CAP STRIKE	15YR	USCPST15	5.520
18) USD CAP STRIKE	20YR	USCPST20	5.570

Australia 61 2 9777 8600 Brazil 5511 3048 4500 Europe 44 20 7330 7500 Germany 49 69 920410
Hong Kong 852 2977 6000 Japan 81 3 3201 8900 Singapore 65 6212 1000 U.S. 1 212 318 2000 Copyright 2007 Bloomberg L.P.
G639-436-0 06-Jun-2007 10:46:36

Fig. 10.11. Market quoted implied volatilities for caps. Used with permission of Bloomberg LP.

Calibrating the LIBOR market model

In Section 10.5.1, a cap with reset dates at times $T_1, \ldots, T_M$ and payoffs at times $T_2, \ldots, T_{M+1}$ is characterized as a portfolio of M caplets that are call

01 N2N121 Equity **TTSV**

200<Go> to view in Launchpad

11:36 **Tullett Prebon Information** PAGE 1 / 3

USD Swaption Volatilities

Option	Swap:	1Y		2Y		3Y		4Y	
1 Month	1)	21.50	11)	22.45	21)	20.95	31)	19.95	11:05
3 Month	2)	20.00	12)	22.00	22)	19.95	32)	19.20	9:19
6 Month	3)	19.90	13)	21.00	23)	19.70	33)	18.95	10:34
1 Year	4)	21.25	14)	20.65	24)	19.50	34)	18.75	9:26
2 Year	5)	19.85	15)	19.25	25)	18.55	35)	17.95	8:27
3 Year	6)	18.65	16)	18.20	26)	17.75	36)	17.30	8:31
4 Year	7)	17.75	17)	17.45	27)	17.05	37)	16.80	8:32
5 Year	8)	17.30	18)	16.95	28)	16.70	38)	16.40	8:32
7 Year	9)	16.35	19)	15.90	29)	15.75	39)	15.50	8:32
10 Year	10)	14.95	20)	14.65	30)	14.55	40)	14.45	8:32

Page forward for 5, 7 10 & 12 Year

Fig. 10.12. Market quoted implied volatilities for swaptions. Used with permission of Bloomberg LP.

options on zero-coupon bonds with payoffs on the calls occurring at the time they are calculated. The LIBOR market model assumes that for the ith caplet the instantaneous forward rate $F_i(t) = F(t, T_i, T_{i+1})$ follows a driftless GBM under $\mathbb{P}_{T_i}$; see (10.41), in which $\nu_i(t)$ is a deterministic function. To calibrate the LIBOR market model to the caplet volatilities σ_i^2 in (10.43), the functions $\nu_i(u)$ in (10.43) are chosen to have the form $\nu_i(u) = \nu(u, T_i)$. Let

$$\nu_{i,k}^2 = \int_{T_{k-1}}^{T_k} \nu^2(u, T_i)du, \qquad 1 \le k \le i,$$

with T_0 equal to the current time t. Then (10.43) can be written as $(T_i - t)\sigma_i^2 = \sum_{k=1}^{i} \nu_{i,k}^2$. Rebonato (2002) introduces three building blocks for $\nu(u, T)$: a purely time-dependent component $\xi(u)$, a purely forward-rate-specific component $\eta(T)$, and a time-homogeneous component $\zeta(T - u)$, so that $\nu(u, T) = \xi(u)\eta(T)\rho(T - u)$. The purely time-dependent component $\xi(u)$ specifies that, when economic news arrives, all forward rates share the same "responsiveness" to the shocks, regardless of their maturity. The purely forward-rate-specific component $\eta(T)$ describes the effect of different forward rates on the instantaneous volatility. The time-homogeneous component $\zeta(T - u)$ specifies the time-to-maturity effect of a forward rate. Rebonato (2002) also recommends using parametric functions for these three building blocks; e.g., he proposes to use the following functional form

$$\zeta(T-u) = \big[\alpha + \beta(T-u)\big]e^{-\gamma(T-u)} + \delta$$

for $\zeta(T-u)$, with certain constraints on the parameters α, β, γ, and δ. Therefore $\nu_{i,k}^2$ can be parameterized as $\nu_{i,k}^2(\boldsymbol{\theta})$, and $\boldsymbol{\theta}$ can be estimated by the least squares criterion

$$\min_{\boldsymbol{\theta}} \sum_{i=1}^{M} \left[(T_i - t)\sigma_i^2 - \sum_{k=1}^{i} \nu_{i,k}^2(\boldsymbol{\theta}) \right]^2. \tag{10.50}$$

10.6.2 Econometric approach to fitting term-structure models

Instead of using daily prices (expressed in terms of implied volatilities from Black's model) of interest rate derivatives to refit term-structure models, the econometric approach uses the multivariate time series of bond yields to fit term-structure models and test them. For one-factor short-rate models, the continuous-time models of r_t are replaced by their discrete-time approximations, and short-maturity T-bill yields are chosen as proxies for r_t. Section 9.7.3, which describes the application of the generalized method of moments by Chan et al. (1992) to estimate the parameters of different short-rate models and to test the validity of these models based on one-month Treasury bill yields from June 1964 to December 1989, provides an illustration of the econometric approach. Subsequent work by Aït-Sahalia (1996) and Stanton (1997) introduced nonparametric methods for estimating the drift and volatility functions μ, σ of the short-rate process $dr_t = \mu(r_t)dt + \sigma(r_t)dw_t$ by making use of its stationary distribution.

Unit-root nonstationarity and regime-switching models

The empirical analysis of multivariate time series of bond yields in Section 10.3, however, shows the inadequacy of the one-factor short-rate models in the preceding paragraph. Moreover, contrary to the assumed stationary distribution, the empirical analysis has demonstrated unit-root nonstationarity and cointegration in U.S. Treasury rates. An alternative to interpreting long memory from the observed unit-root nonstationarity is that the short-rate process may have undergone regime switching among different regimes. Empirical research has indeed provided strong evidence of regime switching in U.S. short-term interest rates (see Hamilton, 1988; Driffill, 1992; Albert and Chib, 1993; Cai, 1994; Gray 1996; Bekaert, Hodrick, and Marshall 2001). Ang and Bekaert (2002a, b) find that regime-switching models replicate certain nonlinear patterns of the nonparametrically estimated drift and volatility functions of short rates and have better out-of-sample forecasts than single-regime models. Various macroeconomic and political events (such as the monetary experiment of the late 1970s, wars involving the United States, and the October 1987 stock market crash) have been used to explain regime switching in U.S. interest rates.

10.6.3 Volatility smiles and a substantive-empirical approach

As pointed out by Rebonato (2004, p. 708), the timing of the wide acceptance of the LIBOR market model (LMM) turned out to be "almost ironic" because it coincided with progressively marked smiles in the implied volatility curves. Moreover, according to Rebonato, "as the interest rate volatility surfaces underwent dramatic and sudden shape changes in 1998 [a tumultuous year in several markets], the hedges suggested by the LMMs calibrated to normal volatility regimes often proved dramatically wrong," leading to heavy trading losses. One approach to address these volatility smiles is to introduce jump components into the lognormal process for the forward rates, as in Glasserman and Kou (2003). Another approach is to use the CEV model described in Section 8.3.2. Derman (2004, p. 249) has remarked on the difficulties in addressing the smile dynamics in a satisfactory manner: "Though we know much more about the *theories* of the smile, we are still on a darkling plain regarding what's correct. A decade of speaking with traders and theorists has made me wonder what 'correct' means. If you are a theorist you must never forget that you are traveling through lawless roads where the local inhabitants don't respect your principles. The more I look at the conflict between markets and theories, the more limitations of models in the financial and human world become apparent to me."

It has been noted in Sections 10.6.1 and 10.6.2 that the calibration approach uses daily implied volatilities of interest rate derivatives while the econometric approach uses time series of bond yields for different maturities. Whereas the calibration approach is targeted toward pricing and hedging in interest rate markets, the econometric approach aims at fitting and testing term-structure models for forecasting future movements of interest rates, inflation rates, and other economic variables that determine the prices of financial assets; see Campbell, Lo, and MacKinlay (1997, pp. 418–424). The possibility of combining both kinds of data (bond yields and implied volatilities of caps, swaptions, and other interest rate derivatives) and combining the calibration and econometric approaches into a substantive-empirical approach like that in Section 8.3.5 for equity options is discussed by Lai, Pong, and Xing (2007).

Exercises

10.1. The file bonds_dec2006.txt contains the data for the 31 U.S. Treasury bonds and Treasury notes used in the illustrative example in Section 10.2.2. Instead of Svensson's model used in that example, apply the following methods to estimate the yield curve:
(a) smoothing cubic splines;
(b) exponential splines;
(c) the Nelson-Siegel model (10.17);
(d) the extended Vasicek model of the instantaneous forward rate

$$f(0,t) = \beta_0 - \beta_1 \frac{1 - e^{-at}}{at} + \beta_2 \frac{(1 - e^{-at})^2}{4at}, \tag{10.51}$$

in which a, β_0, β_1, and β_2 are unknown parameters.

10.2. The file d_12libor_8707.txt contains the 1-month, 2-month, ..., 12-month LIBOR (in U.S. dollars) from January 2, 1990 to March 22, 2007. The data are obtained from http://www.Economagic.com.
(a) Plot the twelve LIBOR time series. Do the series look stationary? Is differencing needed to obtain (weak) stationarity?
(b) Perform PCA of these rates (or their successive differences in the case of nonstationarity) and examine the contributions of the first three principal components to the overall variability.

10.3. The file m_swap_0006.txt contains monthly U.S. 1-year, 2-year, 3-year, and 5-year swap rates from July 2000 to May 2006. The data are obtained from http://www.Economagic.com.
(a) Plot the time series of these rates.
(b) Perform a unit-root test for each rate.
(c) Perform cointegration analysis of these rates, including the cointegration test for the number of cointegration vectors and maximum likelihood estimation.
(d) Fit an ARIMA-GARCH model to the 1-year swap rate.

10.4. Assuming that the underlying term structure model is the Vasicek model (10.25) and that the daily yield curve is given, one can calibrate the model parameters (k, θ, σ) and estimate today's instantaneous interest rate by minimizing

$$\sum_{i=1}^{n} [R(T_i) - R(T_i; \kappa, \theta, \sigma)]^2,$$

where $R(T_i; \kappa, \theta, \sigma)$ is given by (10.29) with $t = 0$. Consider the years to maturity and yields of 31 U.S. Treasury bonds and Treasury notes on December 20, 2006 contained in the file bonds_yield_dec2006.txt.

(a) Use these data to calibrate the model parameters (κ, θ, σ) and instantaneous interest rates on December 20, 2006.

(b) With the fitted model and the initial short-term interest rate 4.75%, calculate the price of a 2.5-year European call option, with strike price $99, on a bond that matures in three years, pays coupons semiannually at 5% rate and has par value $100.

(c) Consider the same problem as in (a) and (b), but use the CIR model (10.27).

10.5. Consider the short-rate model

$$dr_t = \big[\alpha_0(t) + \alpha_1(t)r_t\big]dt + \beta(t)\sqrt{r_t}dw_t, \qquad (10.52)$$

where $\{w_t\}$ is a standard Brownian motion and $\alpha_0(t)$, $\alpha_1(t)$, and $\beta(t)$ are unspecified smooth functions. Assume that r_t is observed at discrete time points $t_0 < \cdots < t_n$. Let $x_i = r_{t_i}$, $\delta_i = t_i - t_{i-1}$, $y_i = x_i - x_{i-1}$, and $z_i = x_i^2 - x_{i-1}^2$. The discrete-time approximation to (10.52) can be written as

$$y_i \approx \big[\alpha_0(t_i) + \alpha_1(t_i)x_i\big]\delta_i + \beta(t_i)\sqrt{x_i\delta_i}\epsilon_i,$$

in which ϵ_i are i.i.d. standard normal random variables. Dividing both sides by δ_i yields

$$\frac{y_i}{\delta_i} \approx \alpha_0(t_i) + \alpha_1(t_i)x_i + \beta(t_i)\sqrt{\frac{x_i}{\delta_i}}\epsilon_i. \qquad (10.53)$$

(a) In view of (10.53), one way to estimate $\alpha_0(t)$ and $\alpha_1(t)$ is to use locally weighted running-mean smoothers (see Sections 7.2.1 and 7.2.2). Specifically, for time period t, choose $a_0 = \alpha_0(t)$ and $a_1 = \alpha_1(t)$ to minimize

$$\sum_{i=1}^{n} \frac{\delta_i}{x_i}\Big(\frac{y_i}{\delta_i} - a_0 - a_1 x_i\Big)^2 K\Big(\frac{t_i - t}{\lambda}\Big) \qquad (10.54)$$

over a_0 and a_1, where K is a smooth kernel function and $\lambda > 0$ the bandwidth. Apply this method to estimate $\alpha_0(t)$, $\alpha_1(t)$, and $\beta(t)$ from the weekly U.S. 3-month T-bill rates from January 1990 to June 2007 contained in the file `w_3mtcm9007.txt`. The data are obtained from the Federal Reserve Bank of St. Louis. Take $\delta_i = 1/52$ (i.e., 1-week), $\lambda = 0.25$ and $K =$ Epanechnikov kernel. Plot the estimated functions (after estimating them over a grid of points).

(b) Application of Ito's formula (see Appendix A) to r_t^2 and the stochastic differential equation (10.52) for r_t yields the discrete-time approximation

$$z_i \approx \left\{ \left[2\alpha_0(t_i) + \beta^2(t_i)\right] x_i + 2\alpha_1(t_i) x_i^2 \right\} \delta_i + 2\beta(t_i)\sqrt{x_i^3 \delta_i} \epsilon_i^*,$$

where ϵ_i^* are i.i.d. standard normal random variables; equivalently,

$$\frac{z_i}{\delta_i} \approx \left[2\alpha_0(t_i) + \beta^2(t_i)\right] x_i + 2\alpha_1(t_i) x_i^2 + 2\beta(t_i)\sqrt{\frac{x_i^3}{\delta_i}} \epsilon_i^*. \qquad (10.55)$$

Hence one can estimate $\beta(t)$ by choosing $b = \beta(t)$ that minimizes

$$\sum_{i=1}^{n} \frac{\delta_i}{x_i^3} \left(\frac{z_i}{\delta_i} - (2\widehat{a}_0 + b^2) x_i - 2\widehat{a}_1 x_i^2 \right)^2 K\left(\frac{t_i - t}{\lambda} \right) \qquad (10.56)$$

over b, where $(\widehat{a}_0, \widehat{a}_1)$ is the minimizer of (10.54). Apply this method to estimate $\beta(t)$ from the data in (a), again taking $\delta_i = 1/52$, $\lambda = 0.25$, and K = Epanechnikov kernel. Plot the estimated function (after estimating $\beta(t)$ at a grid of values of t).

11

Statistical Trading Strategies

The finance theories underlying Chapters 8 and 10 assume the absence of arbitrage, leading to pricing models that are martingales after adjustments for the market price of risk. Since the martingale models preclude making risk-adjusted profits via trading strategies, these theories imply that the derivatives markets would only attract *hedgers*, who use derivatives to reduce the risk they face from future movements of stock or bond prices. However, as pointed out by Hull (2006, Chapter 1), derivatives markets have also attracted speculators and arbitrageurs who try to take advantage of the discrepancies between the arbitrage-free theories and the actual market prices. Hedge funds have now become big users of derivatives for all three purposes, namely hedging, speculation, and arbitrage.

Arbitrage opportunities are not expected to last long after they arise because they attract other arbitrageurs and because of the forces of supply and demand; see Hull (2006, p. 14). "Statistical arbitrage," which has become an important activity of many hedge fund managers, uses statistical learning from market prices and trading patterns to identify arbitrage opportunities in various financial markets, evaluates the profits and risks of possible arbitrage positions, and then uses statistical and economic analyses to devise suitable trading strategies. Such arbitrage activity seems to be at odds with the *efficient market hypothesis* (EMH), which basically rules out arbitrage in efficient markets. An important tenet of the EMH is that prices in an efficient market fully reflect all available information, implying that it is impossible to make profits by trading on the basis of an information set; see Samuelson (1965) and Fama (1970). The taxonomy of information sets, due to Roberts (1967), has led to three forms of market efficiency: (i) the *weak* form, in which the information set includes only the history of prices or returns, (ii) the *semistrong* form, in which the information set includes all publicly available information known to all market participants, and (iii) the *strong* form, in which the information set includes all private information of any market participant. Empirical work on the EMH has difficulties in implementation and interpretation, particularly in specifying the information set and defining normal returns against which abnormal (superior) returns can be measured. Campbell, Lo, and MacKinlay (1997, p. 24) suggest that it is more useful to use the EMH as a benchmark, or "a frictionless ideal" relative to which efficiency of a market is measured, than the "all-or-nothing view taken by much of the traditional market-efficiency literature" that focuses on empirical testing of this ideal.

Statistical trading strategies incorporate model uncertainties and additional risks over what absence of arbitrage in economic theory typically assumes. Moreover, perfect efficiency is an unrealistic benchmark in practice because, as pointed out by Grossman and Stiglitz (1980), superior returns have to exist, even in theory, to compensate investors for their information-gathering and information-processing expenses. This chapter considers different aspects of statistical trading strategies. Section 11.1 reviews some *technical trading rules* based on graphical patterns (or *technical analysis*) of the time series of market prices. In particular, it considers a popular Wall Street investment strategy called *pairs trading*. For this trading strategy, it describes the underlying idea of cointegrated time series and reviews recent empirical work on the strategy, to illustrate how statistical methods can help to uncover arbitrage opportunities and devise quantitative trading rules to take advantage of these opportunities. In addition, it considers another class of popular technical trading rules called *momentum strategies*. It also describes data-snooping checks and statistical methods for evaluating the profitability of technical trading rules. In addition, it considers trading rules that are based more on domain knowledge (including *fundamental analysis* of companies, macroeconomic trends, and behavioral patterns of market players) than graphical patterns.

Statistical learning of market patterns can proceed with different levels of resolution. As pointed out in Section 3.1.2, the highest resolution can be obtained from transaction-by-transaction or trade-by-trade data in securities markets. In Section 11.2, we describe statistical models and methods to study market microstructure. It illustrates these statistical methods with intraday transactions of IBM stock from January 2 to March 31, 2003 and gives a brief introduction to real-time trading, which has become popular for hedge funds and investment banks.

Although the Markowitz, CAPM, and Black-Scholes theories in Chapters 3 and 8 assume the absence of market friction and in particular no transaction costs, transaction costs are an important consideration in the design and evaluation of statistical trading strategies. Section 11.3 gives an introduction to estimation and analysis of transaction costs and discusses how transaction costs and the dynamic nature of trading have introduced challenges to the development of statistical trading strategies.

It should be noted that this chapter only considers trading in equity markets. The high-frequency data in Section 11.2 are related to stocks and stock indices, and the dynamic trading in Section 11.3 is related to portfolios of stocks like those considered earlier in Chapter 3 for the single-period setting without transaction costs. The technical trading rules in Section 11.1 are based on time series of stock prices. Even for these relatively well-understood markets, many issues and open problems still remain unresolved, as shown in Section 11.3. Statistical arbitrage is widely used by hedge funds in the derivatives markets (as noted in the first paragraph of this chapter), foreign exchange markets, and interest rate markets, where these issues become more complex and the unresolved methodological problems are even harder. The recent paper by Duarte, Longstaff, and Yu (2007) on fixed-income arbitrage and the monograph by Dacorogna, Gencay, Muller (2001) on real-time trading in foreign exchange markets provide insights and references for statistical trading in these markets.

11.1 Technical analysis, trading strategies, and data-snooping checks

11.1.1 Technical analysis

Technical analysis, also known as "charting," attempts to extract information from the "chart," or past market prices, for patterns and trends that suggest profitable trading rules, which are often called *technical trading rules*. Technical analysis considers (at each time t) asset prices $P_t, P_{t-1}, \ldots, P_{t-m+1}$ within a moving window of width m. Their average corresponds to the running-mean smoother in Section 7.2.1. Moving average trading rules are based on comparing the current price with the current moving average. For example, "in an uptrend, long commitments are retained as long as the price trend remains above the moving average," as noted in the summary of technical trading rules by Sullivan, Timmerman and White (1999). A related idea is the *Bollinger band*. Let $\bar{P}_t^{(m)} = m^{-1}\sum_{i=0}^{m-1} P_{t-i}$ denote the moving average and

$$\widehat{P}_t^{(m)} = \frac{1}{\sum_{i=1}^{m} i}\sum_{i=0}^{m-1}(m-i)P_{t-i}, \qquad \sigma_t^{(m)} = \left\{\frac{1}{m-1}\sum_{i=0}^{m-1}(P_{t-i}-\bar{P}_t^{(m)})^2\right\}^{1/2}.$$

The upper Bollinger band $r_t^+ = \widehat{P}_t^{(m)} + 2\sigma_t^{(m)}$ and the lower Bollinger band $r_t^- = \widehat{P}_t^{(m)} - 2\sigma_t^{(m)}$ were introduced by John Bollinger in the 1980s to define "high" and "low" for a stock price, as illustrated in Figure 11.1. Instead of the running mean, some technical trading rules use the running maximum (i.e., $\max_{0\le i<m} P_{t-i}$) and the running minimum (i.e., $\min_{0\le i<m} P_{t-i}$).

One such rule is to sell (or buy) when P_{t+1} exceeds $\max_{0 \le i < m} P_{t-i}$ (or falls below $\min_{0 \le i < m} P_{t-i}$), as the former case represents "resistance" (i.e., the price level at which the asset price seems to have difficulty rising further), while the latter case represents "support" (i.e., the price level below which the asset price seems less likely to fall).

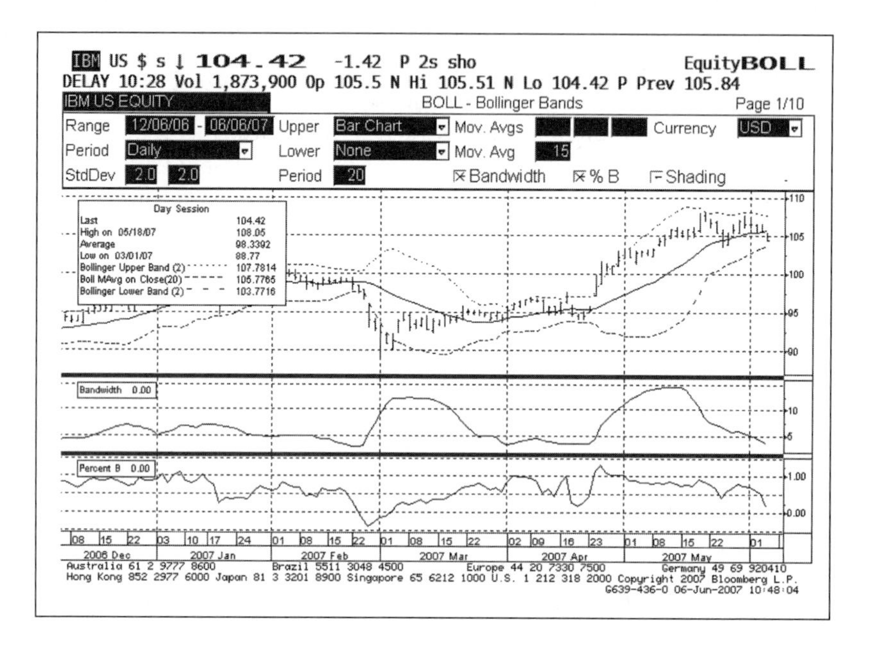

Fig. 11.1. Bollinger bands for IBM stock from December 6, 2007 to June 6, 2007. Used with permission of Bloomberg LP.

Lo, Mamaysky, and Wang (2000) have noted that conventional "charting" is only a graphical form of a more general framework of technical analysis and have proposed a statistical approach to "technical pattern recognition using nonparametric kernel regression." The nonparametric regression techniques they consider are basically those in Section 7.2 dealing with a univariate regressor (which is time t) and using kernel smoothing to estimate $E(P_t)$. The estimated regression function, which is restricted to a given window of data suggests patterns within that time span; see Exercise 11.1.

Other statistical methods besides nonparametric regression are also useful for this general framework of technical analysis. Time series modeling and forecasting are clearly helpful. Whereas the kernel regression approach proposed by Lo, Mamaysky, and Wang (2000) focuses on patterns of past data with the hope that such patterns can be extrapolated to the future, a suitably chosen time series model can address prediction more directly. Another useful statistical method for technical analysis is change-point detection, as pointed out by Shiryaev (2002). In fact, the running maximum and running minimum discussed earlier are related to the CUSUM (cumulative sum) detection rules considered by Shiryaev (2002). These CUSUM rules are based on the detection statistics $\max_{k \le t} S_k - S_t$ or $S_t - \min_{k \le t} S_k$, where S_t is the cumulative sum of the log returns. Lai (1995) has given a general theory of quickest detection by using scan statistics, which are more general than the running maxima/minima in CUSUM rules.

11.1.2 Momentum and contrarian strategies

Trading strategies "to beat the market" date back to the inception of trading in financial instruments. Broadly speaking, there are two classes of strategies that are diametrically opposed in philosophy and execution, the *momentum strategy* which relies on price continuations (or "momentum" in asset prices), and the *contrarian strategy* which is based on price reversals. The momentum strategy exploits the price momentum by buying past winners and selling past losers, while the contrarian strategy makes use of the price reversal by buying past losers and selling past winners.

Contrarian strategies have been found to be profitable for short-term (1 week to 1 month) horizons, and such profitability has been used as evidence against the EMH; see Jegadeesh (1990). Empirical analysis of the cross-correlation matrix of a sample of NYSE (New York Stock Exchange) and AMEX (American Stock Exchange) stocks from July 10, 1962 to December 27, 1994 has shown that short-term contrarian profits may be due to the presence of positive cross-correlations among stock returns; see Section 2.8.3 of Campbell, Lo, and MacKinlay (1997). De Bondt and Thaler (1985, 1987) give another explanation via *behavioral finance*, noting that investors tend to overreact to information in the short run. Momentum strategies have been found to be profitable for medium-term (3- to 12-month) holding periods. Jegadeesh and Titman (1993) give empirical evidence that firms with high returns over the past 3 to 12 months continue to outperform firms with low returns over the sample period in their empirical study. Moskowitz and Grinblatt (1999) have found that momentum strategies are significantly less profitable after controlling for industry momentum. On the other hand, industry momentum investment strategies, which sell stocks from past losing industries and buy stocks from past winning industries, appear highly profitable, even after controlling for size, individual stock momentum, and the cross-sectional dispersion in the mean returns.

11.1.3 Pairs trading strategies

In the mid-1980s, a group of "quants" at Morgan Stanley used sophisticated statistical methods to develop high-tech trading programs executable through automated trading systems. The objective of these quantitative-based strategies is to uncover arbitrage opportunities in equity markets and then buy undervalued assets and sell overvalued ones. *Statistical arbitrage pairs* attempt to uncover arbitrage opportunities by identifying pairs of similar securities, whose prices tend to move together. Therefore, if their prices differ markedly, one of the securities may be undervalued, the other may be overvalued, or both situations may occur. With the expectation that such mispricing will correct itself in the future, one may be able to make a profit by opening a short position for the relatively overvalued security and a long position for the relatively undervalued security. The magnitude of the spread between the two securities indicates the extent of the mispricing and the range of potential profit.

Selecting pairs

A key step in developing a pairs trading strategy is selecting the two stocks for pairs trading. Gatev, Goetzmann, and Rouwenhorst (2006, p. 803) illustrate how this can be done by using stock price data. They form pairs over a 12-month period to be used for trading in the next 6-month period. The first step is to screen out stocks from the CRSP (Center for Research in Security Prices of the University of Chicago, `http://www.crsp.com`) daily files that have one or more days of no trading so that the remaining stocks are liquid and do not have missing daily closing prices. The next step is to compute the normalized price series for each stock, in which the normalizing factor is a cumulative total returns index of the stock over the pairs-formation period. Then a matching partner is chosen for each stock by finding the stock that minimizes the sum of squared deviations between the two normalized price series. The authors point out that this approach "best approximates the description of how traders themselves choose pairs" in their interviews with pairs traders.

An alternative or supplement to this empirical approach is to use fundamentals of the firms to select two stocks that have almost the same risk factor exposures. The spread in their returns is calculated to determine if there is mispricing. Gatev, Goetzmann, and Rouwenhorst (2006) also consider a hybrid of this approach and the empirical approach in the preceding paragraph. They restrict both stocks to belong to the same broad industry categories defined by S&P: Utilities, Transportation, Financials, and Industrials. The minimum squared deviations criterion is again used to match stocks within each of these groups.

Trading design

Having selected two stocks for pairs trading, one has to decide on a trading rule. The rules typically open a long-short position when the pair's prices have diverged by a certain amount and close the position when the spread becomes small. Gatev, Goetzmann, and Rouwenhorst (2006, p. 804) illustrate this idea by "following practice," opening a position in a pair "when prices diverge by more than two historical standard deviations, as estimated (from the spreads) during the pairs formation period," and unwinding the position "at the next crossing of the prices."

An empirical study of the performance of pairs trading

Gatev, Goetzmann, and Rouwenhorst (2006) examine the risk and return characteristics of pairs trading with daily data over the period 1962 through 2002. Their pairs-formation period is 1 year, and the trading period is 6 months following the formation period, as described above: "Once we have paired up all liquid stocks in the formation period, we study the top 5 to 20 pairs with the smallest historical distance measure, in addition to the 20 pairs after the top 100 (pairs 101–120)." Because of the length of the period in the empirical study, some stocks were delisted from CRSP during the period, and the authors used the following convention to take care of this: "If a stock in a pair is delisted from CRSP, we close the position in that pair, using the delisting return, or the last available price. We report the payoffs by going one dollar short in the higher-priced stock and one dollar long in the lower-priced stock." The daily returns on the long and short positions are calculated as the value-weighted returns $\sum_{i \in \Pi} w_{i,t} r_{i,t} / \sum_{i \in \Pi} w_{i,t}$, where Π denotes the portfolio of pairs and $w_{i,t}$ denotes the weights that reflect the payoffs being reinvested during the trading interval. The empirical results of Gatev, Goetzmann, and Rouwenhorst, based on the time series of overlapping 6-month (trading period) returns over the 474 months in the sample, suggest that pairs trading is profitable and that there are diversification benefits from combining multiple pairs in a portfolio; the pairs trading profits typically exceed conservative estimates of transaction costs. Although the raw returns of pairs trading have fallen in recent years, the results of the study show that the risk-adjusted returns have continued to persist despite increased hedge fund activity, which competes away the profits.

Connections to nonstationarity and cointegrated prices

As an illustration, Gatev, Goetzmann, and Rouwenhorst (2006, pp. 803–806) consider the daily normalized prices of two stocks, Kennecott and Uniroyal, in the 6-month pairs trading period from August 1963 to January 1964. The prices generally tended to move together over the trading interval, as shown in their Figure 1. In particular, "the position first opens in the seventh trading day of the period and then remains open until day 36. Over that interval, the spread actually first increased significantly before convergence. The prices remain close during the period and cross frequently. The pair opens five times during the period, however not always in the same direction." They relate such co-movements of prices to the cointegrated prices literature, referring to Bossaerts' (1988) empirical evidence that certain asset prices are cointegrated and Bossaerts and Green's (1989) equilibrium asset-pricing model with nonstationary common factors. They point out that "this interpretation does not imply that the market is inefficient; rather it says that certain assets are weakly redundant, so that any deviation of their price from a linear combination of the prices of other assets is expected to be temporary and reverting." Recall that normalized prices are used for pairs trading instead of actual prices, and therefore the unit-root nonstationarity pertains to the normalized price time series.

A stochastic spread model

Elliott, Van der Hoek, and Malcolm (2005) propose to use a linear state-space model for the observed spread process $\{y_t\}$ between the paired stocks:

$$y_t = x_t + \nu w_t, \qquad x_{t+1} - x_t = a - bx_t + \sigma \epsilon_{t+1}, \tag{11.1}$$

where w_t, ϵ_t are i.i.d. standard normal random variables, ν, σ, b are positive constants, and x_t is an unobservable state price. This model captures the mean reversion property (since $b > 0$) that underlies pairs trading. Letting $\mathcal{Y}_t = (y_1, \ldots, y_t)$, we can use the recursive formulas in Section 5.3.1 to compute the Kalman filter $\widehat{x}_t = E(x_t|\mathcal{Y}_t)$ and the one-step-ahead minimum-variance forecast $\widehat{x}_{t|t-1} = E(x_t|\mathcal{Y}_{t-1})$. The parameters ν, σ, a, b of the linear state-space model can be estimated from historical data by maximum likelihood using arguments similar to those in the second paragraph of Section 5.3.2.

11.1.4 Empirical testing of the profitability of trading strategies

Data snooping

The issue of data-snooping bias has been discussed in Section 3.3.4 in connection with empirical studies of CAPM. Jensen and Bennington (1970) recognized the potential bias of data snooping in evaluating the performance of technical trading rules, pointing out that "given enough computer time, we are sure that we can find a mechanical trading rule which 'works' on a table of random numbers, provided of course that we are allowed to test the rule on the same table of numbers which we used to discover the rule." The last part of the quote is often referred to as "data snooping," and there are two approaches to circumventing the data-snooping bias. The first approach avoids reusing the same dataset by testing a model with a different but comparable dataset or by using different subsamples for training and testing. In particular, using the same model with the training data through the end of 1998 and the testing data over the 1999–2002 period, Gatev, Goetzmann, and Rouwenhorst (2006) conclude that "not only does this additional four-year sample suggest that the results were not simply an artifact of the earlier sample period, over which pairs trading was known to be popular, but it also suggests that the public dissemination of the results has apparently not affected the general risk and return characteristics of the strategy." On the other hand, such large out-of-sample datasets are often difficult to find. A second approach, described below, is to evaluate the true significance level of the multiple hypothesis tests for the profitability claims after fitting the same data to multiple models (trading strategies).

White's reality check and Hansen's SPA test

Let f_k denote the excess return of the kth trading rule over a benchmark and $\psi_k = E(f_k)$ for $k = 1, \ldots, M$. The null hypothesis is that there does not exist a superior model in the collection of M models (rules):

$$H_0 : \max_{1 \le k \le M} \psi_k \le 0. \tag{11.2}$$

Rejecting (11.2) indicates that there is at least one model that outperforms the benchmark. An obvious test statistic for (11.2) is the maximum of the normalized sample average of $f_{k,i}$, i.e.,

$$\bar{V}_n = \max_{1 \le k \le M} \sqrt{n} \bar{f}_k, \tag{11.3}$$

where $\bar{f}_k = \sum_{i=1}^n f_{k,i}/n$, with $f_{k,i}$ being the ith observation of f_k, and $f_{k,1}, \ldots, f_{k,n}$ are the computed returns in a sample of n past prices for the kth trading rule. Letting $\{f_{k,1}^*(b), \ldots, f_{k,n}^*(b)\}$ denote the bth bootstrap sample and $\bar{f}_k^*(b) = \sum_{i=1}^n f_{k,i}^*(b)/n$, White (2000) proposed to approximate the sampling distribution of $\max_{1 \le k \le M} \sqrt{n}(\bar{f}_k - \psi_k)$ by the empirical distribution of

$$\bar{V}_n^*(b) = \max_{1 \le k \le M} \sqrt{n} \big[\bar{f}_k^*(b) - \bar{f}_k \big], \qquad b = 1, \ldots, B. \tag{11.4}$$

To test (11.2), he proposed to compute

$$\widehat{p} = \left\{ \# \text{ of bootstrap samples with } \bar{V}_n^*(b) > \bar{V}_n \right\} \Big/ B \tag{11.5}$$

and to reject (11.2) if $\widehat{p} < \alpha$, where α is some specified significance level, e.g., 0.05. The rationale underlying this bootstrap test is that $\widehat{p}$ is a bootstrap estimate of the p-value of the test statistic (11.3); i.e., the probability that (11.4) exceeds its observed value under the extremal point $\psi_1 = \cdots = \psi_M = 0$

of the null hypothesis. Sullivan, Timmerman and White (1999) and White (2000) applied this bootstrap test to a number of technical trading rules and found that no trading rule significantly outperformed the Dow Jones Industrial Average or S&P 500 index or S&P 500 futures as the benchmark.

Hansen (2005) pointed out two problems with the preceding bootstrap test, which is called *White's reality check*. First, the average returns $\bar{f}_k$ are not studentized; see Section 1.6.3. Second, despite the fact that the null hypothesis (11.2) consists of an infinite number of parameter values, the distribution of White's reality check is based on the "least favorable" null $\psi_1 = \cdots = \psi_M = 0$. To address these problems with White's reality check, Hansen proposed a "superior predictive ability" (SPA) test that replaces $\sqrt{n}\bar{f}_k$ in (11.3) by its studentized version

$$\widetilde{V}_n = \left(\max_{1 \le k \le M} \frac{\sqrt{n}\bar{f}_k}{\widehat{\sigma}_k} \right)_+, \tag{11.6}$$

where $x_+ = \max(x, 0)$ and $\widehat{\sigma}_k$ is a consistent estimator of the standard deviation of $\sqrt{n}\bar{f}_k$. The SPA test also uses a different method to bootstrap the distribution of $\widetilde{V}_n$. Defining $Z^*_{k,i}(b) = f^*_{k,i}(b) - \bar{f}_k \cdot \mathbf{1}_{\{\bar{f}_k \ge -n^{-1/4}\widehat{\sigma}_k/4\}}$ and letting $\bar{Z}^*_k(b)$ denote the sample average of the bth bootstrapped sample $\{Z^*_{k,i}(b)\}_{i=1,\dots,n}$, Hansen (2005) used $\widetilde{V}_n$ in place of $\bar{V}_n$ and the empirical distribution of

$$\widetilde{V}^*_n(b) = \left(\max_{1 \le k \le M} \frac{\sqrt{n}\bar{Z}^*_k(b)}{\widehat{\sigma}^*_k} \right)_+, \quad b = 1, \dots, B, \tag{11.7}$$

instead of that of $\bar{V}^*_n(b)$ in (11.5) to compute $\widehat{p}$. Hansen actually used $\widehat{\sigma}_k$ in (11.7), but as noted by Romano and Wolf (2005, p. 1252), the proper bootstrap version of studentized statistics should use $\widehat{\sigma}^*_k$ instead. Based on the SPA test, Hansen, Lunde, and Nason (2005) found some significant effects, in contrast with the result of Sullivan, Timmerman, and White (1999) based on White's reality check.

Stepwise multiple hypothesis testing

Romano and Wolf (2005) make use of recent advances in multiple hypothesis testing to improve White's reality check and obtain a more comprehensive test of the profitability of trading strategies. The basic idea of multiple testing is to test for the kth strategy

$$H_k : \psi_k \le 0 \text{ versus } H'_k : \psi_k > 0, \text{ for every } 1 \le k \le M, \tag{11.8}$$

subject to a constraint on the *familywise error rate* (FWE), which is the probability of falsely rejecting at least one of the true null hypotheses. More specifically, if P is the true probability measure, let $I_0(P) \subset \{1, \dots, M\}$ denote the set of indices of the true null hypotheses (i.e., $k \in I_0(P)$ if and only if $\psi_k \leq 0$) and define

$$\mathrm{FWE}_P = P\{\text{reject } H_k \text{ for some } k \in I_0(P)\}. \tag{11.9}$$

If $I_0(P) = \emptyset$, $\mathrm{FWE}_P = 0$ by definition. A multiple testing method is said to *control the FWE* (or to have *strong control of the FWE*) *at level* α if $\mathrm{FWE}_P \leq \alpha$ for any P. It is said to *asymptotically control the FWE at level* α as the sample size n becomes infinite if $\limsup_{n\to\infty} \mathrm{FWE}_P \leq \alpha$ for any P.

Romano and Wolf (2005) propose the following *stepdown* bootstrap test and show that it asymptotically controls the FWE at level α. The test statistic for H_k is $T_k = \sqrt{n}\bar{f}_k/\widehat{\sigma}_k$. Relabel the T_k in decreasing order of magnitude as $T_{r(1)} \geq \cdots \geq T_{r(M)}$. For the bth bootstrap sample, let $T_k^*(b) = \sqrt{n}(\bar{f}_k^*(b) - \bar{f}_k)/\widehat{\sigma}_k^*, 1 \leq b \leq B$. The stepwise testing procedure rejects $H_{r(j)}$ sequentially. Suppose R_{j-1} hypotheses are rejected in the first $j-1$ steps (with $R_0 = 0$). At the jth stage, reject $H_{r(\nu)}$ if $T_{r(\nu)} < \widehat{d}_j$, where $\widehat{d}_j$ is the αth quantile of the empirical distribution of $\{\max_{R_{j-1}+1 \leq k \leq M} T_{r(k)}^*(b) : 1 \leq b \leq B\}$, or more precisely,

$$\widehat{d}_j = \inf\left\{x : \left(\#\text{ of bootstrap samples with } \max_{R_{j-1}+1 \leq k \leq M} T_{r(k)}^*(b) \leq x\right)\Big/ B \geq 1-\alpha\right\}.$$

If no further null hypotheses are rejected, stop the stepwise procedure. Otherwise, denote by R_j the total number of hypotheses rejected so far and proceed to stage $j+1$.

Section 9 of Romano and Wolf (2005) applies this stepwise test to compare the performance of $M = 105$ hedge funds with a risk-free investment as the benchmark based on their $n = 147$ monthly returns from January 1992 to March 2004 from the CISDM (Center for International Securities and Derivatives Markets) database. The test identifies six funds that outperform the risk-free rate in the first step and a seventh fund in the next and final step.

11.1.5 Value investing and knowledge-based trading strategies

Whereas technical trading rules are based on statistical patterns of past data windows of stock prices with the hope that such patterns can be extrapolated to the future, an alternative approach is to use much more extensive information to build forecasting models for the future and to use the forecasts for finding and evaluating investment portfolios. Value investors such as Warren Buffett look for stocks that are selling at a deep discount to the company's "intrinsic value," which is derived from the previous values of predicted cash flows of the company over a future time horizon. The predictions are based on "fundamental analysis" of the company, its business sector, and macroeconomic trends. Since there is risk in predicting the future, Buffett talks about favoring companies with wide "economic moats" in the sense that they have competitive advantages that make forecasting safer. His top ten stock picks account for most of the portfolio of Berkshire Hathaway, his publicly traded investment company. Until he identifies a good investment opportunity in which he is confident of future success, he would rather hold cash as a strategic asset. On the other hand, as Lim (2007) points out, there are times when Buffett makes "side bets," such as in currency-related derivative contracts, and the secret of his exceptionally successful investment record may not be just about buying stocks with low price-to-earnings ratios and letting them ride but also in his flexibility and willingness to see value in all sorts of assets.

Because of uncertainties in the forecasts of the future, one has to be prepared for the inherent risk of a trading strategy when the future does not turn out to be what one expects. To a certain extent, statistical forecasts already incorporate such uncertainties in the probability distributions of the forecasts, although these distributions often are not used because of their high dimensionality, involving many economic variables if one wants to perform a thorough statistical analysis. Chapter 12 will describe Monte Carlo methods that work directly with these joint distributions for risk assessment.

As pointed out in Lahart's (2007) commentary on the August 2007 tumble of the stock market following default and other difficulties of subprime mortgage loans, many quantitative hedge funds, "which use statistical models to find winning trading strategies," reported heavy losses because "they all owned many of the same stocks and their models told them all to sell at the same time, driving down the share prices, hurting everyone in the process." Landon (2007) notes that the "pack mentality among hedge funds," using similar strategies for investing in more risky areas such as mortgage-backed securities and credit derivatives to obtain higher returns, "fuels market volatility." This suggests that the knowledge base of trading strategies should contain lessons from past market failures and the behavioral patterns of market players, so that warning signals can be provided for the trading strategy to respond adequately before trouble sets in.

11.2 High-frequency data, market microstructure, and associated trading strategies

High-frequency data are the direct information from markets, in which a unit of information is called a *tick*. Attached to every tick are a *quote* (which may be a price, foreign exchange rate, or interest rate) and a *time stamp* (date and time). Other information, such as transaction volume, is also available in some markets. High-frequency financial data are important for studying market microstructure and in particular price formation processes and trading mechanisms. As pointed out by O'Hara (1995, pp. 3–4), there are two traditional approaches to the study of price formation in economics. One involves the analysis of equilibrium prices; it focuses on solving for a market-clearing price but ignores how exactly this market clearing has been achieved. The other approach involves how traders' demands and supplies are aggregated to find a market-clearing price; e.g., by a "Walrasian auctioneer" or by "market participants playing roles far remote from the passive one of the auctioneer." Domain knowledge in market microstructure and statistical methods for high-frequency data are also important for the development of real-time (also called "algorithmic") trading strategies. High-frequency financial data have certain distinctive features and challenges not shared by traditional (low-frequency) time series data. These features are described below together with their institutional background, statistical modeling and analysis, and implications on market microstructure and real-time trading.

11.2.1 Institutional background and stylized facts about transaction data

Research on market microstructure has often focused on exchanges (and in particular the NYSE) or on OTC (over-the-counter) dealer markets such as NASDAQ. Hasbrouck (2007) gives in his appendix an overview of the NYSE, NASDAQ, and recent developments in U.S. equity trading systems. He points out that most of today's securities markets feature an electronic *limit order book*. A "limit order" is an order that specifies buy or sell, a quantity, and an acceptable price. In such markets, orders arrive sequentially and the price limit of a newly arrived order is compared with those of orders already in the system to see if there is a match. If there is a match, the trade occurs at the price set by the first order. For example, consider the orders A, B, C that arrive sequentially:

A: Buy 200 shares at \$25.50 (per share);

B: Sell 200 shares at \$30.00;

C: Buy 100 shares at \$32.00.

There is no match between B and A, as the price of \$30 is not acceptable to the buyer, so no trade takes place. Since there is an overlap in the acceptable prices of B and C, a trade occurs (for 100 shares) at \$30. The set of unexecuted limit orders held by the system constitutes the "book."

Unexecuted limit orders can be canceled or modified, so the limit order book can change rapidly over time. A system of priority rules governs the sequence in which orders are executed. Price is the top priority; for example, a buy order priced at \$100 is executed before that priced at \$99. Time is a secondary priority. Hidden orders lose priority to visible orders; a trader seeking to buy or sell a large amount (relative to typical quantities posted to the book) may designate the order as "hidden" so that it is not made visible to other market participants. In a limit order market, buyers and sellers interact directly, using brokers mainly as conduits for their orders and to hide their identities.

The data from a limit order book are very detailed and are available in real time by data providers such as Reuters and Bloomberg. Archived historical data are also sold by exchanges such as the NYSE, whose TAQ (Trades and Quotes) database is a rich source of transaction data. These historical data consist of the calendar time t_i, measured in seconds from midnight, at which the ith transaction of an asset takes place, together with the transaction price, the prevailing bid and ask quotes, and the transaction quantity. We next describe some stylized facts about transaction data.

Nonsynchronicity and negative lag-1 autocorrelations

As noted above, transactions such as stock trades on an exchange do not occur at equally spaced time intervals. This results in the nonsynchronicity of the observed asset prices, i.e., the recorded prices of an asset in a transaction database do not form an equally spaced time series. Price nonsynchronicity has made empirical testing of the EMH very difficult. For example, evidence against the EMH has been put forth in the form of significant negative lag-1 autocorrelations of asset returns. In the top panel of Figure 11.2, the autocorrelation function of returns of IBM stock measured at one-minute intervals on January 3, 2003 is plotted against the lags; the returns are computed using the *previous tick interpolation* (which takes the most recent value), similar to the daily closing prices quoted in the financial press. The plot shows significant negative autocorrelations at lag 1, whereas the weak form of the EMH implies that the returns over equally spaced intervals should be uncorrelated. However, because the returns are computed using the previous tick interpolation, negative lag-1 autocorrelations such as those in Figure 11.2 may be spurious, as explained by Miller, Muthuswamy, and Whaley (1994) and the nonsynchronous trading models in Section 11.2.2.

Multiple transactions within a second

Even more complicated than price nonsynchronicity is the occurrence of multiple transactions with different prices at the same time in the transaction database, for which the transaction times are measured to the nearest second. In periods of heavy trading, a second may be too long a timescale.

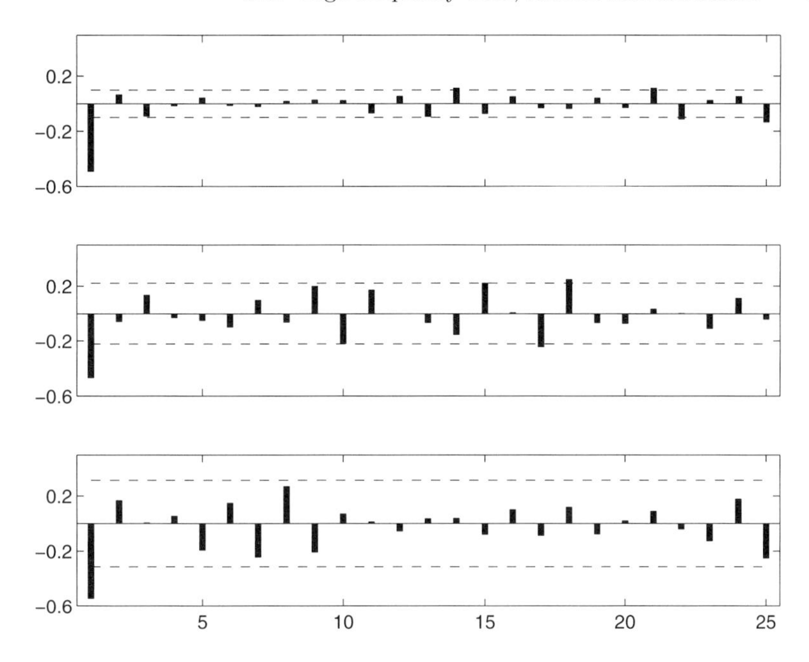

Fig. 11.2. ACFs of price changes for 1-, 5-, and 10-minute intervals (top, middle, and bottom panels, respectively) for IBM stock trading data on January 3, 2003. The horizontal dashed lines represent the rejection boundaries of a 5%-level test of zero autocorrelation at indicated lag.

Discrete (tick-valued) prices

Transaction prices are quoted in discrete units or *ticks*. On the NYSE, the tick size was \$0.125 before June 24, 1997 and \$0.0625 afterward until January 29, 2001, when all NYSE and AMEX stocks started to trade in decimals. For futures contracts on the S&P 500 index, the tick size is \$0.05. Because of the discrete transaction prices, the observed price changes can only take a few distinct values. For example, in 1994, 97.4% of all transactions occurred with no change or a one-tick price change. This also leads to "price clustering," which is the tendency for transaction prices to cluster around certain values; see Campbell, Lo, and MacKinlay (1997, pp. 109–114).

Intraday periodicity

Intraday periodicity/seasonality of trading intensity has been noted by Anderson and Bollerslev (1997), Hasbrouck (1999), and others. On stock exchanges, the seasonality is called *U-shaped* because transactions tend to be heaviest near the beginning and close of trading hours and lightest around the lunch

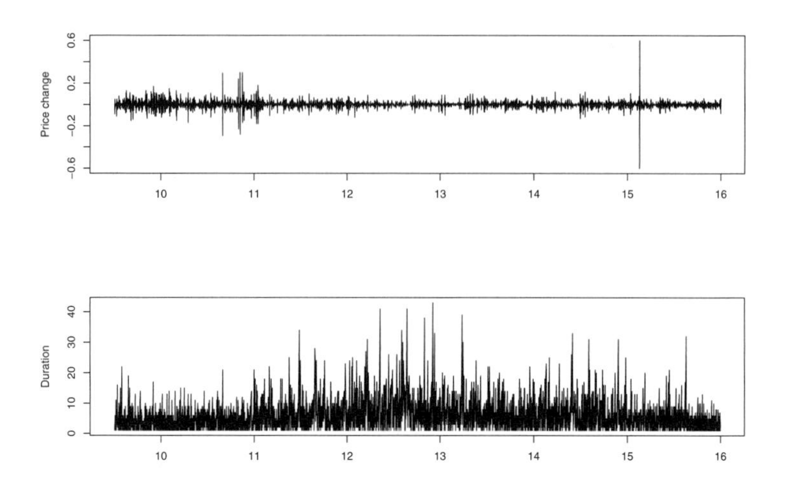

Fig. 11.3. Trading data of IBM stock on January 3, 2003. Top panel: time plot of price changes in consecutive trades. Bottom panel: time plot of duration (in seconds) between trades.

hour. Consequently, the durations between transactions display a daily periodic pattern. The Eurodollar market exhibits a similar U-shaped daily pattern but with the first half of the trading day (when the European markets are still open) more active than the second half (when the European markets have already closed).

Bid–ask spread

An important feature sought by investors in an organized financial market is liquidity, which is the ability to buy or sell significant quantities of a security quickly, anonymously, and with relatively little price impact. To maintain liquidity, many organized exchanges use market makers, who stand ready to buy or sell whenever investors ask them. In return for providing liquidity, market makers are granted the right to post different prices for purchases and sales. They buy at the bid price P_b and sell at a higher ask price P_a. Their compensation is the *bid–ask spread* $P_a - P_b$. The bid–ask spread quotes have discrete values; typical values for equities are 1 or 2 ticks but can exceed 5 ticks in times of low market activity, and foreign exchange spreads are expressed as integers in basis points. Their seemingly small size (see Figure 11.3 for IBM stock) belies the potential importance of bid–ask spreads in determining the time series properties of asset returns. For example, Keim (1989) shows that the "January effect" (that smaller-capitalization stocks seem to outperform larger-capitalization stocks over the few days around the turn of the year) may be partially attributable to closing prices, which are recorded at the bid price at the end of the year but at the ask price at the beginning of the year.

11.2.2 Bid–ask bounce and nonsynchronous trading models

Models that reconcile the EMH with the observed negative lag-1 autocorrelations have been developed in the financial economics literature. Two such models are described below, one based on the bid–ask spread and the other on nonsynchronous trading.

Roll's (1984) model of bid, ask, and transaction prices

Let p_t^* denote the underlying price (or log price, to be consistent with the framework introduced in Section 3.1) at time t of an asset, and suppose that p_t^* is a random walk with i.i.d. zero-mean increments ϵ_t whose common variance is denoted by σ^2, as in EMH. Suppose all trades, buy or sell, are conducted through dealers who charge a cost $c > 0$ per trade and therefore set the bid–ask spread to be $2c$. Then at time t, if there is a trade at transaction price p_t, it may be expressed as $p_t = p_t^* + q_t c$, where q_t is a trade direction indicator, with -1 for a seller-initiated transaction and 1 for a buyer-initiated one. Assume that buys and sells are equally likely and serially independent and that ϵ_t is independent of the buy or sell decision (i.e., that q_t are i.i.d. and independent of $\{\epsilon_i\}$). Let $\Delta p_t = p_t - p_{t-1}$ denote the observed price change at time t. This model, called the *Roll model*, has two parameters c and σ^2, which are related to the estimable quantities $\gamma_0 = \mathrm{Var}(\Delta p_t)$ and $\gamma_1 = \mathrm{Cov}(\Delta p_t, \Delta p_{t-1})$ by

$$\gamma_0 = E(\Delta p_t)^2 = 2c^2 + \sigma^2, \tag{11.10}$$

$$\gamma_1 = E(\Delta p_{t-1} \Delta p_t) = -c^2, \tag{11.11}$$

recalling that $E(\epsilon_t) = 0$ and $\mathrm{Var}(\epsilon_t) = \sigma^2$; see Exercise 11.2. Moreover, $\mathrm{Cov}(\Delta p_t, \Delta p_{t-j}) = 0$ for $j \geq 2$. In the Roll model, it follows from (11.10) and (11.11) that

$$\mathrm{Corr}(\Delta p_t, \Delta p_{t-1}) = \frac{\gamma_1}{\gamma_0} = -\frac{c^2}{2c^2 + \sigma^2} < 0, \tag{11.12}$$

explaining the *bid–ask bounce*, which refers to the connection between the bid–ask spread and the negative lag-1 autocorrelation in an asset return.

Lo and MacKinlay's (1990) model of nonsynchronous trading

Roll's model ignores the nontrading phenomenon in nonsynchronous trading. A stochastic model of nontrading, proposed by Lo and MacKinlay (1990), assumes that for each period the probability that a given security is not traded is π that is time-invariant and independent of the (continuously compounded) returns r_t, which are assumed to be i.i.d. with mean μ and variance σ^2 as in an efficient market. If no trade occurs in period t, the closing price of the period is set to the previous period's closing price and therefore $r_t^{\mathrm{o}} = 0$, where r_t^{o} is the observed return. On the other hand, if the security is traded in period t, its observed return r_t^{o} is the sum of the virtual returns in period t and in all prior consecutive periods in which no trade occurred; i.e., $r_t^{\mathrm{o}} = r_t + r_{t-1} + \cdots + r_{t-k_t}$, where $k_t = \max\{k : \text{no trade occurred in the periods } t-1, t-2, \ldots, t-k\}$. Hence

$$P\{r_t^{\mathrm{o}} = 0\} = \pi, \quad P\{r_t^{\mathrm{o}} = r_t\} = (1-\pi)^2, \quad P\left\{r_t^{\mathrm{o}} = \sum_{i=0}^{k} r_{t-i}\right\} = (1-\pi)^2\pi^k \tag{11.13}$$

for $k \geq 1$. It then follows that

$$E(r_t^{\mathrm{o}}) = \mu, \qquad \mathrm{Var}(r_t^{\mathrm{o}}) = \sigma^2 + 2\pi\mu^2/(1-\pi). \tag{11.14}$$

Similarly, it can be shown that

$$\begin{gathered} P\{r_t^{\mathrm{o}} r_{t-1}^{\mathrm{o}} = 0\} = 2\pi - \pi^2, \quad P\{r_t^{\mathrm{o}} r_{t-1}^{\mathrm{o}} = r_t r_{t-1}\} = (1-\pi)^3, \\ P\{r_t^{\mathrm{o}} r_{t-1}^{\mathrm{o}} = r_t\big(\textstyle\sum_{i=1}^{k} r_{t-i}\big)\} = (1-\pi)^3\pi^{k-1}, \end{gathered} \tag{11.15}$$

$$\mathrm{Cov}(r_t^{\mathrm{o}}, r_{t-1}^{\mathrm{o}}) = -\pi\mu^2. \tag{11.16}$$

Hence the lag-1 autocorrelation of r_t^{o} is

$$\mathrm{Corr}(r_t^{\mathrm{o}}, r_{t-1}^{\mathrm{o}}) = -\frac{(1-\pi)\pi\mu^2}{(1-\pi)\sigma^2 + 2\pi\mu^2} < 0,$$

showing that nonsynchronous trading can cause “spurious” negative lag-1 autocorrelations of the observed returns in efficient markets.

11.2.3 Modeling time intervals between trades

Durations, which are time intervals between trades, contain useful information about intraday market activities. For example, relatively long durations indicate lack of trading activity and provide no new information about market prices. The first step in modeling the durations $\Delta t_l = t_l - t_{l-1}$, where t_l is the time of the lth trade on a trading day, is to model the intraday periodicity described in Section 11.2.1. This is usually carried out by applying nonparametric regression methods (such as smoothing splines and kernel smoothers, as described in Section 7.2) to transaction data from the past few months, yielding a smooth estimate $\widehat{f}(t)$ of the expected duration $f(t)$ of a trade that occurs at time t of a typical trading day. The next step is to build a time series model for the adjusted durations $\Delta t_l^* = \Delta t_l / \widehat{f}(t_l)$. Since $\Delta t_l / f(t_l)$ has mean 1, a natural way is to express Δt_l^* as $\psi_l \epsilon_l$, modeling the ϵ_l as i.i.d. positive random variables with common mean 1 and the ψ_l as some function of past data and certain parameters, similar to the GARCH(h, k) model. This is the idea behind the *autoregressive conditional duration* (ACD) models introduced by Engle and Russell (1998), who assume that ϵ_l has either a standard exponential distribution (with density function e^{-t}) or a standardized Weibull distribution and that the ψ_l are defined recursively in the following:

$$\Delta t_l^* = \psi_l \epsilon_l, \qquad \psi_l = \omega + \sum_{j=1}^{k} \gamma_j \Delta t_{l-j}^* + \sum_{i=1}^{h} \beta_i \psi_{l-i}. \tag{11.17}$$

The Weibull distribution and its standardized form

The *Weibull distribution* with shape parameter α and scale parameter λ is a continuous distribution with density function

$$f_{\alpha,\lambda}(x) = \begin{cases} \alpha \lambda^\alpha x^{\alpha-1} e^{-(\lambda x)^\alpha} & \text{if } x \geq 0, \\ 0 & \text{if } x < 0. \end{cases} \tag{11.18}$$

The mean and variance of a random variable X having a Weibull distribution are

$$E(X) = \frac{1}{\lambda} \Gamma\left(1 + \frac{1}{\alpha}\right), \quad \mathrm{Var}(X) = \frac{1}{\lambda^2} \left\{ \Gamma\left(1 + \frac{2}{\alpha}\right) - \left[\Gamma\left(1 + \frac{1}{\alpha}\right)\right]^2 \right\},$$

and the distribution function is

$$F_{\alpha,\lambda}(x) = 1 - e^{-(\lambda x)^\alpha}, \quad x \geq 0. \tag{11.19}$$

Note that the case $\alpha = 1$ corresponds to an exponential distribution.

Let $Y = X/E(X)$ so that Y has mean 1. The distribution of Y does not depend on λ and is called *standardized Weibull* with shape parameter α. From (11.19), it follows that Y has the density function

$$f_\alpha(y) = \alpha \left[\Gamma\left(1 + \frac{1}{\alpha}\right)\right]^\alpha y^{\alpha-1} \exp\left\{ - \left[\Gamma\left(1 + \frac{1}{\alpha}\right) y\right]^\alpha \right\}, \quad y \geq 0. \tag{11.20}$$

EACD and WACD models and their parameter estimates

Let $x_l = \Delta t_l^*$. Analogous to (6.14), we can express the ACD model (11.17) in ARMA form as

$$x_t = \omega + \sum_{j=1}^{\max(h,k)} (\gamma_j + \beta_j) x_{t-j} + \eta_t - \sum_{i=1}^{h} \beta_i \psi_{t-i}, \tag{11.21}$$

where $\eta_t = x_t - \psi_t$ and $\beta_i = 0$ if $i > h$, and $\gamma_j = 0$ if $j > k$. Since the γ_j and β_i are usually assumed to be nonnegative, (11.21) is weakly stationary if $\sum_{j=1}^{k} \gamma_j + \sum_{i=1}^{h} \beta_i < 1$, as in (6.15). Moreover, under this assumption, the unconditional mean of x_t is

$$E(x_t) = \frac{\omega}{1 - \sum_{j=1}^{k} \gamma_j - \sum_{i=1}^{h} \beta_i}, \tag{11.22}$$

as in (6.16). When the ϵ_t are standard exponential, the ACD model is called an EACD(h,k) model. The model is called WACD(h,k) if the ϵ_t are standardized Weibull. Making use of the joint density function of the i.i.d. $\epsilon_t = x_t/\psi_t$, we can derive the likelihood function of $\boldsymbol{\theta} = (\omega, \gamma_1, \ldots, \gamma_k, \beta_1, \ldots, \beta_h)$ in the case of EACD(h,k) and $\boldsymbol{\theta} = (\alpha, \omega, \gamma_1, \ldots, \gamma_k, \beta_1, \ldots, \beta_h)$ in the case of WACD(h,k). The details are straightforward modifications (making use of e^{-y} or (11.20) in place of the standard normal density) of those in Section 6.3.2 for the GARCH(h,k) model. We can use numerical optimization algorithms described in Section 4.1.1 to maximize the likelihood function and thereby compute the MLE.

For the EACD(1,1) model, let $u_l = d_l \sqrt{x_l}$, where d_l is independent of x_l and is equal to -1 and 1 with probability 0.5. Note that the expected value of u_l is 0 and its variance is ψ_l. Regarding d_l as if it were standard normal would give the same likelihood function as that of the GARCH model in Section 6.3.2. Since this likelihood function only involves u_l^2 (which is equal to the observed x_l), GARCH software can be used to compute the "quasi-maximum likelihood estimator," which is still consistent and asymptotically normal, analogous to the QML estimator for the stochastic volatility model in Section 9.6.1.

An example: Fitting EACD to IBM transaction data

Figure 11.4 gives the histogram of the transaction durations for IBM stock during a 3-month period beginning on January 2, 2003 and ending on March 31, 2003. Figure 11.5 gives the kernel estimate (using a bandwidth of 0.5) of the duration density function; see Section 7.2.2. These data are obtained from the TAQ database of the NYSE via Wharton Research Data Services (`http://wrds.wharton.upenn.edu`). The dataset contains detailed information about each transaction occurring on the NYSE during regular trading hours, from 9:30 a.m. to 4 p.m. EST. There are a total of 409,398 executed limit orders. The minimum duration is 1 second, the maximum duration is 96 seconds, the mean duration is 3.48 seconds, and the standard deviation is 3.45 seconds. As pointed out in Section 11.2.1, the transaction intensity tends to be highest near the beginning and close of the trading day and lowest around the lunch hour. Replacing intensity by duration gives the shape of $\widehat{f}(t)$ in Figure 11.5, where $\widehat{f}$ is the kernel estimate based on these intraday transaction data using a bandwidth of 0.3. Figure 11.6 plots the autocorrelation and partial autocorrelation functions of the durations Δt_i and the adjusted durations Δt_i^*. The ACFs and PACFs of the durations and adjusted durations are positive; see Figure 11.6. Moreover, the Ljung-Box test of $\rho_1 = \cdots = \rho_{50} = 0$ applied to the adjusted durations has a p-value less than 2.2×10^{-16}.

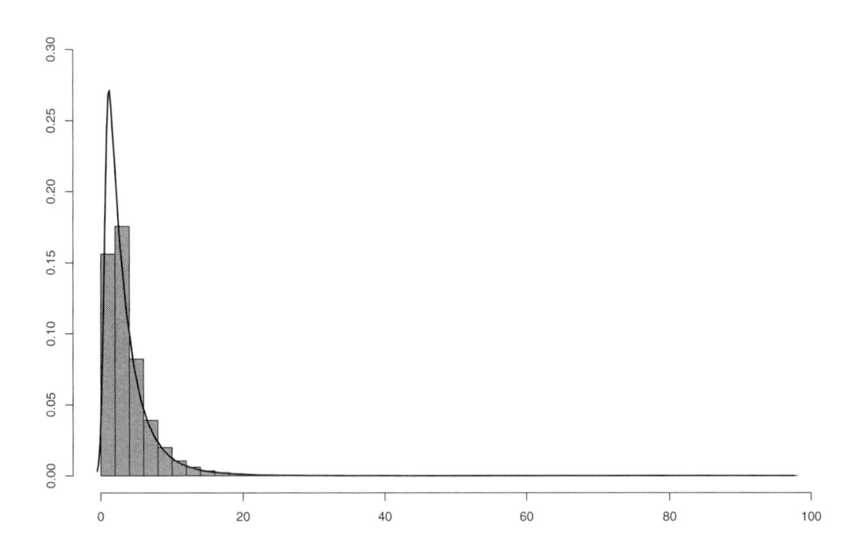

Fig. 11.4. Histogram and density function of transaction durations for IBM stock

To fit the EACD(1, 1) model $x_i := \Delta t_i^* = \psi_i \epsilon_i$, $\psi_i = \omega + \gamma x_{i-1} + \beta \psi_{i-1}$ to these duration data, we use `garchfit` in the `MATLAB` toolbox `garch` to evaluate the quasi-maximum likelihood estimates of the parameters, giving

$$\widehat{\omega} = 0.0011, \qquad \widehat{\gamma} = 0.00446, \qquad \widehat{\beta} = 0.9945.$$

Since $\widehat{\gamma} + \widehat{\beta} = 0.999$, the time series of durations exhibits long memory.

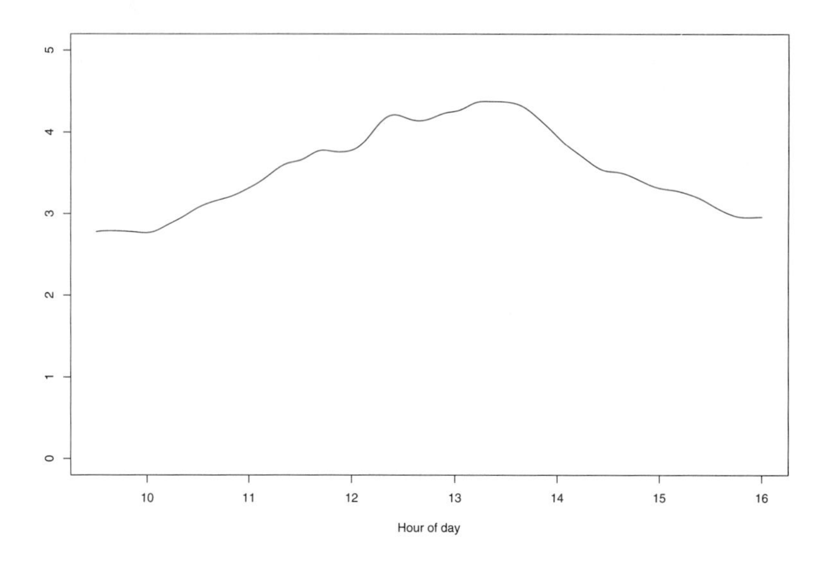

Fig. 11.5. Kernel estimate of the regression function of transaction duration on the time of trading day.

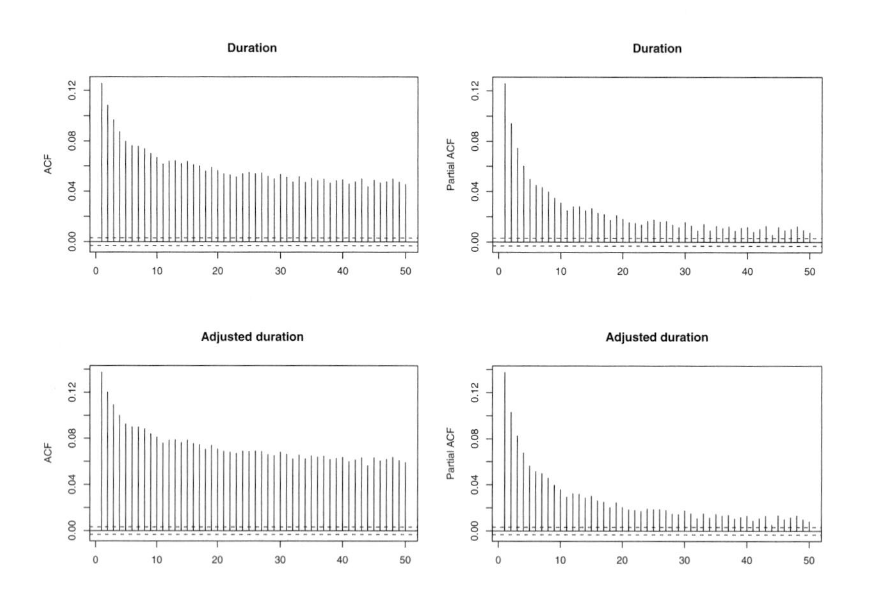

Fig. 11.6. ACFs (left) and PACFs (right) of durations (top panels) and adjusted durations (bottom panels). The dashed lines represent rejection boundaries of 5%-level test of zero ACF or PACF at indicated lag.

11.2.4 Inference on underlying price process

The discreteness of transaction price quotes and the concentration of transactions on zero- or one-tick changes (see Section 11.2.1) make it difficult to estimate the virtual price process p_t^* (see Section 11.2.2). Of particular interest for prediction is the relationship between virtual price changes and observed explanatory variables. Two statistical models of such a relationship have been proposed. Moreover, there has been much recent interest in estimating the volatility of p_t^* from high-frequency data.

The ordered probit model

Let p_t^* be the price (or log-price) of an asset at time t. These prices are not observable; the observations are the transaction price (or its logarithm) p_{t_i} at the ith transaction and the associated vector $\mathbf{x}_i$ of explanatory variables at t_{i-1}. Letting $s_1, \ldots, s_k$ be the set of possible values of the observed price changes $y_i := p_{t_i} - p_{t_{i-1}}$, Hausman, Lo, and MacKinlay (1992) propose a regression model for the *virtual price change* $y_i^* := p_{t_i}^* - p_{t_{i-1}}^*$ of the form

$$y_i^* = \boldsymbol{\beta}^T \mathbf{x}_i + \epsilon_i, \qquad i = 1, \ldots, n. \tag{11.23}$$

It relates the virtual price change to the explanatory variables and uses the transformation that relates the unobservable y_i^* to the observed price changes:

$$y_i = s_j \qquad \text{if } \alpha_{j-1} < y_i^* \le \alpha_j, \quad j = 1, \ldots, k, \tag{11.24}$$

where α_j are real numbers satisfying $-\infty = \alpha_0 < \alpha_1 < \cdots < \alpha_{k-1} < \alpha_k = \infty$. The ϵ_i in (11.23) are assumed to be conditionally independent $N(0, g^2(\mathbf{w}_i, \boldsymbol{\gamma}))$ given $\{\mathbf{x}_1, \ldots, \mathbf{x}_n\}$, where g is a given positive function, $\boldsymbol{\gamma}$ is an unknown parameter vector, and $\mathbf{w}_i$ may contain the adjusted duration Δt_i^*, the bid–ask spread at time t_{i-1}, and other explanatory variables. Combining (11.23) and (11.24) then yields the distribution of the observed y_i:

$$\begin{aligned} P(y_i = s_j | \mathbf{x}_i, \mathbf{w}_i) &= P\big(\alpha_{j-1} < \boldsymbol{\beta}^T \mathbf{x}_i + \epsilon_i \le \alpha_j \big| \mathbf{x}_i, \mathbf{w}_i\big) \\ &= \Phi\left[\frac{\alpha_j - \boldsymbol{\beta}^T \mathbf{x}_i}{g(\mathbf{w}_i, \boldsymbol{\gamma})}\right] - \Phi\left[\frac{\alpha_{j-1} - \boldsymbol{\beta}^T \mathbf{x}_i}{g(\mathbf{w}_i, \boldsymbol{\gamma})}\right], \end{aligned} \tag{11.25}$$

where Φ is the standard normal distribution function. This model, called the *ordered probit model*, contains parameters $\boldsymbol{\gamma}, \alpha_1, \ldots, \alpha_{k-1}$ and $\boldsymbol{\beta}$, which can be estimated by maximizing the likelihood function formed from (11.25). Details are given by Hausman, Lo, and MacKinlay (1992), who estimate the MLEs of the model parameters from the 1988 transaction data of IBM stock with $\mathbf{w}_i^T = (\Delta t_i^*, \text{bid-ask spread at } t_{i-1})$, $g^2(\mathbf{w}_i, \boldsymbol{\gamma}) = 1 + \boldsymbol{\gamma}^T \mathbf{w}_i$ with $\gamma_1 \ge 0, \gamma_2 \ge 0,$

$\mathbf{x}_i$ consisting of $\Delta t_i^*, y_{i-1}, y_{i-2}, y_{i-3}$ and several other market variables, and $k = 9$ categories of virtual price changes.

The Rydberg-Shephard model

Rydberg and Shephard (2003) model the observed price changes y_i directly without involving the virtual price change. They work with the actual transaction price p_{t_i} (rather than its logarithm) and exploit the size S_i (in ticks) of the price change at the ith trade. Let $A_i = 1$ if there is a price change at the ith trade and $A_i = 0$ otherwise. Conditional on $A_i = 1$, let

$$D_i = \begin{cases} 1 & \text{if price increases at the } i\text{th trade,} \\ -1 & \text{if price decreases at the } i\text{th trade.} \end{cases}$$

Let $\mathcal{F}_i$ denote the information set available at the ith transaction. We can express the conditional density of y_i given $\mathcal{F}_{i-1}$ as

$$\begin{aligned} f(y|\mathcal{F}_{i-1}) = P(A_i = 1|\mathcal{F}_{i-1})P(D_i = 1|A_i = 1, \mathcal{F}_{i-1}) \\ \times P(S_i = y|D_i = 1, A_i = 1, \mathcal{F}_{i-1}) \end{aligned} \tag{11.26}$$

for $y > 0$, replacing $D_i = 1$ by $D_i = -1$ in (11.26) for $y < 0$ and noting that $P(y_i = 0|\mathcal{F}_{i-1}) = P(A_i = 0|\mathcal{F}_{i-1})$. The binary nature of A_i and D_i suggests using logistic regression models in Section 4.1.2 to model the factors $P(A_i = 1|\mathcal{F}_{i-1})$ and $P(D_i = 1|A_i = 1, \mathcal{F}_{i-1})$ in (11.26). Specifically, let

$$\operatorname{logit}[P(A_i = 1|\mathcal{F}_{i-1})] = \boldsymbol{\beta}^T \mathbf{x}_i, \tag{11.27}$$

$$\operatorname{logit}[P(D_i = 1|A_i = 1, \mathcal{F}_{i-1})] = \boldsymbol{\gamma}^T \mathbf{z}_i, \tag{11.28}$$

where $\mathbf{x}_i$, $\mathbf{z}_i$ and $\mathbf{w}_i$ are finite-dimensional vectors consisting of elements of $\mathcal{F}_{i-1}$ and $\boldsymbol{\beta}$ and $\boldsymbol{\gamma}$ are parameter vectors. A simple probability model for $(S_i - 1)|A_i, D_i, \mathcal{F}_{i-1}$ is the geometric distribution with density function $\lambda_i(1 - \lambda_i)^k$, $k = 0, 1, \ldots$, whose parameter λ_i is related to $\mathbf{w}_i$ by

$$\operatorname{logit}(\lambda_i) = \begin{cases} \boldsymbol{\theta}^T \mathbf{w}_i & \text{if } D_i = 1, \\ \boldsymbol{\alpha}^T \mathbf{w}_i & \text{if } D_i = -1. \end{cases} \tag{11.29}$$

The log-likelihood function in the Rydberg-Shephard model is therefore of the form $\sum_{i=1}^n \log f(y_i|\mathcal{F}_{i-1})$, in which the parameters $\boldsymbol{\beta}, \boldsymbol{\gamma}, \boldsymbol{\theta}, \boldsymbol{\alpha}$ and the function f are defined by (11.26)–(11.29). Maximum likelihood can be used to estimate the unknown parameters.

Realized volatility

The *quadratic variation* at time t of a log-price process P_t, denoted by $[P]_t$, is defined by

$$[P]_t = \text{p-lim}_{||\Pi||\to 0} \sum_{i=1}^{n} (P_{t(i)} - P_{t(i-1)})^2, \tag{11.30}$$

where $\Pi = \{t(0), \ldots, t(n)\}$ is a partition of $[0, t]$ with $0 = t(0) < \cdots < t(n) = t$, $||\Pi|| = \max_{1 \le k \le n} |t(i) - t(i-1)|$, and $\text{p-lim}_{||\Pi||\to 0} \to 0$ signifies convergence in probability to the same limit for all partitions Π with mesh size $||\Pi||$ approaching 0. The limit in (11.30) exists for square integrable semimartingales (i.e., $EP_t^2 < \infty$ and $P_t = V_t + M_t$, where M_t is a martingale and V_t has bounded variation on bounded time intervals); see Protter (2004). The quantity $RV^{(n)} := \sum_{i=1}^{n} (P_{t(i)} - P_{t(i-1)})^2$ is called the *realized variance* of P_t that is sampled at times $t(0), \ldots, t(n)$, and $\sqrt{RV^{(n)}}$ is called the *realized volatility*. For high-frequency data, although the realized variance of the observed log-price process p_t provides an accurate estimate of $[p]_t$, how to estimate the quadratic variation $[p^*]_t$ of the unobserved underlying log-price process p_t^* is a much harder problem. Under the assumption that p_t is equal to p_t^* plus i.i.d. noise with mean 0 and finite fourth moment, some recent progress has been made in the development of consistent estimators of $[p^*]_t$ by making use of filtering for counting process (Zeng, 2003), or subsampling estimates (Zhang, Mykland and Aït-Sahalia, 2005), or range-based estimates (Christensen and Podolskij, 2007), or kernel-based estimates (Barndorff-Nielsen et al., 2006).

11.2.5 Real-time trading systems

As pointed out by Dacorogna, Gencay, and Muller (2001, Chapter 11), a real-time trading system should incorporate the user's investment and risk profiles to transform available price quotes (e.g., from the limit order book) into actual trading recommendations. It should not change recommendations too rapidly but should support stop-loss around the clock. The system defines a stop-loss price when a trading position is entered. If the most recent price moves adversely beyond the stop-loss price, the stop-loss detector is triggered and causes a deal to close the current open position. Its goal is to prevent excessive loss when the market moves in an unexpected direction. The system should also evaluate the performance of its trading recommendations by using simulated data from various price generation processes and historical data that have not been used to select the strategies; see Sections 12.3 and 12.4 in the next chapter for related statistical methods in risk management.

Such trading systems have been developed from a broad class of technical analysis (pattern recognition) methods and time series forecasting tools, together with the experience and expertise in the financial assets being traded and the trading costs. Understanding the market microstructure is important in designing execution strategies; see Chapter 15 of Hasbrouck (2007), in which it is shown how orders should be split and timed when a trader wants to purchase a large quantity of a security.

11.3 Transaction costs and dynamic trading

11.3.1 Estimation and analysis of transaction costs

As noted by Grinold and Kahn (2000, p. 445), transaction costs "often appear to be unimportant [because] who cares about a 1 or 2 or even 5 percent cost if you expect the stock to double," but are actually important because "they can be the investment management version of death by a thousand cuts." Whereas the portfolio theory in Chapter 3 focused on the portfolio construction problem that does not involve transaction costs, trading is a portfolio optimization problem in the presence of transaction costs that are incurred in carrying out the trades. Transaction costs, however, are difficult to measure and include the following components: (i) commissions, which are the charge per share paid to the broker for executing the trade and are the easiest component to measure, (ii) the bid–ask spread, which is uncertain until one actually trades, and (iii) the trade's market impact, which is unobservable and is assessed by statistical/econometric models of how a trade alters the market; e.g., the BARRA model described in Grinold and Kahn (2000, pp. 449–454) and BARRA (1997).

11.3.2 Heterogeneous trading objectives and strategies

A source of controversy between proponents of "profitable trading models" and their academic critics are the "linguistic barriers" between the two camps, as noted by Lo, Mamaysky, and Wang (2000) who also suggest ways to "bridge this gulf." We add here that one such barrier lies in the implicit (unstated) objective function(s) of the proposed trading model instead of clearly defined objectives (such as maximizing expected returns subject to a specified level of risk in the Markowitz mean-variance portfolio theory) in academic finance. Another such barrier is that whereas academic finance assumes a homogeneous market in which all agents have rational expectations on the "real market value" based on the same model of future price dynamics and optimize the same objective function, the "profitable trading modelers" consider a heterogeneous market in which agents have different objectives, different time horizons for their objectives, and different statistical models and information-processing systems for forecasting future prices and market movements.

To illustrate this point, we cite Grinold and Kahn (2000, p. 1): "From the perspective of the financial economist, active portfolio management appears to be a mundane consideration, if not an entirely dubious proposition. Modern financial economics, with its theories of market efficiency, inspired the move over the past decade away from active management (trying to beat the market) to passive management (trying to match the market)." Indeed the market consists not only of active portfolio managers and passive ones but also others whose strategies cannot be classified into either category. As noted by Grinold and Kahn (2000, pp. 88–89), an active portfolio manager aims to perform well relative to some benchmark portfolio, which may not be the market portfolio as in CAPM (see Section 3.3.1) and is determined by institutional investment on which the performance of the portfolio manager is judged. With this objective in mind and with CAPM providing *consensus expected returns* that serve as a standard of comparison, the active portfolio manager would choose a portfolio with beta near 1 and try to maximize the risk-adjusted alpha of the portfolio; see Grinold and Kahn (2000, pp. 119–127). The risk-adjusted alpha is $\alpha_P - \lambda\sigma_P^2 - TC$, in which α_P is the alpha of the portfolio (see Section 3.3.2), λ is an "active risk aversion" parameter, σ_P is the standard deviation of the return, and TC is the transaction cost (Grinold and Kahn, 2000, pp. 49, 393). Note that λ and TC can vary from agent to agent. Moreover, the asset classes used to form the portfolios can differ among agents, who also have different planning horizons, risk estimators, and forecasts of future alphas of the asset.

11.3.3 Multiperiod trading and dynamic strategies

Dynamic trading involves not only statistical analysis of market information and forecasting future market variables but also the design and implementation of trading strategies. In the area of active portfolio management, Grinold and Kahn (2000, p. 574) point out that an important and difficult open problem is "when should we trade, given the dynamics of returns, risks, and costs over time." Note that Markowitz's efficient frontier and CAPM are single-period theories. Although multiperiod portfolio selection has been studied by several authors since Markowitz's seminal work (e.g., Mossin, 1968; Samuelson, 1969; Hakansson, 1971), "no analytical result, comparable to those in the single-period model, has been reported on mean-variance efficient frontiers" and "researches on multiperiod portfolio selections have been dominated by those of maximizing expected utility functions of the terminal wealth," leading to myopic optimal policies, as noted by Zhou and Li (2000). The inclusion of transaction costs makes the dynamic optimization problem much more difficult, and not much progress has been made. On the other hand, there has been substantial progress in incorporating transaction costs in the infinite-horizon, utility-based, consumption-investment problem introduced by Merton (1971);

see Magill and Constantinides (1976), Davis and Norman (1990), Dumas and Luciano (1991), and Shreve and Soner (1994).

The papers cited above on multiperiod portfolio theory assume the prices are lognormal (i.e., having geometric Brownian motion with known drift and volatility parameters). In practice, these parameters are unknown and have to be estimated from current and past data; moreover, they may also change over the time horizon for which the multiperiod trading strategy is planned. Yin and Zhou (2004) have therefore incorporated regime switching in their extension of Markowitz's mean-variance portfolio selection to continuous-time models. Their formulation, however, still assumes known model parameters and fully observable parameter switches, as in Markowitz's framework. As noted in Section 4.4, simple plug-in implementations that replace the unknown parameters by their sample estimates in Markowitz's efficient frontier do not work well, and a better approach is to use a Bayesian reformulation of Markowitz's optimization problem in the presence of unknown parameters. Bayesian modeling also allows the agent's prior information and subjective judgment to be included in the development of statistical trading strategies. An important function of these models and associated forecasts is to detect profitable trading opportunities. After detecting opportunities and devising strategies to take advantage of them, more conservative (and "objective") statistical methods, such as those described in Section 11.1.4, can be used for evaluating the strategies to see if they should be adopted.

Exercises

11.1. Whereas the investment theories in Chapter 3 assume that the log returns are i.i.d., technical trading rules are based on the implicit assumption that the returns have certain detectable patterns which can reveal profitable trading strategies. Sections II.A, B and III of Lo, Mamaysky, and Wang (2000, abbreviated by LMW) describe five "technical patterns" that are often used in technical analysis, and introduce kernel regression methods to identify the patterns. In addition, statistical tests are proposed for testing whether these technical patterns are significantly informative. To illustrate these and other methods, consider the file `d_logret_5stocks.txt` which contains the monthly log returns of five stocks from January 1991 to December 2006. The five stocks are Apple Inc., International Business Machines Corp., Cisco Systems, Inc., Sun Microsystems Inc. and Hewlett-Packard Co. Following LMW, we divide the 16-year period into four non-overlapping 4-year subperiods.

(a) For each stock and each subperioid, identify technical patterns using Section II.A, B of LMW and using the Bollinger band described in

Section 11.1.1. You can use generalized cross-validation instead of cross-validation that is used by LMW.

(b) Section III of LMW proposes goodness-of-fit tests to compare the quantiles of the conditional distribution of the returns, conditioned on the occurrence of a technical pattern, with the unconditional empirical distribution of the returns. Instead of carrying out these goodness-of-fit tests, compare the conditional and unconditional distributions by examining their Q-Q plot and thereby assess whether the technical pattern is informative.

11.2. Prove (11.10) and (11.11).

11.3. Prove (11.13), (11.14), (11.15), and (11.16).

11.4. Let p_t^* be the underlying price of an asset at time t. Then the transaction price at time t, if there is a trade, may be expressed as $p_t = p_t^* + q_t c$; see Section 11.2.2 for details. Assume that $\Delta p_t^* = p_t^* - p_{t-1}^*$ are i.i.d. standard normal random variables, and that the bid–ask spread is 2 ticks. Compute the lag-1 autocorrelation of the price change series $\Delta p_t = p_t - p_{t-1}$ when the tick size is \$0.125.

11.5. The file `ibm_intratrade_200306.txt` contains the transaction data on the New York Stock Exchange for IBM stock in June 2003. The data are obtained from Wharton Research Data Services.

(a) Let x_i denote the number of trades in the ith 5-minute interval. Ignoring the time gaps between trading days, this gives the time series x_t of the number of trades on IBM stock in 5-minute intervals on the NYSE in June, 2003. Plot the time series and its ACF. Determine if there are intraday period patterns in the series.

(b) Using the last transaction price in the ith 5-minute interval as the stock price in that interval, plot the time series y_t of 5-minute log returns during the period and the corresponding ACF.

(c) Consider the bivariate time series (x_t, y_t). How does y_t vary with x_t? Are there intraday periodic patterns in (x_t, y_t)?

(d) Tabulate the relative frequencies of price changes in multiples of the tick size \$0.0625.

11.6. For the IBM stock transaction data on June 2, 2003 in Exercise 7.5, fit an EACD(1, 1) model to the adjusted durations $\Delta t_i^* = \Delta t_i / \widehat{f}(t_i)$.

12

Statistical Methods in Risk Management

As noted in Chapters 3 and 11, the theory of investment involves two fundamental concepts, namely expected returns and risk. In these chapters, we have conveniently measured risk by historic or model-based volatility in considering the optimal trade-off between maximizing expected returns and minimizing risk. In this chapter, we consider other risk measures and risk management from the corporate/regulatory perspective in order to provide safeguards (via capital reserves, hedging instruments, etc.) against extreme downside price movements, thereby protecting the company and its investors should these rare events occur.

To illustrate the regulatory background underlying the statistical methods for risk management described in the subsequent sections, we give a brief description of the Basel Accord, which was developed in 1988 by the Basel Committee on Banking Supervision and later adopted by the central banks of the G10 countries. The accord represents the first regulatory step toward risk management by setting minimum capital requirements for commercial banks to manage their risk. One such requirement is 8% as the minimum *solvency ratio*, which is the ratio of capital to risk-weighted assets. Another requirement is an upper bound of 20 for the *asset-to-capital multiple*, which is obtained by dividing the bank's total assets by its total capital. In Section 12.1, we introduce the concept of *market risk* and describe the 1996 amendment of the accord that allows banks to use *internal models* to calculate their capital requirements for the market risk. Statistical methods are important for the development of these internal models, which are used to capture the magnitude of the market risks of all trading positions. In particular, internal models are required to compute the maximum loss over ten trading days at a one-sided 99% confidence level. The model parameters have to be updated at least once every three months or more frequently. A number of problems with the 1996 amendment subsequently emerged, and the Basel Committee provided further revisions to address them in 1999 and 2001. The new Basel II Accord, which was adopted by all G10 countries and many others in 2007, extends the risk capital requirements to a broader range of risk types, which are described in Section 12.1, and incorporates traditional banking positions, asset management, and brokerage subsidiaries in the solvency ratios; see `www.bis.org` for details.

Besides outlining important types of financial risks, Section 12.1 also describes several measures of market risk, in particular *Value at Risk* and *Expected Shortfall*. Section 12.2 introduces statistical methods and models for these risk measures in the case of a portfolio of assets. Section 12.3 considers the more complicated case of nonlinear financial instruments such as derivatives and describes simulation-based approaches besides commonly used linear and quadratic approximations. The Basel Committee requires backtesting of internal market risk models, which is considered in Section 12.2.4, and stress testing, which is considered in Section 12.4. In connection with stress testing, extreme value theory and computationally tractable Monte Carlo methods are introduced in Section 12.4.

12.1 Financial risks and measures of market risk

12.1.1 Types of financial risks

Financial risks can be broadly classified into several categories, namely market risk, credit risk, liquidity risk, operational risk, and legal risk. *Market risk* is the risk of loss arising from changes in the value of tradable or traded assets. *Credit risk* is the risk of loss due to the failure of the counterparty to pay the promised obligation. *Liquidity risk* is the risk of loss arising from the inability either to meet payment obligations (funding liquidity risk) or to liquidate positions with little price impact (asset liquidity risk). *Operational risk* is the risk of loss caused by inadequate or failed internal processes, people and systems, or external events. *Legal risk* is the risk of loss arising from uncertainty about the enforceability of contracts.

As this chapter focuses primarily on market risk, we briefly describe the other risks. The loss encompassed in credit risk includes the *exposure*, which is the amount at risk, reduced by the *recovery rate*, which is the proportion paid back. It can occur before the actual default of the counterparty. As pointed out by Jorion (2001, pp. 16–17), "changes in market prices of debt, due to changes in credit ratings or in the market's perception of default, also can be viewed as credit risk," and "the methodological advances in quantifying market risk are now being extended to credit risk." Chapter 20 of Hull (2006) describes some of these extensions. Asset liquidity risk arises when a transaction cannot be conducted at prevailing market prices due to the size of the trading position and/or the nature of the asset (e.g., over-the-counter financial derivatives). Funding liquidity risk refers to the inability to meet payment obligations, which may lead to involuntary selling of assets at depressed prices. The loss ascribed to operational risk may be due to using a flawed model to value trading positions (called "model risk") or technical errors, inadequate supervision of trade execution, fraud, or accident. As pointed out by Jorion (2001, p. 19), "traders using a conventional option pricing model, for instance, could be exposed to model risk if the model is misspecified or if the model parameters are erroneous," and "to guard against model risk, models must be subjected to independent evaluation using market prices, when available, or objective out-of-sample evaluations."

12.1.2 Internal models for capital requirements

While the original Basel Accord focused primarily on the (credit) risks associated with the issuer, the 1996 amendment sought to give more coverage to market risk. In the amendment, there are certain regulatory requirements on the internal models from which the banks calculate their capital requirements. One requirement is that the risk management group in charge of the development and execution of these models should be independent of the business units it monitors and should report directly to senior management. Another requirement is that besides calculating the regulatory capital requirements, these models should be fully integrated into the bank's risk measurement and management, and backtesting and stress testing should be performed on their performance on a regular basis.

As pointed out by Dowd (2005, pp. 9–10), work on internal models by a number of major financial institutions had already started in the late 1970s not only to support their risk management but also to develop the models into a system that could be sold to other financial institutions and corporations. In particular, the RiskMetrics system, to which we have already referred in Section 6.2, originated when the chairman of J.P. Morgan asked his staff to give him, after the close of trading each day, a one-page report indicating risk and potential losses over the next 24 hours across the bank's entire trading portfolio. This led to the development of a system to measure the risks of all trading positions and to aggregate the risks into a single risk measure, called *Value at Risk* (VaR). In 1994, J.P. Morgan made the RiskMetrics system public and the necessary data freely available on the Internet, which led to subsequent improvements of the approach and widespread adoption of VaR systems by financial institutions.

12.1.3 VaR and other measures of market risk

Value at Risk is one of the most important and widely used risk management statistics. It measures the maximum loss of a financial institution's position due to market movements over a given holding period with a given level of confidence. More precisely, VaR is the quantile of the distribution of projected gains and losses over the holding period. Let X denote the change in value of the financial position from the beginning to the end of the holding period. A $100(1-\alpha)\%$ VaR of a long (L) position over the period is defined as the αth quantile of X:

$$\text{VaR}_L = \inf\{x : P(X \le x) \ge \alpha\}. \tag{12.1}$$

Since the holder of a long position suffers a loss when $X < 0$ and since (12.1) is typically negative when α is small, the negative sign signifies a loss for a long position. On the other hand, the holder of a short (S) position suffers a loss when $X > 0$, so the $100(1-\alpha)\%$ VaR of a short position over the holding period is the $(1-\alpha)$ quantile of X:

$$\text{VaR}_S = \inf\{x : P(X > x) \le \alpha\}. \tag{12.2}$$

For small α, VaR_S is typically positive, and the positive sign signifies a loss for a short position.

Coherent risk measures

A *risk measure* is a number $\rho(X)$ associated with the distribution of the random variable X in the preceding paragraph. In particular, the loss associated with VaR is a risk measure with $\rho(X) = (1-\alpha)$th (or absolute value of αth) quantile of X for a short (or long) position. Although VaR has become a standard risk measure for financial risk management due to its interpretability for capital requirements in entering financial positions and the availability of Web-based systems (e.g., the VaR Calculator provided by State Street) to perform VaR calculations, it has been criticized for disregarding losses beyond the VaR level. To remedy this difficulty with VaR, Artzner et al. (1997, 1999) propose four axioms that risk measures need to satisfy in order to be "coherent." Let X_A and X_B denote the change in value of financial positions A and B, respectively, over a time horizon. The risk measure $\rho(\cdot)$ is said to be *coherent* if it satisfies the following properties:

(i) *Monotonicity:* $X_B \le X_A \Rightarrow \rho(X_B) \le \rho(X_A)$.
(ii) *Subadditivity:* $\rho(X_A + X_B) \le \rho(X_A) + \rho(X_B)$.
(iii) *Positive homogeneity:* $\rho(bX_A) = b\rho(X_A)$ for $b > 0$.
(iv) *Translational invariance:* $\rho(X_A + a) = \rho(X_A) - a$ for nonrandom a.

Expected shortfall

When the distribution function F of X is continuous, the *expected shortfall* (ES), proposed by Artzner et al. (1997), is defined as the conditional expectation of loss given that the loss falls beyond the $100(1-\alpha)\%$ VaR level or, more precisely,

$$\mathrm{ES} = \begin{cases} -\int_0^{\alpha} F^{-1}(x)dx/\alpha = -E\big[X|X \leq \mathrm{VaR}_L\big] & \text{(long)}, \\ \int_{1-\alpha}^{1} F^{-1}(x)dx/\alpha = E\big[X|X \geq \mathrm{VaR}_S\big] & \text{(short)}. \end{cases} \tag{12.3}$$

For extensions of ES to general F, see Acerbi (2004, Chapter 2). Whereas the loss associated with VaR is not a coherent risk measure, ES is coherent; see Acerbi (2004).

12.2 Statistical models for VaR and ES

12.2.1 The Gaussian convention and the t-modification

The classical framework of i.i.d. normal returns in Section 3.1.2 provides a convenient framework for VaR and ES calculations. First, the quantiles are the easily calculated standard normal quantiles after subtracting the mean from X and dividing by its standard deviation. Second, it is easy to aggregate over time and over assets. If the daily log return has mean μ and variance σ^2, then the k-day VaR is a quantile of the $N(k\mu, k\sigma^2)$-distribution. For a portfolio consisting of p assets with daily returns $r_1, \ldots, r_p$ and corresponding weights $w_1, \ldots, w_p$, the return of the portfolio is $r = \sum_{i=1}^{p} w_i r_i$, and the mean and variance of the portfolio return are given by

$$\mu = \sum_{i=1}^{p} w_i \mu_i, \qquad \sigma^2 = \sum_{i=1}^{p} w_i^2 \sigma_i^2 + 2 \sum_{i=1}^{p} \sum_{j=i+1}^{p} w_i w_j \sigma_{ij}, \tag{12.4}$$

where μ_i $(1 \leq i \leq p)$ are the mean returns and σ_i^2 and σ_{ij} $(1 \leq i, j \leq p)$ are the variances and covariances of these assets; see Section 3.2.1. If all asset returns are jointly normally distributed and hence $r \sim N(\mu, \sigma^2)$, then the $100(1-\alpha)\%$ VaR of a long position over k days is

$$\mathrm{VaR}_L(k, \alpha) = k\mu - \sqrt{k} z_{1-\alpha} \sigma, \tag{12.5}$$

and the $100(1-\alpha)\%$ ES of a long position over k days is

$$\mathrm{ES}_L(k, \alpha) = -k\mu + \sqrt{k} \sigma \phi(z_{1-\alpha})/\alpha, \tag{12.6}$$

where $z_{1-\alpha}$ is the $(1-\alpha)$th quantile of the standard normal distribution and $\phi(\cdot)$ is the density function of the standard normal distribution, for which $z_\alpha = -z_{1-\alpha}$. Similarly, for a short position,

$$\mathrm{VaR}(k, \alpha) = k\mu + \sqrt{k} z_{1-\alpha} \sigma, \qquad \mathrm{ES}(k, \alpha) = k\mu + \sqrt{k} \sigma \phi(z_{1-\alpha})/\alpha.$$

In practice, both μ and σ are unknown and have to be estimated from past data. For VaR calculations, the normal distribution is often replaced by a Student t-distribution that has fatter tails as in Section 6.3.

12.2.2 Applications of PCA and an example

The covariances σ_{ij} in (12.4) are the essential ingredients of portfolio risk. Since they are unknown in practice, the task of estimating them from limited historical data becomes increasingly difficult with an increasing number of assets. A widely used method to overcome this "curse of dimensionality" in VaR models is to perform principal component analysis (PCA) to determine whether σ^2 in (12.4) can be approximated by the sum of variances of a relatively small number of principal components.

As an illustration, consider the seven weekly U.S. Treasury rates with maturities 1, 2, 3, 5, 7, 10, and 20 years from October 1, 1993 to March 23, 2007 in Section 10.3; see Figure 10.9 and Table 10.3, in which the factor loadings and standard deviations of factor scores using the covariance matrix of interest rate changes are given. The first three principal components account for about 99% of the variance in the data. Consider a portfolio with the exposures to interest rate moves shown in Table 12.1, in which a 1-basis-point change in the 1-year (or 2-year, ..., or 20-year) rate causes the portfolio value to increase by \$8 (or 4, ..., or 2) million; 1 basis point = 0.01%. From the results in Table 10.3(a), it follows that the exposure (measured in millions of dollars) to the first principal component f_1 of the covariance matrix of interest rate changes is

$$8 \times 0.279 + 4 \times 0.384 - 8 \times 0.416 - 3 \times 0.423 + 2 \times 0.412 - 0.384 + 2 \times 0.324 = 0.259,$$

and the exposure to the second principal component f_2 is

$$8 \times 0.590 + 4 \times 0.398 - 8 \times 0.239 + 3 \times 0.044 - 2 \times 0.240 + 0.363 - 2 \times 0.495 = 3.427.$$

Table 12.1. Change in portfolio value for 1-basis-point rate change.

Maturity (year)	1	2	3	5	7	10	20
Change (\$$10^6$)	8	4	−8	−3	2	−1	2

The change in the portfolio value can therefore be represented to a good approximation by $\Delta P = 0.259 f_1 + 3.427 f_2$, and the standard deviation of ΔP is therefore $\sigma_P = \sqrt{(0.259)^2\sigma_1^2 + (3.427)^2\sigma_2^2} = 24.61$, where σ_1 and σ_2 are the standard deviations of f_1 and f_2, respectively. Hence, the 99% 1-week VaR is $24.61 \times 2.33 = 57.34$ (million dollars), assuming a normal distribution for ΔP.

12.2.3 Time series models

We have assumed in the preceding sections that the returns are i.i.d. as in Chapter 3. However, since risk may vary with time and since the available data are time series of past returns, better VaR models can be built by incorporating the time series properties of the returns data. In particular, the time series models of Chapter 5 can be used to model the mean returns over time, and the conditional heteroskedastic models of Chapter 6 can be used to model the time-varying volatilities, as in the ARMA(p,q)-GARCH(h,k) model (6.34). The model can be estimated by maximum likelihood as indicated in Section 6.4.2 and implemented by the `garchfit` function in the `GARCH` toolbox of `MATLAB`. The conditional distribution of r_{t+1} given the information available at time t is $N(\widehat{r}_{t+1|t}, \widehat{\sigma}^2_{t+1|t})$, where $\widehat{r}_{t+1|t}$ and $\widehat{\sigma}^2_{t+1|t}$ are given by (6.35) and (6.36). Hence the $100(1-\alpha)\%$ 1-day VaR and ES of a long position are $\widehat{r}_{t+1|t} - \widehat{\sigma}_{t+1|t} z_{1-\alpha}$ and $\widehat{r}_{t+1|t} - \widehat{\sigma}_{t+1|t}\phi(z_{1-\alpha})/\alpha$, respectively, assuming normal ϵ_t.

For a k-day holding period, we aggregate over time by $r_t[k] = r_{t+1} + \cdots + r_{t+k}$ as in Section 12.2.1. Therefore $E\big(r_t[k]\big|\mathcal{F}_t\big) = \widehat{r}_{t+1|t} + \cdots + \widehat{r}_{t+k|t}$, and the same argument as that leading to (6.37) in Section 6.4.1 can be used to show that

$$
\begin{aligned}
r_t[k] - E\big(r_t[k]\big|\mathcal{F}_t\big) &= \sum_{h=1}^{k}(r_{t+h} - \widehat{r}_{t+h|t}) = \sum_{h=1}^{k}\sum_{i=1}^{h}\psi_{h-i}u_{t+i},\\
\mathrm{Var}\Big(\big\{r_t[k] - E(r_t[k]\big|\mathcal{F}_t)\big\}\Big|\mathcal{F}_t\Big) &= \mathrm{Var}\left[\sum_{h=1}^{k}\sum_{i=1}^{h}\psi_{h-i}u_{t+i}\middle|\mathcal{F}_t\right]\\
&= \mathrm{Var}\left\{\left[\sum_{i=1}^{k}\left(\sum_{h=i}^{k}\psi_{h-i}\right)u_{t+i}\right]\middle|\mathcal{F}_t\right\}\\
&= \sum_{i=1}^{k}\left(\sum_{h=i}^{k}\psi_{h-i}\right)^2 E\big(\sigma^2_{t+i}\big|\mathcal{F}_t\big).
\end{aligned}
\tag{12.7}
$$

12.2.4 Backtesting VaR models

Backtesting is a statistical procedure for testing whether the projected losses under a VaR model differ significantly from the actual losses. It played a basic role in the Basel Committee's decision to allow internal VaR models for capital requirements in the 1996 amendment, which outlines the framework of backtests. Suppose a financial institution provides $100(1-\alpha)\%$ 1-day VaR values for a total of n days. The backtest counts the number of times, denoted by τ, the actual loss exceeds the previous day's VaR. If the internal VaR model is accurate, the probability of such exceedances is α, and therefore the expected number of exceedances is $n\alpha$. If the number of exceedances is significantly larger than $n\alpha$, the VaR model is considered to be underestimating risk and the regulator may impose penalties for allocating insufficient capital to cover the risk. Statistical significance of the number of exceedances is measured by the p-value of a binomial test that assumes the occurrences of exceedance to be independent Bernoulli trials so that the number of occurrences follows a binomial(n, α) distribution. Using the normal approximation to the binomial distribution, the test statistic is

$$Z = (\tau - n\alpha) / \sqrt{n\alpha(1-\alpha)}.$$

The backtesting procedure adopted by the Basel Committee consists of recording the number of exceedances of the 99% daily VaR over the last year with $n = 250$ trading days. Up to four exceedances is considered acceptable (p-value > 0.83, binomial test), putting the bank in a "green" zone that has no penalty. For $5 \leq \tau \leq 9$, the bank falls into a "yellow" zone in which the penalty is up to the regulator, depending on the circumstances that cause the exceedances. For $\tau \geq 10$, the bank falls into a "red" zone and receives an automatic penalty. The penalty is in the form of an increase in the *multiplicative factor* in setting the bank's market capital charge. The general market capital charge is set at the larger of the previous day's VaR or the average VaR of the last 60 business days times the multiplicative factor determined by local regulators, subject to a minimum value of 3. The multiplicative factor is increased progressively from the yellow zone to the red zone; see Jorion (2001, pp. 137–138).

12.3 Measuring risk for nonlinear portfolios

Whereas asset prices are traditionally modeled by geometric Brownian motion, the prices of derivatives such as options are nonlinear functions of the underlying asset prices, and fixed-income securities depend nonlinearly on interest rates with different maturities; see Chapters 8 and 10. There are two approaches to measuring the risk of a portfolio that includes nonlinear financial instruments. One involves local valuation, approximating the nonlinear functions by linear or quadratic functions by using Taylor expansions. The other approach is to use Monte Carlo simulations for full valuation.

12.3.1 Local valuation via Taylor expansions

Delta-normal valuation

To begin, consider a financial instrument whose value is a function $V(S)$ of a risk factor S. Its *delta* refers to the first derivative $V'(S_0)$, denoted by Δ_0, at the current value S_0 of the risk factor. The change ΔV in its value is related to the change ΔS in S by the linear approximation $\Delta V \approx V'(S_0)\Delta S$. The delta-normal approach to VaR and ES assumes ΔS to be normal $N(0, \sigma^2)$ so that ΔV can be approximated by $N(0, \Delta_0^2\sigma^2)$. As an illustration, consider a bond portfolio that has price P, duration D, and yield to maturity y. The change ΔP in the bond price relative to a change Δy in yield can be described by $\Delta P \approx -DP\Delta y$, and the standard deviation σ of ΔP can be approximated by $\sigma \approx DP\sigma_y$, where σ_y is the standard deviation of changes in the level of the yield. The VaR and ES of the portfolio can then be calculated under the assumption of a normal distribution for Δy. Note that this duration approach is only a first-order approximation and makes the restrictive assumption that the change in yield is the same for all maturities.

Delta-gamma approximations and the Greeks

Let $V = f(t, \mathbf{S})$ be the value of a portfolio depending on the prices $S_1, \dots, S_m$ (the components of the vector $\mathbf{S}$) of m underlying assets at time t. The change ΔV in the portfolio's value can be approximated by the Taylor expansion

$$\Delta V \approx \frac{\partial V}{\partial t}\Delta t + \left(\frac{\partial V}{\partial S_1}, \dots, \frac{\partial V}{\partial S_m}\right)^T \Delta\mathbf{S} + \frac{1}{2}\Delta\mathbf{S}^T \left(\frac{\partial^2 V}{\partial S_i \partial S_j}\right)_{1\le i,j\le m} \Delta\mathbf{S}. \tag{12.8}$$

The first and second derivatives of V with respect to the components of $\mathbf{S}$ are called the *delta* and the *gamma* of the portfolio, respectively, and $\partial V/\partial t$ is called the portfolio's *theta*. In particular, the delta of a European call option is given by (8.8), and it follows from the Black-Scholes formula (8.6) that the call option's theta and gamma are given by

$$\Theta = -\frac{S_t \phi(d_1)\sigma e^{-q(T-t)}}{2\sqrt{T-t}} + qS_t\Phi(d_1)e^{-q(T-t)} - rKe^{-r(T-t)}\Phi(d_2), \tag{12.9}$$

$$\Gamma = \frac{\phi(d_1)}{S_t\sigma\sqrt{T-t}}. \tag{12.10}$$

Note that the Black-Scholes PDE (8.5) for a European call or put can be expressed as

$$\Theta + (r-q)S\Delta + \frac{1}{2}\sigma^2 S^2 \Gamma = rV. \tag{12.11}$$

A portfolio is called *delta-neutral* if its delta is zero. Other "Greeks" include the *vega* $\partial V/\partial\sigma$, which is the rate of change with respect to the volatility σ of the underlying asset, and the *rho* $\partial V/\partial r$, which is the rate of change with respect to the interest rate.

12.3.2 Full valuation via Monte Carlo

The delta-normal approach to VaR and ES can be refined by using second-order approximations that involve gamma besides delta. However, the approximation is still local, and the formulas involving gradient vectors and Hessian matrices may be complicated to implement, especially for portfolios with a large number of underlying assets and their derivatives. Section 9.1.5 of Jorion (2001) uses the collapse of Barings Bank in 1995 to illustrate the dangers of using these local approximations, which fail to warn of large losses for portfolios that have relatively small delta-normal VaR.

Instead of relying on local approximations, the *full valuation* approach evaluates the distribution of ΔV by Monte Carlo simulations, generating a large number of trajectories of $(\mathbf{S}_0, \mathbf{S}_t)$ and therefore also of the realizations of $\Delta V = V(t, \mathbf{S}_t) - V(0, \mathbf{S}_0)$, so that the empirical distribution of these realizations can be used to estimate the distribution of ΔV. Not only does this approach obviate the need to use questionable first-order approximations but it can also dispense with the assumption of normal $\Delta\mathbf{S}$ in the delta-normal approach.

What distribution for $\Delta\mathbf{S}$ should one use in simulating the distribution of ΔV? The *historical simulation* approach samples $\Delta\mathbf{S}$ from the empirical distribution of price changes over a given period of time in the immediate past, e.g., daily price changes over the past year. The approach is relatively simple to implement if historical data are collected in-house for daily marking to market. But it can be cumbersome for large portfolios with complicated data structures. Although it makes no distributional assumptions on $\Delta\mathbf{S}$ such as normality and uses historical data to estimate the distribution nonparametrically, it makes the implicit assumption that the historical values of $\Delta\mathbf{S}$ are independent realizations of the same underlying distribution and ignores the time series aspects of past data.

An alternative to simulating $\Delta\mathbf{S}$ for the basic market variables from historical data is to simulate them from a stochastic model whose parameters are estimated from historical data. For example, one can use a time series model of the type in Section 12.2.3 for each component of the vector $\Delta\mathbf{S}$. For the joint distribution of these components, an often-used method is that of Gaussian or t-copula functions, described below. Another method is to use PCA to reduce the correlation matrix of $\Delta\mathbf{S}$ to a relatively small number of factors, as in Section 12.2.2.

12.3.3 Multivariate copula functions

Let $\mathbf{X} = (X_1, \ldots, X_m)^T$ be a continuous random vector with given marginal distribution functions F_i for X_i. Since $U_i = F_i(X_i)$ is uniformly distributed, we can specify the joint distribution of $\mathbf{X}$ by a *copula function*, which is the joint distribution function of m uniform random variables. Specifically, the copula function $C(u_1, \ldots, u_m)$ defined on $0 \le u_i \le 1$ $(1 \le i \le m)$ can be used to provide a joint distribution function of $\mathbf{X}$ via

$$\begin{aligned} C\big(F_1(x_1), \ldots, F_m(x_m)\big) &= P\big\{F_1(X_1) \le F_1(x_1), \ldots, F_m(X_m) \le F_m(x_m)\big\} \\ &= P\big\{X_1 \le x_1, \ldots, X_m \le x_m\big\}. \end{aligned} \tag{12.12}$$

Definition 12.1. Let $\Phi_{\boldsymbol{\rho}}$ be the distribution function of an m-variate normal distribution with zero means, unit variances, and correlation matrix $\boldsymbol{\rho}$, and let Φ denote the standard normal distribution function. Then the copula function

$$C(u_1, \ldots, u_m) = \Phi_{\boldsymbol{\rho}}\big(\Phi^{-1}(u_1), \ldots, \Phi^{-1}(u_m)\big), \quad 0 \le u_1, \ldots, u_m \le 1, \tag{12.13}$$

is called a *Gaussian copula* function.

In particular, a bivariate Gaussian copula with correlation coefficient ρ is given by

$$C_G(u, v) = \int_{-\infty}^{\Phi^{-1}(u)} \int_{-\infty}^{\Phi^{-1}(v)} \frac{1}{2\pi\sqrt{1-\rho^2}} \exp\left(-\frac{s^2 + t^2 - 2\rho st}{2(1-\rho^2)}\right) ds dt. \tag{12.14}$$

For a random vector $\mathbf{X}$ with given marginal distribution function F_i, the Gaussian copula approach specifies the dependence structure of $Z_i := \Phi^{-1}(F_i(X_i))$ by specifying the pairwise correlations between the jointly normal Z_i that are marginally standard normal. This joint modeling approach is widely used in the credit derivatives market; see Li (2000).

t-copulas

A variant of the Gaussian copula, called Student's t-copula, is often used to accommodate heavy tails of asset returns. For the univariate Student t-distribution with k degrees of freedom that has the density function (2.29), its distribution function has the form

$$T_k(x) = \int_{-\infty}^{x} \frac{\Gamma((k+1)/2}{\sqrt{\pi k}\Gamma(k/2)} \left(1 + \frac{s^2}{k}\right)^{-(k+1)/2} ds. \tag{12.15}$$

Correspondingly, for the m-variate Student t-distribution with k degrees of freedom that has the density function f given by (2.34), its distribution function can be written as

$$T_{\boldsymbol{\Sigma},k}(\mathbf{x}) = \int_{-\infty}^{x_1} \cdots \int_{-\infty}^{x_m} f(\mathbf{t}) dt_1 \ldots dt_m. \tag{12.16}$$

Definition 12.2. Let $T_{\boldsymbol{\Sigma},k}$, given by (12.16), be the distribution function of an m-variate Student t-distribution with covariance matrix $\boldsymbol{\Sigma}$ and k degrees of freedom, and let T_k, given by (12.15), be the distribution function of a univariate Student t-distribution with k degrees of freedom and variance 1. Then the copula function

$$C(u_1, \ldots, u_m) = T_{\boldsymbol{\Sigma},k}(T_k^{-1}(u_1), \ldots, T_k^{-1}(u_m)),\ 0 \le u_1, \ldots, u_m \le 1, \tag{12.17}$$

is called a multivariate *Student t-copula* function.

In particular, a bivariate Student's t-copula with k degrees of freedom and correlation coefficient ρ (so that $\boldsymbol{\Sigma}$ has diagonal elements equal to 1 and off-diagonal elements equal to ρ) is given by

$$C_T(u, v) = \int_{-\infty}^{T_k^{-1}(u)} \int_{-\infty}^{T_k^{-1}(v)} \frac{1}{2\pi\sqrt{1-\rho^2}} \left(1 + \frac{s^2 + t^2 - 2\rho st}{k(1-\rho^2)}\right)^{-\frac{k+2}{2}} dtds. \tag{12.18}$$

12.3.4 Variance reduction techniques

To compute $E(X)$ by Monte Carlo, the usual idea is to simulate i.i.d. $X_1, \ldots, X_n$, and then estimate $E(X)$ by the sample mean $\bar{X} = \sum_{i=1}^n X_i/n$. The sample mean is unbiased with variance $\mathrm{Var}(X)/n$.

Antithetic variables

Direct Monto Carlo evaluation of the expectation μ of a random variable X (e.g., X = return of a portfolio, or X = payoff of an option, or $X = \mathbf{1}_A$) involves generating B independent replicates $X_1, \ldots, X_B$ of X so that $B^{-1}\sum_{i=1}^B X_i$ can be used to estimate $E(X)$ with standard error $\sqrt{\mathrm{Var}(X)/B}$. The method of *antithetic variables* provides variance reduction by simulating m i.i.d. pairs $(X_i, \widetilde{X}_i)$ such that X_i and $\widetilde{X}_i$ have the same distribution as X and $\mathrm{Cov}(X_i, \widetilde{X}_i) < 0$. For example, if X has a symmetric distribution (such as normal or Student-t), then one can choose $\widetilde{X}_i = -X_i$. These "antithetic pairs" can be used to estimate μ by $\widehat{\mu} = (2m)^{-1}\sum_{i=1}^n (X_i + \widetilde{X}_i)$, which is unbiased and has variance

$$\begin{aligned} m^{-1}\mathrm{Var}\Big((X_1 + \widetilde{X}_1)/2\Big) &= (2m)^{-1}\{\mathrm{Var}(X_1) + \mathrm{Cov}(X_1, \widetilde{X}_1)\} \\ &< (2m)^{-1}\mathrm{Var}(X_1). \end{aligned}$$

Control variates

The method of *control variates* involves simulating m i.i.d. pairs (X_i, Y_i) in which Y_i has *known* expectation EY. It estimates $\mu := EX$ by

$$\widehat{\mu} = m^{-1} \sum_{i=1}^{m} \{X_i - b(Y_i - E(Y))\}, \tag{12.19}$$

which is unbiased. The optimal choice of b is $b^* = \text{Cov}(X, Y)/\text{Var}(Y)$, for which $\text{Var}(\widehat{\mu})/\text{Var}(m^{-1}\sum_{i=1}^{m} X_i) = 1 - \rho^2$, where $\rho = \text{Corr}(X, Y)$. The values $\text{Var}(Y)$ and $\text{Cov}(X, Y)$ can be estimated from the simulated samples $\{Y_1, \ldots, Y_m\}$ and $\{(X_1, Y_1), \ldots, (X_m, Y_m)\}$, and these estimates can then be used to form the optimal b in (12.19).

Stratified sampling

Instead of sampling from a large sample space, sampling representative values from its strata can reduce the variance of the Monte Carlo estimate of $Eh(\mathbf{X})$. Suppose the sample space is the union of κ disjoint events, which we label for convenience as $1, \ldots, \kappa$. Letting p_k be the probability of event k, we can write $Eh(\mathbf{X}) = \sum_{k=1}^{\kappa} p_k E_k h(\mathbf{X})$, where E_k denotes the conditional expectation of $h(\mathbf{X})$ given that event k has occurred. *Stratified sampling* generates B i.i.d. pairs $(K_i, \mathbf{X}_i)$ by first generating K_i from the distribution $P(K_i = k) = p_k$, $k = 1, \ldots, \kappa$, and then generating $\mathbf{X}_i$ from its conditional distribution given K_i. These stratified samples are used to estimate $\mu := E(h(\mathbf{X}))$ by

$$\widehat{\boldsymbol{\mu}} = \sum_{k=1}^{\kappa} p_k \left\{ \sum_{i=1}^{B} h(\mathbf{X}_i) \mathbf{1}_{\{K_i = k\}} \right\} \Big/ \left(\sum_{i=1}^{B} \mathbf{1}_{\{K_i = k\}} \right).$$

Importance sampling

Instead of sampling the $\mathbf{X}_i$ from the distribution P, *importance sampling* samples i.i.d. $\mathbf{X}_i$ from some other distribution Q and estimates $\mu := Eh(\mathbf{X})$ by

$$\widehat{\mu} = \frac{1}{B} \sum_{i=1}^{B} h(\mathbf{X}_i) R(\mathbf{X}_i), \tag{12.20}$$

where $R(\mathbf{X}_i)$ is the likelihood ratio of P with respect to Q. In particular, $R(\mathbf{X}_i) = f(\mathbf{X}_i)/g(\mathbf{X}_i)$ in the case where the density function of $\mathbf{X}$ is f under P and g under Q such that $g = 0 \Rightarrow f = 0$.

12.4 Stress testing and extreme value theory

12.4.1 Stress testing

Whereas risk measures such as VaR or ES consider normal market conditions to evaluate potential losses, extraordinary situations that can cause severe losses and bankruptcy also need to be incorporated into risk assessment and management. *Stress testing*, the goal of which is to identify and manage situations that can cause extraordinary losses, is required by the Basel Committee as one of the conditions to be satisfied when using internal models. However, unlike those for backtesting, the guidelines for stress testing are vague. Stress testing uses extreme scenarios that might occur given the current economic environment and global uncertainties. A simple method to create scenarios is to move key market variables sequentially by a large amount, ignoring correlations. While this method has often been used, the financial industry has recently realized that identification of extreme scenarios should be driven by the particular portfolio at hand. For example, a highly leveraged portfolio with a long position in corporate bonds offset by a short position in Treasuries could suffer sharp losses if there is an unusual decrease in the correlations between the rates of corporate and Treasury bonds. A recent *Wall Street Journal* article by Smith and Pulliam (2007) says, "This leveraging binge has regulators and others worried. ... When markets turn bad, leverage can create a snowball effect. Lenders and derivatives dealers demand that investors provide them with more collateral — the stocks, cash or other assets they pledge to cover potential losses. Sometimes, investors dump stocks and bonds to raise cash. Prices drop more, losses accelerate, and more selling ensues. Some Wall Street analysts have taken to referring to a nightmare version of this scenario as 'The Great Unwind'."

12.4.2 Extraordinary losses and extreme value theory

Whereas VaR typically uses recent historical data and assumes a Gaussian or Student t-distribution for normal market conditions, stress testing considers situations that are either absent or rare from historical data. Extreme value theory provides two classes of statistical models for rare returns. They are the generalized extreme value (GEV) and the generalized Pareto distributions. Consider a sample of successive returns $r_1, \dots, r_n$. The *order statistics* $r_{(1)}, \dots, r_{(n)}$ of the sample are obtained by rearranging the sample values in increasing order of magnitude; i.e., $r_{(1)} \le r_{(2)} \le \cdots \le r_{(n)}$. Thus, $r_{(1)} = \min_{1 \le i \le n} r_i$ and $r_{(n)} = \max_{1 \le i \le n} r_i$. The GEV distribution arises from the asymptotic distribution of $r_{(1)}$ as $n \to \infty$, which is related to an extraordinary loss for a long position. A simple sign change yields a similar asymptotic distribution of $r_{(n)}$ as $n \to \infty$, which is related to an extraordinary loss for a short position.

To begin, assume that r_t are i.i.d. Then $P(r_{(1)} \geq x) = P(r_t \geq x$ for all $1 \leq t \leq n) = P^n(r_1 \geq x)$. This result can be used to show that if there exist constants $\alpha_n > 0$ and β_n such that $(r_{(1)} - \beta_n)/\alpha_n$ converges in distribution, then the limiting distribution function F must be of the form

$$F(x) = \begin{cases} 1 - \exp[-(1+cx)^{1/c}] & \text{if } c \neq 0, \\ 1 - \exp[-\exp(x)] & \text{if } c = 0, \end{cases} \tag{12.21}$$

for $x < -1/c$ if $c < 0$ and for $x > -1/c$ for $c > 0$. The case $c = 0$ in (12.21) is referred to as the *Gumbel* family, while $c < 0$ corresponds to the *Fréchet* family and $c > 0$ the *Weibull* family. The parameter c is called the *shape parameter* of the GEV distribution, $-1/c$ is called the *tail index*, and β_n and α_n are called the *location* and *scale* parameters, respectively. Extreme value theory also gives necessary and sufficient conditions on the left tail of the distribution of r_1 for the asymptotic distribution of $r_{(1)}$ to belong to one of the three GEV types. It has been extended to serially dependent r_t when the dependence is weak, as in the case of stationary sequences with square summable ACFs; see Leadbetter, Lindgren, and Rootzén (1983).

Maximum likelihood estimation

For a given sample, there is only a single minimum or maximum, and (α_n, β_n, c) cannot be estimated with only one observation ($r_{(1)}$ or $r_{(n)}$). A method that has been used to circumvent this difficulty is to divide the sample into subsamples and to apply extreme value theory to the subsamples. Specifically, divide a sample of size T into k nonoverlapping subsamples, each with n observations:

$$\{r_1, \dots, r_n\}, \{r_{n+1}, \dots, r_{2n}\}, \dots, \{r_{(k-1)n+1}, \dots, r_{kn}\},$$

assuming for simplicity that $T = nk$. Let $M_{n,i} = \min\{r_{(i-1)n+j} : 1 \leq j \leq n\}$ for $i = 1, \dots, k$. For sufficiently large n, we can apply extreme value theory to each subsample to conclude that the subsample minima $M_{n,i}$, $i = 1, \dots, k$, can be regarded as a subsample of k observations from a GEV distribution after renormalization; i.e., $(M_{n,i} - \beta)/\alpha$ has distribution function (12.21). In this way, the parameters α, β, c can be estimated by maximizing the likelihood function

$$L(\alpha, \beta, c) = \prod_{i=1}^{k} \left\{ \frac{1}{\alpha} f\left(\frac{M_{n,i} - \beta}{\alpha} \right) \right\}, \tag{12.22}$$

where $f(x)$ is the derivative of the distribution function $F(x)$ in (12.21). Under certain regularity conditions, the MLE $(\widehat{\alpha}, \widehat{\beta}, \widehat{c})$ is consistent and $(\widehat{\alpha}, \widehat{\beta})$ is asymptotically normal; see Embrechts, Klüppelberg, and Mikosch (1997). The `evir` package in `R` can be used to perform likelihood inference on the parameters α, β, and c. The MLEs can be computed by using the `R` function `gev`.

An example

Table 12.2 gives the MLEs, computed by the `R` function `gev`, of the parameters of the GEV distribution fitted to the subsamples of daily log returns of the NASDAQ 100 index from March 11, 1999 to April 20, 2007, using $n = 20, 40$ (days) for each subsample period; standard errors are given in parentheses.

Table 12.2. MLE of GEV parameters estimated from subsample minima and subsample maxima of daily log returns.

Subsample Size	c	100α	100β
	Subsample minima		
$n = 20$	−0.285 (0.130)	0.590 (0.050)	−1.026 (0.073)
$n = 40$	−0.272 (0.189)	0.653 (0.082)	−1.270 (0.115)
	Subsample maxima		
$n = 20$	−0.386 (0.122)	0.547 (0.039)	1.053 (0.062)
$n = 40$	−0.456 (0.198)	1.254 (0.114)	0.665 (0.084)

Excess losses over a threshold and generalized Pareto distribution

Instead of subsamples of prespecified sizes, an alternative approach is to use exceedances of the returns over some prespecified threshold. Consider a sample of successive returns $r_1, \ldots, r_n$ and a short financial position for them. If there exist constants $\alpha_n > 0$ and β_n such that $(r_{(n)} - \beta_n)/\alpha_n$ converges in distribution, then the limiting distribution function F has the form

$$F(x) = \begin{cases} \exp[-(1-cx)^{1/c}] & \text{if } c \neq 0, \\ \exp[-\exp(x)] & \text{if } c = 0, \end{cases} \tag{12.23}$$

for $1 - cx > 0$ if $c \neq 0$. In this case, for η_n so chosen that $\alpha_n - c(\eta_n - \beta_n) \to \psi > 0$, $P\{r_i \leq x + \eta_n | r_i > \eta_n\}$ can be shown to coverage to

$$G_{c,\psi}(x) = \begin{cases} 1 - \left(1 - cx/\psi\right)^{1/c} & \text{if } c \neq 0, \\ 1 - \exp(-x/\psi) & \text{if } c = 0, \end{cases} \tag{12.24}$$

in which $\psi := \alpha - c(\eta - \beta) > 0$, $x > 0$ when $c \leq 0$ and $0 < x \leq \psi/c$ when $c > 0$. The function $G_{c,\psi}$ is the distribution function of the *generalized Pareto distribution*. The parameters ψ and c can be estimated from the subsample of returns r_i that exceed η_n by the method of maximum likelihood; see Embrechts, Klüppelberg, and Mikosch (1997). The `pot` package in `R` and the function `gpfit` in `MATLAB` can be used to perform likelihood inference on the parameters ψ and c.

12.4.3 Scenario analysis and Monte Carlo simulations

Scenario analysis involves evaluating the gain or loss on a portfolio over a specified period under a variety of scenarios. The scenarios cover a wide range of possibilities and can involve a chain of events, as in the subprime mortgage market meltdown in 2007. Monte Carlo simulations are best suited to the inherent complexities of these scenarios. To handle rare events in scenario analysis, importance sampling is particularly effective, changing the physical measure under which the event rarely occurs to a new measure under which the event is no longer rare; see Glasserman and Li (2005) and Chan and Lai (2007) for recent developments of the importance sampling approach. It is important to be comprehensive in generating scenarios. One approach is to examine 10 or 20 years of past data and to choose the most extreme events as scenarios; see Hull (2006, p. 733). As pointed out by Browning (2007), how some companies survived while others failed in these past extreme events also provides useful lessons in coping with these adversities should they occur.

Exercises

12.1. Explain, with examples, why the 99% daily VaR for short positions is not a coherent risk measure.

12.2. The file `intel_d_logret.txt` contains daily log returns of Intel stock from July 9, 1986 to June 29, 2007. Compute the 99% 1-day and 10-day VaR for a long position of $1 million using the following methods (as internal models in Section 12.1.2):
- (a) GARCH$(1,1)$ model with standard normal ϵ_t;
- (b) EGARCH$(1,1)$ model with standard normal ϵ_t;
- (c) ARMA$(1,1)$-GARCH$(1,1)$ model with ϵ_t having the standardized Student t-distribution whose degrees of freedom are to be estimated from the data;

(d) the GEV distribution for extreme (negative) returns with subperiod length of 20 trading days.

12.3. Consider a European call option with parameters as follows: current stock price S_0, strike K, risk-free rate r, volatility rate σ, and time to maturity T years. Assuming a geometric Brownian motion for the stock price process S_t, use the delta-normal valuation to compute the 95% VaR and ES over a horizon of 5 days for (a) a short position and (b) a long position.

12.4. Consider the bivariate Gaussian copula function $C(u,v)$ given by (12.14). Prove the following properties:

(a) $C(u,0) = C(0,v) = 0$, $C(u,1) = u$, $C(1,v) = v$.

(b) $C(u,v) = \int_0^u \int_0^v \frac{\partial^2 C(u,v)}{\partial u \partial v} dudv$.

(c) $C(x,y) = \int_0^x \Phi\left(\frac{\Phi^{-1}(y) - \rho\Phi^{-1}(u)}{\sqrt{1-\rho^2}}\right) du$.

12.5. The bivariate t-copula function with correlation coefficient ρ and ν degrees of freedom is given by (12.18). A method of simulating a sample (X,Y) whose distribution function is the t-copula function consists of the following two steps:

Step 1. Simulate independent random variables U, V that are uniformly distributed in $[0,1]$.

Step 2. Set $X = U$ and evaluate

$$Y := t_\nu\left(t_{\nu+1}^{-1}(V)\sqrt{\frac{(1-\rho^2)\left[\nu + t_\nu^{-1}(U)^2\right]}{\nu+1}} + \rho t_\nu^{-1}(U)\right).$$

(a) Write a program to simulate bivariate t-copulas; you can use `MATLAB` or `R` functions. (*Hint*: The function `runif` in `R` and the function `rand` in `MATLAB` generate $U[0,1]$ random variables. The functions `dt` and `qt` in `R` or the functions `tcdf` and `tinv` in `MATLAB` can be used to compute $t_\nu(\cdot)$ and $t_\nu^{-1}(\cdot)$.)

(b) Use your program in (a) to simulate 1000 samples of t-copulas for $(\rho,\nu) = (0.3, 3)$ and $(\rho,\nu) = (0.6, 7)$. Compare the histograms of the samples for these two (ρ,ν) values.

12.6. A group of investors is currently investing in a portfolio of a market index and three stocks. The group wants to develop a model for assessing the market risk of the portfolio by using data in the past year (264 trading days) on the daily profit (or loss) M for the market index and S_1, S_2, S_3 for the three stocks. These P/L (profit/loss) data are contained in the file `d_risk_profitloss.txt`, in which each number represents a daily profit (with the negative sign denoting a loss) in units of $100,000. Suppose you are a quantitative analyst whom the group has asked for help in building a one-factor model.

(a) The simplest way to begin is to assume that (M, S_1, S_2, S_3) has a multivariate normal distribution. Estimate by maximum likelihood the mean $(\mu_M, \mu_1, \mu_2, \mu_3)^T$, the variances $\sigma_M^2, \sigma_1^2, \sigma_2^2, \sigma_3^2$, and the correlation matrix of this normal vector.

(b) Assuming the Gaussian one-factor model in (a), calculate the 99% 1-day VaR and ES of the portfolio $M + S_1 + S_2 + S_3$.

(c) Let $m = (M - \mu_M)/\sigma_M$ and $s_i = (S_i - \mu_i)/\sigma_i$ be the standardized daily P/L. The one-factor model assumes that $s_i = r_i \epsilon_i + \sqrt{1 - r_i^2}\, m$, where ϵ_1, ϵ_2, and ϵ_3 are independent and represent the idiosyncratic components of s_i (see Section 3.3.1). Show that in the present Gaussian case, this one-factor model is equivalent to the assumption that s_1, s_2, and s_3 are conditionally independent given m.

(d) While still keeping normal marginal distributions, you proceed to relax the jointly normal assumption in the one-factor model by using bivariate copulas instead, in view of (c). First use $\widetilde{m} := \Phi(m)$, $\widetilde{s}_i := \Phi(s_i)$ as uniform random variables after replacing the unknown parameters $\mu_M, \sigma_M, \mu_1, \sigma_1, \ldots, \mu_3, \sigma_3$ by their estimates in (a). Then fit three bivariate t-copula models to $(\widetilde{m}, \widetilde{s}_1)$, $(\widetilde{m}, \widetilde{s}_2)$, and $(\widetilde{m}, \widetilde{s}_3)$ by using maximum likelihood. You can use the `MATLAB` program `tcopula2dfit.m`, or write your program to fit these three models.

(e) Write a program by modifying that in Exercise 12.7 to simulate $(\widetilde{m}, \widetilde{s}_1, \widetilde{s}_2, \widetilde{s}_3)$ under the one-factor model by first simulating $\widetilde{m} \sim U[0, 1]$ and then simulating $\widetilde{s}_1$, $\widetilde{s}_2$, $\widetilde{s}_3$ independently from the three bivariate t-copulas obtained in (d).

(f) Use your program in (e) to generate 10,000 simulated samples of $M + S_1 + S_2 + S_3$, which is the daily P/L of the portfolio. Use these simulations to compute the daily 99% 1-day VaR and ES for the portfolio. Compare the result with those in (b).

(g) Perform diagnostic checks on the assumption of Gaussian marginal distributions by examining the Q-Q plots of m, s_1, s_2, and s_3 and by performing the Jacque-Bera test for each marginal distribution. Discuss how you would relax the assumption of Gaussian marginals in this approach.

Appendix A. Martingale Theory and Central Limit Theorems

A sequence of random variables S_n with finite expectations is called a *martingale* if $E(S_n|\mathcal{F}_{n-1}) = S_{n-1}$, where $\mathcal{F}_m$ is the information set at time m (which consists of all events that have occurred up to time m). A prototypical example is $S_n = \sum_{i=1}^n y_i$, where $y_1, y_2, \ldots$ are independent random variables with zero means. In this special case, if $Ey_i^2 < \infty$ for all i, then $\text{Var}(S_n) = \sum_{i=1}^n E(y_i^2)$; moreover, under the Lindeberg condition in (A.5) below, the central limit theorem (CLT) holds; i.e.,

$$\left(\sum_{i=1}^n y_i\right) \Big/ \left(\sum_{i=1}^n \sigma_i^2\right)^{1/2} \text{ has a limiting } N(0,1) \text{ distribution} \tag{A.1}$$

as $n \to \infty$, where $\sigma_i^2 = E(y_i^2)$. This result can be extended to martingales by using the *conditional variance* $\sum_{i=1}^n E(y_i^2|\mathcal{F}_{i-1})$ in lieu of $\sum_{i=1}^n \sigma_i^2$ in (A.1).

The *martingale central limit theorem* can be stated in a sharper form, called the *functional form*; see Durrett (2005, pp. 409–411). To begin, instead of the martingale S_n, consider the martingale differences $y_n = S_n - S_{n-1}$. More generally, $\{x_{n,m}, \mathcal{F}_{n,m}, 1 \le m \le n\}$ is called a *martingale difference array* if $E(x_{n,m}|\mathcal{F}_{n,m-1}) = 0$ for $1 \le m \le n$, where $\mathcal{F}_{n,k}$ is the information set at time k. The first index n can be regarded as the sample size, with $x_{n,m}$ being the mth observation in the sample. For the special case of sums of zero-mean random variables in (A.1), $x_{n,m} = y_m / \sqrt{\sum_{i=1}^n \sigma_i^2}$. Whereas the standard normal distribution is the distribution limit in (A.1), the functional CLT has Brownian motion as the limiting stochastic process: A stochastic process $\{w_t, t \ge 0\}$ is called *Brownian motion* if

(a) $w_{t_0}, w_{t_1} - w_{t_0}, \ldots, w_{t_k} - w_{t_{k-1}}$ are independent for all $k \ge 1$ and $0 \le t_0 < t_1 < \cdots < t_k$, (A.2)

(b) $w_t - w_s \sim N(0, t-s)$ for $t \ge s \ge 0$, and (A.3)

(c) $P\{w_t \text{ is continuous at } t \text{ for all } t \ge 0\} = 1$. (A.4)

Martingale CLT

Suppose $\{x_{n,m}, \mathcal{F}_{n,m}, 1 \leq m \leq n\}$ is a martingale difference array. Let $S_{n,k} = \sum_{m=1}^{k} x_{n,m}$, $V_{n,k} = \sum_{m=1}^{k} E(x_{n,m}^2 | \mathcal{F}_{n,m-1})$. Suppose that, as $n \to \infty$,

(i) $V_{n,[nt]} \to t$ in probability for every $0 \leq t \leq 1$, and
(ii) $\sum_{m=1}^{n} E(x_{n,m}^2 \mathbf{1}_{\{|x_{n,m}|>\delta\}} \big| \mathcal{F}_{n,m-1}) \to 0$ in probability for every $\delta > 0$.

Then $\{S_{n,[nt]}, 0 \leq t \leq 1\}$ converges in distribution to Brownian motion as $n \to \infty$.

For the precise definition of "convergence in distribution" (also called "weak convergence") of the process $\{S_{n,[nt]}, 0 \leq t \leq 1\}$ (or for its linearly interpolated modification that is continuous), see Durrett (2005, p. 403). The martingale CLT implies that, as $n \to \infty$, $S_{n,n}$ has the limiting $N(0,1)$ distribution (which is the distribution of w_1). It also yields limiting distributions for $\max_{1 \leq k \leq n} S_{n,k}$ or for $n^{-1} \sum_{k=1}^{n} S_{n,k}^2$, which will be described in Appendix C. Condition (ii) in the martingale CLT is often referred to as *conditional Lindeberg*. When $x_{n,1}, \ldots, x_{n,n}$ are independent zero-mean random variables such that $\sum_{m=1}^{n} E(x_{n,m}^2) = 1$, (ii) reduces to the classical Lindeberg condition

$$\lim_{n \to \infty} \sum_{m=1}^{n} E(x_{n,m}^2 \mathbf{1}_{\{|x_{n,m}|>\delta\}}) = 0 \text{ for every } \delta > 0, \tag{A.5}$$

which is necessary and sufficient for $S_{n,n}$ to have a limiting standard normal distribution.

The concept of martingales can be readily extended to continuous-time processes $\{M_t, t \geq 0\}$. We call M_t a martingale if $E(M_t | \mathcal{F}_s) = M_s$ for $t \geq s$, where $\mathcal{F}_s$ is the information set consisting of all events that have occurred up to time s. Since the Brownian motion w_t has independent Gaussian increments, $\{w_t, t \geq 0\}$ is a martingale. The martingale CLT basically says that the martingale $\{S_{n,k} : 1 \leq k \leq n\}$ converges in distribution to the Gaussian martingale, and condition (i) says that the conditional variance of $S_{n,k}$ converges to that of w_t, noting that for $t \geq s$

$$E(w_t^2 | \mathcal{F}_s) = E\{(w_s + w_t - w_s)^2 | \mathcal{F}_s\} = w_s^2 + E(w_t - w_s)^2 = w_s^2 + t - s,$$

by the independent increments property (A.2).

Applications to linear regression

First consider the case of independent ϵ_t with $E\epsilon_t = 0$ and $E\epsilon_t^2 = \sigma^2$ in the regression model

$$y_t = \boldsymbol{\beta}^T \mathbf{x}_t + \epsilon_t, \quad 1 \leq t \leq n, \tag{A.6}$$

in which $\mathbf{x}_t$ are nonrandom variables, as in Section 1.1.4. Then $\widehat{\boldsymbol{\beta}} - \boldsymbol{\beta} = \left(\sum_{t=1}^{n} \mathbf{x}_t \mathbf{x}_t^T\right)^{-1} \sum_{t=1}^{n} \mathbf{x}_t \epsilon_t$; see (2.21). Since the $\mathbf{x}_t$ are nonrandom, $\widehat{\beta}_j - \beta_j$

can be written in the form $\sum_{t=1}^n a_{nt}^{(j)} \epsilon_t$, where $a_{nt}^{(j)}$ are nonrandom constants. Fix j and simply write a_{nt} instead of $a_{nt}^{(j)}$. Since $\mathrm{Var}(\widehat{\beta}_j) = \sigma^2 \sum_{t=1}^n a_{nt}^2 = \sigma^2 c_{jj}$, where c_{jj} is the jth diagonal element of $(\sum_{t=1}^n \mathbf{x}_t \mathbf{x}_t^T)^{-1}$, we can write $\left(\sigma^2 \sum_{t=1}^n a_{nt}^2\right)^{-1/2} (\widehat{\beta}_j - \beta_j) = \sum_{m=1}^n x_{n,m}$, where

$$x_{n,m} = (\epsilon_m/\sigma)\left(a_{nm} \Big/ \sqrt{\sum_{t=1}^n a_{nt}^2}\right). \tag{A.7}$$

The Lindeberg condition (A.5) is satisfied if for some $r > 2$

$$\sup_t E|\epsilon_t|^r < \infty \text{ and } \lim_{n\to\infty} \left(\max_{1\le t\le n} a_{nt}^2\right) \Big/ \left(\sum_{t=1}^n a_{nt}^2\right) = 0, \tag{A.8}$$

noting that

$$E x_{n,m}^2 \mathbf{1}_{\{|x_{n,m}|>\delta\}} \le \left(\frac{a_{nm}^2}{\sum_{t=1}^n a_{nt}^2}\right)^{r/2} \frac{E|\epsilon_m|^r}{\delta^{r-2}\sigma^r}.$$

Hence, under (A.8), $(\widehat{\beta}_j - \beta_j)/\mathrm{se}(\widehat{\beta}_j)$ has a limiting $N(0,1)$ distribution as $n \to \infty$, where $\mathrm{se}(\widehat{\beta}_j) = \sigma\sqrt{c_{jj}}$. Since its limiting normal distribution does not depend on the distributions of the ϵ_t, $(\widehat{\beta}_j - \beta_j)/\mathrm{se}(\widehat{\beta}_j)$ is an approximate pivot, as pointed out in Section 1.6.3.

The martingale CLT can be used to remove the assumption of nonrandom $\mathbf{x}_t$ in the regression model (A.6). Note that the martingale difference assumption (1.47) basically says that the input $\mathbf{x}_t$ at time t can depend on past inputs and outputs $(y_s, \mathbf{x}_s)$ with $s < t$. Variances are now replaced by conditional variances, as in (1.49), and likewise $\sup_t E|\epsilon_t|^r < \infty$ in (A.8) can be replaced by $\sup_t E(|\epsilon_t|^r | \mathbf{x}_t, \epsilon_{t-1}, \ldots, \epsilon_1, \mathbf{x}_1) < \infty$. The assumption (1.49) is crucial to obtain a martingale difference sequence that satisfies condition (i) of the martingale CLT after renormalization of the type (A.7). As will be shown in Appendix C, asymptotic normality may no longer hold in situations where (1.49) fails to hold.

Applications to likelihood theory

The arguments for the case of i.i.d. X_t in (2.37)–(2.43) can be readily extended to the general case considered in (2.45); see Exercise 2.9. The conditional covariance matrix of the martingale $\nabla l_n(\boldsymbol{\theta}_0)$ is

$$\mathbf{I}_n(\boldsymbol{\theta}_0) := -\sum_{t=1}^n E\Big[\nabla^2 \log f_{\boldsymbol{\theta}}(X_t|X_1,\ldots,X_{t-1})\big|_{\boldsymbol{\theta}=\boldsymbol{\theta}_0}\Big| X_1,\ldots,X_{t-1}\Big];$$

see Exercise 2.9(b). To apply the martingale CLT, let $\mathbf{y}_t = \nabla \log f_{\boldsymbol{\theta}}(X_t|X_1, \ldots, X_{t-1})\big|_{\boldsymbol{\theta}=\boldsymbol{\theta}_0}$ and assume that

$$\sup_t E(||\mathbf{y}_t||^r \big| X_1, \ldots, X_{t-1}) < \infty \quad \text{for some } r > 2, \tag{A.9}$$

$$n^{-1}\mathbf{I}_n(\boldsymbol{\theta}_0) \text{ converges to a positive definite nonrandom matrix} \tag{A.10}$$

in probability. Since $\boldsymbol{\theta}_0$ is unknown, we have to replace it by $\widehat{\boldsymbol{\theta}}$. In addition, analogous to (2.42) for the i.i.d. case, it is often more convenient to replace $\mathbf{I}_n(\boldsymbol{\theta}_0)$ by the observed Fisher information matrix $-\sum_{t=1}^n \nabla^2 \log f_{\boldsymbol{\theta}}(X_t|X_1, \ldots, X_{t-1})\big|_{\boldsymbol{\theta}=\widehat{\boldsymbol{\theta}}}$.

The law of large numbers for sums of independent random variables y_i with $E(y_i) = \mu_i$ and $\text{Var}(y_i) = \sigma_i^2$ says that if $\sum_{i=1}^\infty \sigma_i^2 = \infty$, then

$$\left(\sum_{i=1}^n y_i - \sum_{i=1}^n \mu_i\right) \Big/ \sum_{i=1}^n \sigma_i^2 \to 0 \quad \text{as } n \to \infty \text{ with probability 1.} \tag{A.11}$$

This can be extended to general random variables y_i by martingale strong laws of the form

$$\frac{\sum_{i=1}^n \{y_i - E(y_i|\mathcal{F}_{i-1})\}}{\sum_{i=1}^n \text{Var}(y_i|\mathcal{F}_{i-1})} \to 0 \text{ with probability 1 on } \left\{\sum_{i=1}^\infty \text{Var}(y_i|\mathcal{F}_{i-1}) = \infty\right\}.$$

Multivariate Brownian motion, Ito's lemma, and the Feynman-Kac formula

In the applications to likelihood theory and regression models with stochastic regressors, we have actually used the multivariate version of the martingale CLT, in which conditional variance is replaced by the conditional covariance matrix and Brownian motion is replaced by its multivariate counterpart. A *p-dimensional Brownian motion* $\mathbf{w}(t) = (w_1(t), \ldots, w_p(t))^T$ simply puts together p independent Brownian motions $w_i(t)$, $1 \le i \le p$, into a p-dimensional vector. Therefore it has independent increments and continuous sample paths such that $\mathbf{w}_t \sim N(\mathbf{0}, t\mathbf{I})$. The p-dimensional Brownian motion plays a basic role not only in the multivariate CLT but also in stochastic calculus, which extends ordinary calculus (differentiation, integration, and the chain rule) to trajectories that have random components generated by the increments of a p-dimensional Brownian motion. The laws of motion governing these trajectories can be described by a *stochastic differential equation* (SDE) of the form

$$d\mathbf{x}_t = \mathbf{b}(t, \mathbf{x}_t)dt + \boldsymbol{\sigma}(t, \mathbf{x}_t)d\mathbf{w}_t \tag{A.12}$$

in which $\mathbf{x}_t \in \mathbb{R}^p$, $\mathbf{b}(t, \mathbf{x}_t) \in \mathbb{R}^p$, and $\boldsymbol{\sigma}(t, \mathbf{x}_t)$ is a $p \times p$ matrix. In the case $\boldsymbol{\sigma} = \mathbf{0}$, the SDE reduces to an ordinary differential equation and the chain rule of differentiation can be applied to a smooth real-valued function $f(t, \mathbf{x}_t)$, yielding $df(t, \mathbf{x}_t) = \left(\frac{\partial}{\partial t} + \sum_{i=1}^{p} b_i \frac{\partial}{\partial x_i} \right) f(t, \mathbf{x}_t)$, which corresponds to a first-order approximation of f. The essential difference in stochastic calculus is due to the fact that $d\mathbf{w}_t (d\mathbf{w}_t)^T = \mathbf{I}dt$, which suggests that one cannot ignore second-order derivatives of f for terms associated with $d\mathbf{w}_t$ when $\boldsymbol{\sigma}$ is not identically equal to the zero matrix. Letting $\boldsymbol{\sigma}\boldsymbol{\sigma}^T = (a_{ij})_{1 \le i,j \le p}$ and introducing the operator $\mathcal{A}_t$ on twice continuously differentiable real-valued functions u by

$$\mathcal{A}_t u = \sum_{i=1}^{p} b_i(t, x) \frac{\partial u}{\partial x_i} + \frac{1}{2} \sum_{i=1}^{p} \sum_{j=1}^{p} a_{ij}(t, x) \frac{\partial^2 u}{\partial x_i \partial x_j}, \tag{A.13}$$

Ito's lemma provides the chain rule for smooth real-valued functions of the SDE (A.12):

$$df(t, \mathbf{x}_t) = \left(\frac{\partial}{\partial t} + \mathcal{A}_t \right) f(t, \mathbf{x}_t) dt + \left(\boldsymbol{\nabla} f(t, \mathbf{x}_t) \right)^T \boldsymbol{\sigma}(t, \mathbf{x}_t) d\mathbf{w}_t, \tag{A.14}$$

where $\boldsymbol{\nabla}$ denotes the gradient vector $(\partial/\partial x_1, \ldots, \partial/\partial x_p)^T$.

The operator $\mathcal{A}_t$ defined by (A.13) is called the *infinitesimal generator* of the SDE (A.12) and completely characterizes the SDE. The *Feynman-Kac formula* expresses the solution of the PDE

$$\begin{cases} \dfrac{\partial f}{\partial t} + \mathcal{A}_t f = hf & \text{for } (t, \mathbf{x}) \in [0, T) \times \mathbb{R}^q, \\ f(T, x) = g(x) & \text{for } \mathbf{x} \in \mathbb{R}^q, \end{cases} \tag{A.15}$$

in terms of the expectation of the SDE $\mathbf{x}_t$ with infinitesimal generator $\mathcal{A}_t$:

$$f(t, \mathbf{x}) = E\left[g(X_T) e^{-\int_t^T h(u, X_u) du} \middle| \mathbf{x}_t = \mathbf{x} \right]. \tag{A.16}$$

A special case of the Feynman-Kac formula is (10.24), which expresses the price of an interest rate derivative given by the PDE (10.23) in terms of an expectation.

Appendix B. Limit Theorems for Stationary Processes

Birkhoff's ergodic theorem (see Section 6.2 of Durrett (2005)) gives the strong law of large numbers for strictly stationary sequences $\{x_n\}$ with $E|x_1| < \infty$: With probability 1, $n^{-1} \sum_{i=1}^{n} x_i$ converges to a random variable that has mean $\mu := Ex_1$. For covariance stationary sequences, a weak law of large numbers holds: There exists a random variable x_∞ such that $E(x_\infty) = \mu$ and

$$\lim_{n\to\infty} E\left(n^{-1}\sum_{i=1}^{n} x_i - x_\infty\right)^2 = 0;$$

see Doob (1953, pp. 489–490). If the autocovariance function γ_h of $\{x_n\}$ satisfies $\sum_{h=1}^{\infty} |\gamma_h| < \infty$, then we can replace x_∞ by μ since

$$\left\{E\sum_{i=1}^{n}(x_i - \mu)\right\}^2 = n\gamma_0 + 2\{(n-1)\gamma_1 + (n-2)\gamma_2 + \cdots + \gamma_{n-1}\}.$$

Under certain conditions, which are generically called "mixing conditions," sums S_n of zero-mean stationary sequences can be approximated by martingales and the martingale CLT can be applied to prove a corresponding CLT for S_n; see Durrett (2005, pp. 420–424) for details.

Appendix C. Limit Theorems Underlying Unit-Root Tests and Cointegration

Unit-root nonstationary regressors

Let $\mathbf{x}_t$ be a unit-root nonstationary multivariate time series (see Definition 9.3(i)). We show here that assumption (1.49) is violated, resulting in non-normal asymptotic distributions, and even inconsistency in certain cases, of the OLS estimate $\widehat{\boldsymbol{\beta}}$ in the regression model $y_t = \boldsymbol{\beta}^T \mathbf{x}_t + \epsilon_t$, $1 \le t \le n$. To fix the ideas, consider the univariate case for which

$$\widehat{\beta} = \left(\sum_{t=1}^{n} x_t y_t\right) \Big/ \left(\sum_{t=1}^{n} x_t^2\right) = \beta + \left(\sum_{t=1}^{n} x_t \epsilon_t\right) \Big/ \left(\sum_{t=1}^{n} x_t^2\right). \tag{C.1}$$

Since x_t is unit-root nonstationary, $x_t - x_{t-1} = u_t$, where u_t is weakly stationary, and therefore $x_t = x_0 + \sum_{i=1}^{t} u_i$. Suppose $E(u_i) = 0$. Under certain conditions (see Appendices A and B), $\left(\sum_{i=1}^{m} u_i\right)/\sqrt{n}$ behaves like $\theta w_{m/n}$, where $\{w_t, 0 \le t \le 1\}$ is Brownian motion and $\theta^2 = \lim_{n\to\infty} n^{-1}\mathrm{Var}\left(\sum_{i=1}^{m} u_i\right)$. In particular, if u_i are i.i.d. with mean 0 and variance θ^2, then we can apply the martingale CLT in Appendix A. Hence

$$\frac{1}{n^2} \sum_{m=1}^{n} x_t^2 = \frac{\theta^2}{n} \sum_{m=1}^{n} \left(\frac{x_0 + \sum_{i=1}^{m} u_i}{\theta\sqrt{n}}\right)^2 \tag{C.2}$$

converges in distribution to $\int_0^1 w_t^2 dt$ as $n \to \infty$ by the continuous mapping theorem, stated below. This shows that to attain a nonzero limit in (1.49), we have to choose c_n of order n^2, but instead of converging in probability to a nonrandom limit, $n^{-2}\sum_{t=1}^{n} x_t^2$ converges in distribution to the random functional $\int_0^1 w_t^2 dt$ of Brownian motion.

Continuous mapping theorem

As already pointed out in Appendix A, the functional form of the CLT enables one to derive limit distributions of functionals of the martingale. To describe this idea more precisely, the functionals are required to be continuous, and the discrete-time martingale has to be linearly interpolated to form a continuous process. Thus, instead of $S_{n,[nt]}$ in the martingale CLT, we let $S_n(t) = S_{n,m}$ for $t = m/n$ and define $S_n(t)$ by linear interpolation of $S_{n,m}$ and $S_{n,m+1}$ for $m/n \leq t \leq (m+1)/n$. In this way, we have a sequence of continuous processes $\{S_n(t), 0 \leq t \leq 1\}$, which can be viewed as a sequence of random variables taking values in the space $C[0,1]$ of continuous functions $f : [0,1] \to \mathbb{R}$. The space $C[0,1]$ is a metric space with the sup-norm metric $d(f,g) = \max_{0 \leq t \leq 1} |f(t) - g(t)|$.

The *continuous mapping theorem* says that if X_n is a sequence of random variables taking values in a metric space C and converging in distribution to X, then $\psi(X)$ converges in distribution to $\psi(X)$ for every continuous function $\psi : C \to \mathbb{R}$. To apply the continuous mapping theorem to (C.2), we first note that the function $\psi : C[0,1] \to \mathbb{R}$ defined by $\psi(f) = \int_0^1 f^2(t)dt$, $f \in C[0,1]$, is continuous. The next step is to approximate the integral $\int_0^1 f^2(t)dt$ by the Riemann sum $n^{-1} \sum_{m=1}^n f^2(n/m)$.

Spurious regression

Up to now we have only considered the regressor x_t in the model $y_t = \beta x_t + \epsilon_t$ but have not made any assumptions on ϵ_t for the analysis of $\sum_{t=1}^n x_t \epsilon_t$ in (C.1). The model in Section 9.4.3, used by Granger and Newbold to illustrate spurious regression, assumes that $\epsilon_t \sim N(0,t)$ and that $\{\epsilon_i\}$ and $\{x_i\}$ are independent sequences. Let $\xrightarrow{\mathcal{D}}$ denotes convergence in distribution. By the continuous mapping theorem,

$$\frac{1}{n^2} \sum_{m=1}^{n} x_m \epsilon_m = \frac{1}{n} \sum_{m=1}^{n} \frac{x_m}{\sqrt{n}} \frac{\epsilon_t}{\sqrt{n}} \xrightarrow{\mathcal{D}} \int_0^1 w_t \widetilde{w}_t dt, \tag{C.3}$$

where $\widetilde{w}_t$ is a Brownian motion independent of w_t. Applying the continuous mapping theorem to $(n^{-2} \sum_{m=1}^n x_m \epsilon_m, n^{-2} \sum_{m=1}^n x_m^2)$ in (C.1) then yields

$$\widehat{\beta} - \beta \xrightarrow{\mathcal{D}} \left(\int_0^1 w_t \widetilde{w}_t dt \right) \Big/ \left(\int_0^1 w_t^2 dt \right),$$

which shows that $\widehat{\beta}$ cannot converge in probability to β.

OLS in unit-root nonstationary AR(1) models

We next consider the case where $\epsilon_t = u_{t+1}$, which arises in the AR(1) model $x_{t+1} = \beta x_t + u_{t+1}$ with $\beta = 1$ (and hence the AR(1) model is unit-root nonstationary) and $y_t = x_{t+1}$. Assuming $x_0 = 0$, we can write

$$x_n^2 = \left(\sum_{i=1}^n u_i\right)^2 = \sum_{i=1}^n u_i^2 + 2\sum_{i=1}^n x_{i-1}u_i.$$

Hence

$$\frac{1}{n}\sum_{t=0}^{n-1} x_t\epsilon_t = \frac{1}{n}\sum_{t=0}^{n-1} x_t u_{t+1} = \frac{1}{2}\left\{\left(\frac{x_n}{\sqrt{n}}\right)^2 - \frac{1}{n}\sum_{i=1}^n u_i^2\right\} \xrightarrow{\mathcal{D}} \frac{\sigma^2}{2}(w_1^2 - 1) \quad (C.4)$$

since $n^{-1}\sum_{i=1}^n u_i^2$ converges in probability to $E(u_1^2) = \mathrm{Var}(\epsilon_1) = \sigma^2$ by the law of large numbers.

Combining (C.1), (C.4) with $n^{-2}\sum_{t=0}^{n-1} x_t^2 \xrightarrow{\mathcal{D}} \int_0^1 w_t^2 dt$ shows that $\widehat{\beta}$ converges in probability to β and that $\widehat{\sigma}^2 := n^{-1}\sum_{t=1}^n (y_t - \widehat{\beta}x_t)^2$ is a consistent estimate of σ^2. Applying the continuous mapping theorem to (C.1) then yields the following nonnormal distribution of the studentized statistic when $\beta = 1$:

$$\frac{\widehat{\beta} - \beta}{\widehat{\sigma}\Big/\sqrt{\sum_{t=0}^{n-1} x_t^2}} \xrightarrow{\mathcal{D}} \frac{1}{2}\frac{w_1^2 - 1}{\sqrt{\int_0^1 w_t^2 dt}}. \quad (C.5)$$

In the preceding AR(1) model, the mean level is assumed to be 0 so that the regression model does not have an intercept term. When the AR(1) model takes the more general form $x_t = \mu + \beta x_{t-1} + u_t$, μ and β can be estimated consistently by the OLS estimate $(\widehat{\mu}, \widehat{\beta})$ and modification of the preceding argument can be used to show that, under the unit-root hypothesis $\beta = 1$,

$$\frac{\widehat{\beta} - 1}{\widehat{\mathrm{se}}(\widehat{\beta})} \xrightarrow{\mathcal{D}} \frac{1}{2}\frac{w_1^2 - 1 - w_1\int_0^1 w_t dt}{\sqrt{\int_0^1 w_t^2 dt - \left(\int_0^1 w_t dt\right)^2}}, \quad (C.6)$$

where $\widehat{\mathrm{se}}(\widehat{\beta}) = \widehat{\sigma}\Big/\sqrt{\sum_{t=0}^{n-1}(x_t - \bar{x})^2}$, $\bar{x} = n^{-1}\sum_{t=0}^{n-1} x_t$, and $\widehat{\sigma}^2 = n^{-1}\sum_{t=1}^n (x_t - \widehat{\mu} - \widehat{\beta}x_{t-1})^2$.

Limiting distribution of the ADF statistic under the unit-root hypothesis

For the AR(1) model, (C.6) already gives the limiting distribution of the augmented Dickey-Fuller statistic (9.29) under the unit-root null hypothesis.

For the more general AR(p) model written in the form (9.28), it can be shown that, under $\beta_1 = 1$, the OLS estimate $(\widehat{\mu}, \widehat{\beta}_1, \ldots, \widehat{\beta}_p)$ is again consistent, with $\widehat{\beta}_1$ converging to β_1 at rate $1/n$ and with the other estimates converging to the parameter values at rate $1/\sqrt{n}$. Because of this faster rate of convergence for $\widehat{\beta}_1$, the ADF statistic still has the limiting distribution given by (C.6).

Limiting distribution of likelihood ratio statistic in cointegration test

Making use of the functional CLT and the continuous mapping theorem, Johansen (1988) showed that when there are $h(< k)$ cointegration vectors, the likelihood ratio statistic (9.42) converges in distribution, as $n \to \infty$, to

$$\operatorname{tr}\left\{\left[\int_0^1 \mathbf{w}_t d\mathbf{w}_t^T\right]^T \left[\int_0^1 \mathbf{w}_t d\mathbf{w}_t^T\, dt\right]^{-1} \left[\int_0^1 \mathbf{w}_t d\mathbf{w}_t^T\right]\right\}, \tag{C.7}$$

where $\mathbf{w}_t$ is $(k-h)$-dimensional Brownian motion. In the case $h = k-1$, (C.7) reduces to

$$\frac{\left(\int_0^1 w_t dw_t\right)^2}{\int_0^1 w_t^2 dt} = \left\{\frac{(w_1^2 - 1)/2}{\sqrt{\int_0^1 w_t^2 dt}}\right\}^2, \tag{C.8}$$

which is the square of the limiting random variable in (C.5). The equality in (C.8) follows from Ito's lemma (see Appendix A): $dw_t^2 = 2w_t dw_t + dt$.

References

Acerbi, C. (2004). Coherent representations of subjective risk-aversion. In *Risk Measures for the 21st Century*. Szegö, G. (Ed.). Wiley. New York.

Adams, K.J. and Deventer, D.R. (1994). Fitting yield curves and forward rate curves with maximum smoothness. *Journal of Fixed Income*, 4, 52–62.

Aït-Sahalia, Y. (1996). Nonparametric pricing of interest rate derivative securities. *Econometrica*, 64, 527–560.

AitSahlia, F. and Lai, T.L. (2001). Exercise boundaries and efficient approximations to American option prices and hedge parameters. *Journal of Computational Finance*, 4, 85–103.

Akaike, H. (1973). Information theory and an extension of the maximum likelihood principle. In B. N. Petrov and F. Csake (Eds.), *The Second International Symposium on Information Theory*, 267–281, Budapest: Akádemiai, Kiado.

Albert, J.H., and Chib, S. (1993). Bayes inference via Gibbs sampling of autoregressive time series subject to Markov mean and variance shifts. *Journal of Business and Economic Statistics*, 11, 1–15.

Andersen, L. and Brotherton-Ratcliffe, R. (1997). The equity option volatility smile: An implicit finite-difference approach. *Journal of Computational Finance*, 1, 5–37.

Anderson, T.G. and Bollerslev, T. (1997). Heterogeneous information arrivals and return volatility dynamics: uncovering the long-run in high frequency returns. *Journal of Finance*, 52, 975–1005.

Anderson, T.W. (1951). Estimating linear restrictions on regression coefficients for multivariate normal distributions. *The Annals of Mathematical Statistics*, 22, 327–351.

—— (2003). *An Introduction to Multivariate Statistical Analysis*, 3rd ed. Wiley, New York.

Ang, A. and Bekaert, G. (2002a). Short rate non-linearities and regime switches. *Journal of Economic Dynamics and Control*, 26, 1243–1274.

—— (2002b). Regime switches in interest rates. *Journal of Business and Economic Statistics*, 20, 163–182.

Artzner, P., Delaen, F., Eber, J.-M., and Heath, D. (1997). Thinking coherently. *Risk*, 10 (November), 68–71.

—— (1999). Coherent measures of risk. *Mathematical Finance*, 9, 203–228.

Banz, R. (1981). The relation between return and market value of common stocks. *Journal of Financial Economics*, 9, 3–18.

BARRA (1997). *Market Impact Model Handbook*, MSCI Barra, Berkeley, CA.

Barndorff-Nielsen, O.E., Hansen P.R., Lundo, A., and Shephard, N. (2006). Designing realised kernels to measure the ex-post variation of equity prices in the presence of noise. Working paper, Nuffield College, University of Oxford.

Bartlett, M.S. (1937). The statistical conception of mental factors. *The British Journal of Psychology*, 28, 97–104.

—— (1954). A note on multiplying factors for various chi-squared approximations. *Journal of the Royal Statistical Society, Series B*, 16, 296–298.

Basu, S. (1977). The investment performance of common stocks in relation to their price to earnings ratios: A test of the efficient market hypothesis. *Journal of Finance*, 32, 663–682.

Bauwens, L., Laurent, S., and Rombouts, J.V.K. (2006). Multivariate GARCH models: A survey. *Journal of Applied Econometrics*, 21, 79–109.

Bekaert, G., Hodrick, R.J., and Marshall, D.A. (2001). Peso problem explanations for term structure anomalies. *Journal of Monetary Economics*, 48, 241–270.

Bickel, P.J. and Levina, E. (2007). Regularized estimation of large covariance matrices. To appear in *The Annals of Statistics*.

Black, F. (1976). The pricing of commodity contracts. *Journal of Financial Economics*, 3, 167–179.

Black, F. and Scholes, M. (1973). The pricing of options and corporate liabilities. *Journal of Political Economy*, 81, 637–659.

Black, F. and Karasinski, P. (1991). Bond and option pricing when short-term rates are lognormal. *Financial Analysts Journal*, 47, 52–59.

Bollerslev, T. (1986). Generalized autoregressive conditional heteroscedasticity. *Journal of Econometrics*, 31, 307–327.

Bossaerts, P. (1988). Common nonstationary components of asset prices. *Journal of Economic Dynamics and Control*, 12, 347–364.

Bossaerts, P. and Green, R.C. (1989). A general equilibrium model of changing risk premia: Theory and tests. *Review of Financial Studies*, 2, 467–493.

Box, G.E.P. and Cox, D.R. (1964). An analysis of transformation. *Journal of the Royal Statistical Society, Series B*, 26, 211–246.

Brace, A., Gatarek, D., and Musiela, M. (1997). The market model of interest rate dynamics. *Mathematical Finance*, 7, 127–154.

Brigo, D. and Mercurio, F. (2006). *Interest Rate Models: Theory and Practice*, 2nd ed. Springer, New York.

Broadie, M. and Detemple, J. (1996). American option valuation: New bounds, approximations, and a comparison of existing methods. *Review of Financial Studies*, 9, 1211–1250.

Broadie, M., Detemple, J., Ghysels, E., and Torres, O. (2000). Nonparametric estimation of American options' exercise boundaries and call prices. *Journal of Economic Dynamics and Control*, 24, 1829–1857.

Browning, E.S. (2007). Lessons of past may offer clues to market's fate. *The Wall Street Journal*, August 20: A1, A9.

Cai, J. (1994). A Markov model of switching-regime ARCH. *Journal of Business and Economic Statistics*, 12, 309–316.

Campbell, J., Lo, A., and MacKinlay, A.C. (1997). *The Econometrics of Financial Markets*. Princeton University Press, Princeton, NJ.

Carr, P., Jarrow, R., and Myneni, R. (1992). Alternative characterizations of American put options. *Mathematical Finance*, 2, 87–106.

Chan, H.P. and Lai, T.L. (2007). Efficient importance sampling for Monte Carlo evaluation of exceedance probabilities. *The Annals of Applied Probability*, 17, 440–473.

Chan, K.C., Karolyi, C.A., Longstaff, F.A., and Sanders, A.B. (1992). An empirical comparison of alternative models of the short-term interest rate. *Journal of Finance*, 47, 1209–1227.

Chen N., Roll, R., and Ross, S. (1986). Economic forces and the stock market. *Journal of Business*, 59, 383–403.

Christensen, K. and Podolskij, M. (2007). Realized range-based estimation of integrated variance. *Journal of Econometrics*, 141, 323–349.

Connor, G. and Korajczyk, R.A. (1988). Risk and return in an equilibrium APT: Application of a new test methodology. *Journal of Financial Econometrics*, 21, 255–289.

Cook, R.D. (1979). Influential observations in linear regression. *Journal of the American Statistical Association*, 74, 169–174.

Cox, J.C., Ingersoll, J.E., and Ross, S.A. (1985). A theory of the term structure of interest rates. *Econometrica*, 53, 385–407.

Cox, J.C. and Ross, S.A. (1976). The valuation of options for alternative stochastic processes. *Journal of Financial Economics*, 3, 145–166.

Cox, J.C., Ross, S.A., and Rubinstein, M. (1979). Option pricing: A simplified approach. *Journal of Financial Economics*, 7, 229–263.

Dacorogna, M., Gencay, R., and Muller, U. (2001). *An Introduction to High-Frequency Finance*. Academic Press, San Diego, CA.

Davidon, W.C. (1959). Variable metric method for minimization. *Atomic Energy Commission Research and Development Report*, ANL-5990.

Davis, M.H.A. and Norman, A.R. (1990). Portfolio selection with transaction costs. *Mathematics of Operations Research*, 15, 676–713.

De Bondt, W.F.M. and Thaler, R. (1985). Does the stock market overreact? *Journal of Finance*, 40, 793–805.

—— (1987). Further evidence of investor overreaction and stock market seasonality. *Journal of Finance*, 42, 557–581.

Derman, E. (2004). *My Life as a Quant.* Wiley, New York.

Derman, E., and Kani, I. (1994). Riding on a smile. *Risk*, 7, 32–39.

Diebold, F.X. (1986). Modeling the persistence of conditional variances: A comment. *Econometric Reviews*, 5, 51–56.

Doob, J.L. (1953). *Stochastic Processes.* Wiley, New York.

Dowd, K. (2005). *Measuring Market Risk*, 2nd ed. Wiley, New York.

Driffill, J. (1992). Change in regime and the term structure: A note. *Journal of Economic Dynamics and Control*, 16, 165–173.

Duarte, D., Longstaff, F., and Yu, F. (2007). Risk and return in fixed income arbitrage: Nickels in front of steamroller? *Review of Financial Studies*, 20, 769–811.

Dumas, B. and Luciano, E. (1991). An exact solution to a dynamic portfolio choice problem under transaction costs. *Journal of Finance*, 46, 577–595.

Dunsmuir, W. (1979). A central limit theorem for parameter estimation in stationary vector time series and its applications to models for a signal observed with noise. *The Annals of Statistics*, 7, 490–506.

Dupire, B. (1994). Pricing with a smile. *Risk*, 7, 18–20.

Durrett, R. (2005). *Probability: Theory and Examples*, 3rd ed. Duxbury Press, Belmont, CA.

Elliott, R.J., Van der Hoek, J., and Malcolm, W.P. (2005). Pairs trading. *Quantitative Finance*, 5, 271–276.

Embrechts, P., Klüppelberg, C., and Mikosch, T. (1997). *Extremal Events in Finance and Insurance.* Springer-Verlag, Berlin.

Engle, R.F. (1982). Autoregressive conditional heteroscedasticity with estimates of the variance of United Kingdom inflation. *Econometrica*, 50, 987–1007.

Engle, R.F. and Kroner, K.F. (1995). Multivariate simultaneous generalized ARCH. *Econometric Theory*, 11, 122–150.

Engle, R.F. and Patton, A.J. (2001). What good is a volatility model? *Quantitative Finance*, 1, 237–245.

Engle, R.F. and Russell, J.R. (1998). Autoregressive conditional duration: A new model for irregularly spaced transaction data. *Econometrica*, 66, 1127–1162.

Fama, E.F. (1970). Efficient capital markets: A review of theory and empirical work. *Journal of Finance*, 25, 383–417.

Fama, E.F. and French, K.R. (1992). The cross-section of expected stock returns. *Journal of Finance*, 47, 427–465.

—— (1993). Common risk factors in the returns on stocks and bonds. *Journal of Financial Economics*, 33, 3–56.

—— (1996). Multifactor explanations of asset pricing anomalies. *Journal of Finance*, 51, 55–84.

Ferson, W.E. (1989). Changes in expected security returns, risk and the level of interest rates. *Journal of Finance*, 44, 1191–1214.

Ferson, W.E. and Harvey, C.R. (1991). The time variation of economic risk premiums. *Journal of Political Economy*, 99, 385–415.

Fletcher, R., and Powell, M.J.D. (1963). A rapidly convergent descent method for minimization. *Computer Journal*, 6, 163–168.

Frankfurter, G.M., Phillips, H.E., and Seagle, J.P. (1971). Portfolio selection: The effects of uncertain means, variances, and co-variances. *Journal of Financial and Quantitative Analysis*, 6, 1251–1262.

Friedman, J.H. (1991). Multivariate adaptive regression splines (with discussion). *The Annals of Statistics*, 19, 1–141.

Gatev, E., Goetzmann, W.N., and Rouwenhorst, K.G. (2006). Pairs trading: Performance of a relative-value arbitrage rule. *The Review of Financial Studies*, 19, 797–827.

Glasserman, P. and Kou, S.G. (2003). The term structure of simple forward rates with jump risk. *Mathematical Finance*, 13, 383–410.

Glasserman, P. and Li, J. (2005). Importance sampling for portfolio credit risk. *Management Science*, 51, 1643–1656.

Granger, C. and Newbold, P. (1974). Spurious regression in econometrics. *Journal of Econometrics*, 2, 111–120.

Gray, S.F. (1996). Modeling the conditional distribution of interest rates as a regime-switching process. *Journal of Financial Economics*, 42, 27–62.

Green, P. and Silverman, B. (1994). *Nonparametric Regression and Generalized Linear Models*. Chapman and Hall, Glasgow.

Grinold, R.C. and Kahn, R.N. (2000). *Active Portfolio Management*, 2nd ed. McGraw-Hill, New York.

Grossman, S.J. and Stiglitz, J.E. (1980). On the impossibility of informationally efficient markets. *The American Economic Review*, 70, 393–408.

Hakansson, N.H. (1971). Capital growth and the mean-variance approach to portfolio selection. *Journal of Financial and Quantitative Analysis*, 6, 517–557.

Hamilton, J.D. (1988). Rational expectations econometric analysis of changes in regime. *Journal of Economic Dynamics and Control*, 12, 385–423.

—— (1989). A new approach to the economic analysis of nonstationary time series and the business cycle. *Econometrica*, 57, 357–384.

Hamilton, J.D. and Susmel, R. (1994). A conditional heteroskedasticity and change in regimes. *Journal of Econometrics*, 64, 307–333.

Hansen, L.P. (1982). Large sample properties of generalized method of moments estimators. *Econometrica*, 50, 1029–1054.

Hansen, P.R. (2005). A test for superior predictive ability. *Journal of Business and Economic Statistics*, 23, 365–380.

Hansen, P.R., Lunde, A., and Nason, J.M. (2005). Model confidence sets for forecasting models. Federal Reserve Bank of Atlanta Working Paper.

Hasbrouck, J. (1999). The dynamics of discrete bid and ask quotes. *Journal of Finance*, 54, 2109–2141.

—— (2007). *Empirical Market Microstructure: The Institutions, Economics, and Econometrics of Securities Trading.* Oxford University Press, New York.

Hausman, J.A., Lo, A.W., and MacKinlay, A.C. (1992). An ordered probit analysis of transaction stock prices. *Journal of Financial Economics*, 31, 319–379.

Heath, D., Jarrow, R., and Morton, A. (1992). Bond pricing and the term structure of interest rates: A new methodology for contigent claims valuation. *Econometrica*, 60, 77–105.

Hoerl, A.E. and Kennard, R.W. (1970). Ridge regression: Biased estimation for nonorthogonal problems. *Technometrics*, 12, 55–67.

Huang, J.Z., Liu, N., Pourahmadi, M. and Liu, L. (2006). Covariance matrix selection and estimation via penalised normal likelihood. *Biometrika*, 93, 85–98.

Hull, J. (2006). *Options, Futures, and Other Derivatives*, 6th ed. Prentice-Hall, Englewood Cliffs, NJ.

Hull, J. and White, A. (1987). The pricing of options on assets with stochastic volatilities. *Journal of Finance*, 42, 281–300.

—— (1990). Pricing interest rate derivative securities. *Review of Financial Studies*, 3, 573–592.

Hutchinson, J.M., Lo, A.W., and Poggio, T. (1994). A nonparameric approach to pricing and hedging derivative securities via learning networks. *Journal of Finance*, 49, 851–889.

Ing, C.K., Lai, T.L. and Chen, Z. (2007). Consistent estimation of high-dimensional covariance matrices under sparsity constraints. Working paper, Department of Statistics, Stanford University.

Jacka, S.D. (1991). Optimal stopping and the American put. *Mathematical Finance*, 1, 1–14.

Jacquier, E., Polson, N.G., and Rossi, P.E. (1994). Bayesian analysis of stochastic volatility models (with discussion). *Journal of Business and Economic Statistics*, 12, 371–417.

James, W. and Stein, C. (1961). Estimation with quadratic loss. *Proceedings of the Fourth Berkeley Symposium on Mathematical Statistics and Probability*, 1, 311–319.

Jamshidian, F. (1989). An exact bond pricing formula. *Jounal of Finance*, 44, 205–209.

—— (1997). LIBOR and swap market models and measures. *Finance and Stochastics*, 1, 293–330.

Jarrow, R., Ruppert, D., and Yu, Y. (2004). Estimating the interest rate term structure of corporate debt with a semiparametric penalized spline model. *Journal of the American Statistical Association*, 75, 544–554.

Jegadeesh, N. (1990). Evidence of predictable behavior of security returns. *Journal of Finance*, 45, 881–898.

Jegadeesh, N. and Titman, S. (1993). Returns to buying winners and selling losers: Implications for stock market efficiency. *Journal of Finance*, 48, 65–91.

—— (1995). Overreaction, delayed reaction, and contrarian profits. *The Review of Financial Studies*, 8, 973–993.

Jensen, M.C. (1968). The performance of mutual funds in the period 1945–1964. *Journal of Finance*, 23, 389–461.

Jensen, M.C. and Bennington, G. (1970). Random walks and technical theories: Some additional evidence. *Journal of Finance*, 25, 469–482.

Jobson, J.D., and Korkie, B. (1980). Estimation of Markowitz efficient portfolios. *Journal of the American Statistical Association*, 75, 544–554.

Johansen, S. (1988). Statistical analysis of cointegration vectors. *Journal of Economic Dynamics and Control*, 12, 231–254.

—— (1991). Estimation and hypothesis testing of cointegration vectors in Gaussian vector autoregressive models. *Econometrica*, 59, 1551–1580.

Jorion, P. (2001). *Value at Risk*, 2nd ed. McGraw-Hill, New York.

Ju, N. (1998). Pricing an American option by approximating its early exercise boundary as a multipiece exponential function. *The Review of Financial Studies*, 11, 627–646.

Keim, D. (1989). Trading patterns, bid-ask spreads, and estimated security returns: the case of common stocks at calendar turning points. *Journal of Financial Economics*, 25, 75–97.

Kothari, S., Shanken, J., and Sloan, R. (1995). Another look at the cross-section of expected returns. *Journal of Finance*, 50, 185–224.

Lahart, J. (2007). How the 'quant' playbook failed. *The Wall Street Journal*, August 24, C1–C2.

Lai, T.L. (1995). Sequential change-point detection in quality control and dynamical systems. *Journal of the Royal Statistical Society, Series B*, 57, 613–658.

Lai, T.L. and Lim, T.W. (2007). Option hedging theory under transaction costs. Working paper, Department of Statistics, Stanford University.

Lai, T.L., Lim, T.W., and Chen, L. (2007). A new approach to pricing and hedging options with transaction costs. Working paper, Department of Statistics, Stanford University.

Lai, T.L., Liu, H., and Xing, H. (2005). Autoregressive models with piecewise constant volatility and regression parameters. *Statistica Sinica*, 15, 279–301.

Lai, T.L., Pong, C.K., and Xing, H. (2007). A mechanistic-empirical approach to pricing and hedging interest-rate derivatives. Working paper, Department of Statistics, Stanford University.

Lai, T.L. and Wei, C.Z. (1983). Some asymptotic properties of general autoregressive models and strong consistency of least squares estimates of their parameters. *Journal of Multivariate Analysis*, 13, 1–23.

Lai, T.L. and Wong, S. (2004). Valuation of American options via basis functions. *IEEE Transactions on Automatic Control*, 49, 374–385.

—— (2006). Combining domain knowledge and statistical models in time series analysis. In *Time Series and Related Topics*, H.C. Ho, C.K. Ing, and T.L. Lai (Eds.). Institute of Mathematical Statistics, Beachwood, OH, 193–209.

Lai, T.L., and Xing, H. (2006). Structural change as an alternative to long memory in financial time series. In *Advances in Econometrics*, 20, T. Fomby and D. Terrell (Eds.). Elsevier, Amsterdam, 209–228.

—— (2007). Stochastic change-point ARX-GARCH models and their applications to econometric time series. Working paper, Department of Statistics, Stanford University.

Lai, T.L., Xing, H., and Chen, Z. (2007). Mean-variance portfolio optimization when means and variances are unknown. Working paper. Department of Statistics, Stanford University.

Landon, T. (2007). Pack mentality among hedge funds fuels market volatility. *The New York Times*, August 13, C1–C2.

Leadbetter, M.R., Lindgren, G., and Rootzén, H. (1983). *Extremes and Related Properties of Random Sequences and Series.* Springer-Verlag, New York.

Ledoit, P. and Wolf, M. (2003). Improve estimation of the covariance matrix of stock returns with an application to portfolio selection. *Journal of Empirical Finance*, 10, 603–621.

—— (2004). Honey, I shrunk the sample covariance matrix. *Journal of Portfolio Management*, 30, 110–119.

Li, D.X. (2000). On default correlation: A copula function approach. *Journal of Fixed Income*, 9, 43–54.

Lim, P.J. (2007). The oracle nobody knows. *U.S. News & World Report*, August 6, 57.

Lintner, J. (1965). The valuation of risk assets and the selection of risky investments in stock portfolios and capital budgets. *Review of Economics and Statistics*, 47, 13–37.

Lo, A., and MacKinlay, A.C. (1990). Data-snooping biases in tests of financial asset pricing models. *Review of Financial Studies*, 3, 431–468.

Lo, A.W., Mamaysky, H., and Wang, J. (2000). Foundations of technical analysis: Computational algorithms, statistical inference, and empirical implementation. *Journal of Finance*, 40, 1705–1765.

Magill, M.J.P. and Constantinides, G.M. (1976). Portfolio selection with transaction costs. *Journal of Economic Theory*, 13, 245–263.

Mallows, C.L. (1973). Some comments on C_p. *Technometrics*, 15, 661–676.

Merton, R.C. (1971). Optimum consumption and portfolio rules in a continuous-time model. *Journal of Economic Theory*, 3, 373–413.

—— (1973). Theory of rational option pricing, *The Bell Journal of Economics and Management Science*, 4, 141–183.

Michaud, R.O. (1989). *Efficient Asset Management.* Harvard Business School Press, Boston.

Miller, M., Muthuswamy, J., and Whaley, R. (1994). Mean reversion of Standard and Poor's 500 Index basis changes: Arbitrage-induced or statistical illusion. *Journal of Finance*, 49, 479–513.

Moskowitz, T.J. and Grinblatt, M. (1999). Do industries explain momentum? *Journal of Finance*, 54, 1249–1290.

Mossin, J. (1968). Aspects of rational insurance purchasing. *Journal of Political Economy*, 76, 553–568.

Nelson, D.B. (1991). Conditional heteroscedasticity in asset returns: a new approach. *Econometrica*, 59, 347–370.

Nelson, D.B. and Siegel, A. (1987). Parsimonious modelling of yield curves. *Journal of Business*, 60, 473–489.

Newey, W.K. and West, K.D. (1987). A simple, positive definite, heteroskedasticity and autocorrelation consistent covariance matrix. *Econometrica*, 55, 703–708.

O'Hara, M. (1995). *Market Microstructure Theory.* Blackwell, Cambridge, MA.

Perron, P. (1989). The great crash, the oil price shock, and the unit root hypothesis. *Econometrica*, 57, 1361–1401.

Phillips, P.C.B. and Perron, P. (1988). Testing for a unit root in time series regression. *Biometrika*, 75, 335–346.

Press, W.H., Flannery, B.P., Teukolsky, S.A., and Vetterling, W.T. (1992). *Numerical Recipes in C: The Art of Scientific Computing.* Cambridge University Press, Cambridge.

Protter, P. (2004). *Stochastic Integration and Differential Equations.* Springer, New York.

Rappoport, P. and Reichlin, L. (1989). Segmented trends and nonstationary time series. *Economic Journal* (supplement), 99, 168–177.

Rebonato, R. (2002). *Modern Pricing of Interest Rate Derivatives.* Princeton University Press, Princeton, NJ.

—— (2004). Interest-rate term-structure pricing models: A review. *Proceedings of the Royal Society of London, Series A*, 460, 667–728.

Reinsel, G.C. and Velu, R.P. (1998). *Multivariate Reduced Rank Regression: Theory and Applications.* Springer, New York.

Roberts, H. (1967). Statistical versus clinical prediction of the stock market. Unpublished manuscript. Center for Research in Security Prices, University of Chicago.

Roll, R. (1977). A critique of the asset pricing theory's tests: Part I. *Journal of Financial Economics*, 4, 129–176.

—— (1984). Simple implicit measure of the effective bid-ask spread in an efficient market. *Journal of Finance*, 39, 1127–1139.

Roll, R. and Ross, S. (1980). An empirical investigation of the arbitrage pricing theory. *Journal of Finance*, 5, 1073–1103.

Romano, J.P. and Wolf, M. (2005). Stepwise multiple testing as formalized data snooping. *Econometrica*, 73, 1237–1282.

Ross, S.A. (1976). The arbitrage theory of capital asset pricing. *Journal of Economic Theory*, 13, 341–360.

—— (1987). Finance. In *The New Palgrave: A Dictionary of Economics*, J. Eatwell, M. Milgate, and P. Newman (Eds.). Stockton Press, New York, 2, 322–336.

Rubinstein, M. (1994). Implied binomial trees. *Journal of Finance*, 49, 771–818.

Rydberg, T.H. and Shephard, N. (2003). Dynamics of trade-by-trade price movements: Decomposition and models. *Journal of Financial Econometrics*, 1, 2–25.

Samuelson, P.A. (1965). Proof that properly anticipated prices fluctuate randomly. *Industrial Management Review*, 6, 41–49.

—— (1969). Lifetime portfolio selection by dynamic stochastic programming. *Review of Economics and Statistics*, 51, 239–246.

Schwarz, G. (1978). Estimating the dimension of a model. *The Annals of Statistics*, 6, 461–464.

Sharpe, W.F. (1964). Capital asset prices: A theory of market equilibrium under conditions of risk, *Journal of Finance*, 19, 425–442.

Shiryaev, A.N. (2002). Quickest detection problems in the technical analysis of financial data. In *Mathematical Finance–Bachelier Congress 2000*, H. Geman, D. Madan, S. Pliska, and T. Vorst (Eds.). Springer, New York, 487–521.

Shreve, S.E. and Soner, H.M. (1994). Optimal investment and consumption with transaction costs. *The Annals of Applied Probability*, 4, 206–236.

Smith, R. and Pulliam, S. (2007). As funds leverage up, fears of reckoning rise. *The Wall Street Journal*, April 30, pp. A1, A12.

Stambaugh, R. (1982). On the exclusion of assets from tests of the two parameter model. *Journal of Financial Economics*, 10, 235–268.

Stanton, R. (1997). A nonparametric model of term structure dynamics and the market price of interest rate risk. *Journal of Finance*, 52, 1973–2002.

Sullivan, R., Timmerman, A., and White, H. (1999). Data-snooping, technical trading rule performance, and the bootstrap. *Journal of Finance*, 59, 1647–1691.

Svensson, L.E.O. (1994). Estimating and interpreting forward interest rates: Sweden 1992–1994. *NBER Working Paper Series* # 4871.

Vasicek, O. (1977). An equilibrium characterization of the term structure. *Journal of Financial Economics*, 5, 177–188.

Vasicek, O.A. and Fong, H.G. (1982). Term structure modeling using exponential splines. *Journal of Finance*, 37, 339–348.

Venables, W.N. and Ripley, B.D. (2002). *Modern Applied Statistics with S*, 4th ed. Springer, New York.

White, H. (2000). A reality check for data snooping. *Econometrica*, 68, 1097–1126.

Yin, G. and Zhou, X.Y. (2004). Markowitz's mean-variance portfolio selection with regime switching from discrete-time models to their continuous-time limits. *IEEE Transactions on Automatic Control*, 49, 349–360.

Zeng, Y. (2003). A partially observed model for micromovement of asset prices with Bayes estimation via filtering. *Mathematical Finance*, 13, 411–444.

Zhang, L., Mykland, P.A., and Aït-Sahalia, Y. (2005). A tale of two time scales: Determining integrated volatility with noisy high-frequency data. *Journal of the American Statistical Association*, 100, 1394–1411.

Zhou, X. and Li, D. (2000). Continuous-time mean-variance portfolio selection: A stochastic LQ framework. *Applied Mathematics and Optimization*, 42, 19–33.

Index

Printed in the United States of America